ELECTROANALYTICAL CHEMISTRY

Basic Principles and Applications

ELECTROANALYTICAL CHEMISTRY

Basic Principles and Applications

JAMES A. PLAMBECK

Department of Chemistry
University of Alberta

1807 1982

A Wiley-Interscience Publication
JOHN WILEY & SONS
New York   Chichester   Brisbane   Toronto   Singapore

Library of Congress Cataloging in Publication Data:

Plambeck, James A.
    Electroanalytical Chemistry: Basic Principles and Applications

    "A Wiley-Interscience publication."
    Includes bibliographical references and index.
    1. Electrochemical analysis. I. Title.

QD115.P48   543'.0871   82-2803
ISBN 0-471-04608-6   AACR2

Printed in the United States of America

10 9 8 7 6 5 4 3 2 1

To William, Joanna, Cheryl, and Michael—our joys and our distractions

TO THE STUDENT

This textbook of electroanalytical chemistry has been written for the author's half-year course at the University of Alberta, and is appropriate for students at the advanced undergraduate and beginning graduate level. It is possible to use this text with the background of one single but solid year of general university chemistry; additional physical chemistry background is preferable, however, especially in thermodynamics. References to appropriate monographs at the advanced graduate and research level are given at the end of each chapter. These should be used for further study and as sources of references to the research literature. Original research references within the text have been minimized, and many are of historical or educational significance.

I recommend that you proceed through the textbook in its present chapter order. It is reasonable, however, to take up Part 3 prior to Part 2 if you already have some physical chemistry background. Some of the material in Part 1 may be omitted initially, if you are prepared to use the index and refer back to it later. Study of isolated chapters is not recommended, as the chapters have not been written as self-contained units. The chapter study problems are an integral part of the book and should all be completed.

I have emphasized SI units because these follow the quantity calculus. Any physical quantity is expressed as the product of a number and a unit. Mathematical operations should be carried out on both alike. The familiar 1 mmol/liter, or 1 mmol/dm^3, is identical to 1 mol/m^3.

Throughout the text, the symbols for all physical quantities are set in italic or Greek, while the symbols for all units are in roman type. Symbols used only for convenience in writing equations are set in roman type. Tables of symbols are given in Appendix 1.

PREFACE

Electroanalytical chemistry has developed rapidly over the last several decades both in theory and in practice, and further development is likely as the digital revolution reaches further into commercial electroanalytical instrumentation. Numerous excellent monographs and reviews dealing with specific areas of electroanalytical chemistry now exist, but a modern textbook capable of guiding a worker new to the field does not. The purpose of the present work is to cover the full scope of modern electroanalytical chemistry at an introductory level and to guide the student to further reading at the advanced monograph and review level in specific areas. For this reason the ubiquitous Laplace transform and other more mathematical approaches have been avoided, although their results are used. The mathematical treatments will be found in the references cited.

Several areas of current electroanalytical interest have not been included in this book. These areas in which electroanalytical techniques are used together with the methodology of other areas, such as chromatography, electron spin resonance, and absorption spectrophotometry, as well as the specific interfacing of electroanalytical instrumentation to digital equipment, have been omitted for reasons of space.

The author is indebted to H. A. Laitinen for his original introduction to electroanalytical chemistry; to his colleagues at the University of Alberta for helpful comments, especially Byron Kratochvil; to manufacturers of electroanalytical instrumentation for photographs of their equipment; and to his wife, Olga, for proofreading and grammatical assistance. The manuscript was computer typeset by the author and Mrs. L. Ziola using the TEXTFORM system of the University of Alberta. Page masters were prepared by the University of Alberta Printing Services.

James A. Plambeck

Edmonton, Alberta, Canada
January 1982

CONTENTS

ELECTROANALYTICAL CHEMISTRY

Basic Principles and Applications

PART 1

BASIC INFORMATION

CHAPTER
1
ELECTRICAL CONCEPTS

Concepts, measurement of physical quantities related to those concepts, and the units in which the measurements are made are inseparable. The units of the *Système International* (SI), which are emphasized in this text and are increasingly in use worldwide, can serve as a starting point for untangling the skein.

1.1 SI UNITS

The International System of Units, normally abbreviated SI, is an improved metric system adopted by the Eleventh CGPM (General Conference of Weights and Measures) in 1960.

TABLE 1.1   Base Units of SI

Quantity	Unit	Symbol
Length	meter	m
Mass	kilogram	kg
Time	second	s
Temperature	kelvin	K
Amount of substance	mole	mol
Electric current	ampere	A
Luminous intensity	candela	cd

The entire system is constructed from seven base units, each of which represents a single physical quantity as shown in Table 1.1. Each of these quantities is measurable, according to a defined procedure, a specific physical object (mass), or in terms of measurements based upon fundamental properties of matter.

The advantage of the SI is that when any physical quantity is written out in SI base units, or in units derived from them, any mathematical manipulations

TABLE 1.2   Exponents and Prefixes of the SI

Exponent	Prefix	Symbol	Exponent	Prefix	Symbol
18	exa-	E	−18	atto-	a
15	peta-	P	−15	femto-	f
12	tera-	T	−12	pico-	p
9	giga-	G	−9	nano-	n
6	mega-	M	−6	micro-	μ
3	kilo-	k	−3	milli-	m
2	hecto-	h	−2	centi-	c
1	deka-	da	−1	deci-	d

performed with them will follow the quantity calculus. This means that if the entries in any equation are put in with their units and algebraic manipulations are performed upon the units along with the numbers to which they refer, the result will come out with the correct numbers *and units*. This is of considerable advantage in all areas of science. Electrochemists have been among the slower groups to convert completely to the SI system of coherent units. The present text attempts to use SI units exclusively in the body of the text. Problems, however, use the original units of their references, and data in the Appendixes are given in all useful units.

As in earlier versions of the metric system, SI units can be designated as decimal fractions or multiples by the use of appropriate prefixes. The acceptable prefixes are given in Table 1.2. Any prefix can be applied to any base unit except the kilogram. The kilogram takes prefixes as if the base unit were the gram: for example, 10^{-6} kg is written as 1 milligram (mg) rather than 1 microkilogram (μkg). Only positive and negative multiples of 1000 are generally given a prefix name or symbol, and the use of only such multiples is encouraged.

Since in scientific usage physical quantities often vary over many orders of magnitude, it is convenient to use *scientific notation* to express them. In scientific notation, a quantity is expressed as a number containing a decimal point multiplied by 10 to some integer exponent. If the quantity is a physical quantity, it includes a unit as well as the number. In scientific notation the exponent can have any integer value, and the decimal point usually follows the first digit of the number. An alternative approach, called *engineering notation* and available on some pocket calculators, restricts the integer exponents to those divisible by 3; the decimal point then follows any digit of the number. In either notation it is sometimes convenient to add a preceding zero digit to the number, for example,

TABLE 1.3 Derived Units and Symbols of the SI

Derived Quantity	Unit	Unit Symbol	Base Unit Equivalent	General Symbol
Area	square meter	m^2	m^2	A
Force	newton	N	$kg\ m/s^2$	Fa
Pressure	pascal	Pa	$kg/m\ s^2$	p
Energy	joule	J	$kg\ m^2/s^2$	E
Power	watt	W	$kg\ m^2/s^3$	P
Charge	coulomb	C	$A\ s$	Q
Potential	volt	V	$kg\ m^2/s^3A$	E
Resistance	ohm	Ω	$kg\ m^2/s^3A^2$	R
Conductance	siemens	S	$s^3A^2/kg\ m^2$	κ
Capacitance	farad	F	$s^4A^2/kg\ m$	C
Inductance	henry	H	$kg\ m^2/s^2A^2$	L
Frequency	hertz	Hz	s^{-1}	f

anot used in this work.

expressing 463 as 0.463×10^3. The SI prefixes are particularly convenient for engineering notation.

There are many physical quantities that are not those of the base units of the SI. These are measured in units derived from the seven base units. Volume, since it is the cube of a length, can be measured in the cube of the base unit for length, the cubic meter. Density, mass per unit volume, is then measured in kilograms per cubic meter (kg/m^3). Whenever exponents of this type are used with SI prefixes, the exponent applies to the prefix as well; thus nm^2, or square nanometer, is interpreted as $(nm)^2$ rather than $n(m^2)$. Some of these derived units are used sufficiently often that special names and symbols are used for them. These are listed in Table 1.3 with their definitions in terms of base units of the SI; this list, while not exhaustive, does cover most units of electrochemical interest. The most relevant derived units are the *volt*, the *coulomb* , the *joule*, the *watt*, the *ohm*, the *farad*, and the *henry*. The volt is often described as the unit of electromotive force as well as the unit of potential or potential difference.

The *faraday*, or Faraday's constant, is sometimes considered to be another derived unit of charge, the charge carried by one mole of electrons. It is the product of the Avogadro number and the electronic charge and thus an experimentally obtained unit rather than a defined one. Using modern values, one faraday is (6.02252×10^{23} electrons)/mole electrons multiplied by (1.60210×10^{-19} C/electron), yielding $96,487.0 \pm 1.6$ C/mole electrons. This quantity is generally symbolized by F, often written in script form rather than as the italic letter used in this text. It is better to consider the faraday a conversion factor between units rather than a unit in its own right.

1.2   ELECTRICAL CONCEPTS AND UNITS

The basic relationships between electrical units are closely related to their definitions and to the operational significance of the definitions. These relationships are fundamental to electrochemistry as well, and knowledge of them will be assumed throughout the remainder of this work.

1.2.1   Electrical Quantities and Units

Electrical quantities include *current, charge, potential* or *voltage, resistance, capacitance*, and *inductance*. Other quantities are used to express the variation of some of these with time, either absolutely or in relation to each other; some of these are discussed in the following section.

The fundamental electrical unit is the unit of current, the *ampere*, which is a base unit of the SI. The unit of charge, the *coulomb*, is the product of the ampere and the second. The SI unit of force, the *newton*, is derived from the kilogram, meter, and second since force is the product of mass and acceleration. The unit of energy, the *joule*, is simply the newton-meter; the energy/charge quotient, the joule/coulomb, is the unit of potential difference, the *volt*. The joule is therefore the volt-coulomb as well as the newton-meter. The unit of resistance, the *ohm*, is the quotient of the volt and the ampere. The unit of capacitance, the *farad*, is the quotient of the coulomb and the volt. The unit of inductance, the *henry*, is the joule/ampere^2.

In a purely direct current (DC) circuit employing perfect components, only charge, current, voltage, and resistance have meaning. A perfect capacitance acts as an infinitely large resistance, and a perfect inductance acts as an infinitely small resistance, to the passage of direct current.

1.2.2   DC Electrical Relationships

Electrical *resistance* is observed whenever an electrical potential drives an electric current through matter, since the structure of the matter will to a greater or lesser extent impede the flow of electrons. This fact is stated in the most fundamental electrical relationship, *Ohm's law*:

$$E = IR \tag{1.1}$$

Ohm's law serves to define the unit of resistance as well as the quantity itself; one *ohm* (1 Ω) of resistance is the resistance present when one volt of electrical potential is required to drive a current of one ampere.

TABLE 1.4   Characteristics of Pure Materials at 25°C[a]

Material	Dielectric Constant	Conductivity (S/m)	Viscosity (g/m s)	Density (g/cm³)
Acetic acid	6.13	1.12×10^{-6}	1.23	1.049
Acetone	20.7	6×10^{-6}	0.30	0.785
Acetonitrile	36.0	2×10^{-5}	0.34	0.777
Benzene	2.2	5×10^{-5}	0.60	0.874
Cyclohexane	2.0	—	0.90	0.774
Deuterium oxide	77.9	1×10^{-5}	1.09	1.104
Dimethyl-formamide	36.7	2×10^{-5}	0.80	0.944
Dimethyl-sulfoxide	45	3×10^{-6}	1.96	1.096
Dioxane	2.2	5×10^{-5}	1.26	1.028
Ethanol	25	0.14×10^{-6}	1.08	0.785
Ethylenediamine	13	1×10^{-5}	1.54	0.891
Methanol	32.6	1×10^{-5}	0.54	0.787
1-Propanol	20.1	2×10^{-6}	1.93	0.801
2-Propanol	18	9×10^{-6}	2.07	0.781
Propylene carbonate	64.4	1×10^{-4}	2.53	1.19
Pyridine	12	1×10^{-6}	0.88	0.978
Tetrahydrofuran	7.6	4×10^{-8}	0.46	0.880
Water	78.3	6.0×10^{-8}	0.8904	1.000
Vacuum	1.00	0.0	0.0	0.000
Silver	—	62.8×10^{6}	solid	10.50
Copper	—	56.5×10^{6}	solid	8.96
Lead	—	4.5×10^{6}	solid	11.35
Mercury	—	1.044×10^{6}	1.53	13.546
Carbon	—	0.128	solid	2.0
Sulfur	4.0	1.25×10^{-14}	solid	2.0
Polystyrene	2.6	1×10^{-16}	solid	1.06
Glass	7.0	—	solid	2.6

[a]Values from various sources. The SI dynamic viscosity unit of kilograms per meter per second (kg/m s) is equal in cgs units to 10 P (poise), so the values given in this table are in centipoises. Glass values are for borosilicate glass.

It is sometimes more convenient to work with the reciprocal of resistance, which is the *conductance*. The conductance is also given by Ohm's law, now in the form:

$$E = I/G \qquad (1.2)$$

The unit of conductance, the *siemens* (S), is the reciprocal of the ohm. A conductance of one siemens will carry a current of one ampere under a driving potential of one volt.

Resistance and conductance are properties of all matter. If the matter is uniform, then the resistance and conductance are properties only of the nature of the matter and its shape. For uniform conductors, resistance is directly proportional to length and inversely proportional to cross-sectional area. Conductance is inversely proportional to length l and directly proportional to cross-sectional area A:

$$R = \rho l/A \qquad (1.3)$$

$$G = \kappa A/l \qquad (1.4)$$

The proportionality constant ρ is characteristic only of the nature of the matter, since the dependence upon the shape of the conductor has been removed. It is called the *resistivity* and has the units of ohm-meter (Ω-m). *Specific resistance* is an older term for resistivity. The proportionality constant κ, likewise characteristic only of the nature of the matter, is called the *conductivity* and has the units of siemens per meter (S/m). *Specific conductance* is an older term for conductivity. Resistivities and conductivities can be measured for any uniform matter that can conduct an electric current, whether pure elements, pure compounds, mixtures, or solutions (Table 1.4).

In addition to Ohm's law, it is useful to recall that electrical power dissipated along a resistive path is the product of the current flowing through the resistive path and the potential difference driving that current, EI. When the current is in amperes and the driving potential is in volts, the power dissipated is in *watts*; one watt is one joule per second. From Ohm's law the Joule heating of a resistive path by flow of electricity through it is I^2R, since the electrical power dissipated is quantitatively converted into heat. Joule heating is rarely a problem in electroanalytical methods. Low currents are used except in some coulometric applications, and coulometry is not sensitive to small increases in temperature.

By DC electrical relationships is meant those relationships that exist between time-invariant electrical quantities. Any measured electrical quantity such as current or voltage can be divided into two components: a constant or time-invariant part, the *DC component*, and a part that varies with time, the *AC component*. Real quantities may include either or both components.

Resistance in an electrical circuit can be connected in a sequential path or *series circuit*, in which the current must flow through each resistance in turn, or through a branching of alternative paths, that is, a *parallel circuit*. Any resistance path, however complex, can be resolved into these simple equivalent circuits. This is done by application, and when necessary, successive application, of *Kirchhoff's law*, which states that the sum of the currents flowing into any junction of paths must be zero, since there is no net buildup of charge at the junction point. The effect of any purely resistive network seen by an external electrical circuit can thus be expressed as a single resistance value.

In a series circuit the current through the circuit is the same at all points. The total voltage driving the current is obtainable from Ohm's law, the actual current, and the sum of the resistances in the series. The fraction of the total voltage appearing across each resistor is directly proportional to the fraction of the total resistance that the resistor constitutes. Series circuits are used as *voltage dividers* to provide a specific desired lower voltage from a DC voltage source.

In a parallel circuit the same voltage difference appears across the two parallel paths, but the current is proportioned between them, with most of the current taking the path of least resistance. The resistance of a parallel network is necessarily less than that of any of its branches, since the branches are alternative. The *conductance* of the circuit, which is the reciprocal of its resistance, is the sum of the conductances of the alternative paths; therefore, the total resistance R of a parallel circuit is the reciprocal of the sum of the reciprocals of the resistances of its alternative paths:

$$1/R = 1/R_1 + 1/R_2 + 1/R_3 + \ldots$$

The total current flowing in a parallel resistance network is given by Ohm's Law using the applied voltage and the total resistance as calculated above. The fraction of the total current flowing through each resistor is inversely proportional to the fraction of the total resistance that the resistor constitutes. In other words, the fractional current is directly proportional to the fractional conductance.

1.2.3   Electrical Quantities Varying with Time

Any electrical quantity may vary with time, but in electroanalytical practice the quantities of greatest interest are current and voltage. Since the usual manner of measuring a current that varies with time is to pass that current through a fixed pure resistance and thereby obtain a voltage that varies with time exactly as does the current, a discussion of time-varying voltage will suffice.

Time-varying electrical quantities are divided into the two classes of *aperiodic waveforms* and *periodic waveforms*. Aperiodic waveforms are also known as *transient signals*. An aperiodic waveform is a signal that does not repeat itself indefinitely, whereas a periodic waveform does repeat itself indefinitely. Both are significant in analytical electrochemistry.

The simplest periodic waveform is one that varies sinusoidally with time, as shown in Figure 1.1. Sinusoidal displacement of an object with time is called *simple harmonic motion* and is common in nature. A sinusoidal current is

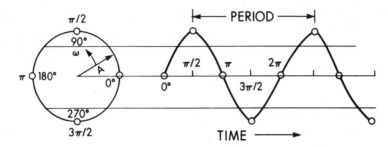

Figure 1.1 Sine wave development from rotating vector. (*A*), amplitude of vector. Units given in degrees and radians.

produced by the rotation of a coil of wire in a uniform magnetic field such as that of a power generator, which is the reason that AC power is obtainable from the mains.

A sine wave is the simplest periodic waveform because, as will be seen later, it is the only periodic waveform that can be characterized by a single *period*, or *frequency*. This figure shows the generation of a sine wave from a rotating vector of magnitude *A*. As the vector rotates counterclockwise with uniform *angular velocity* ω, the sine wave results from projection of the vector on the vertical axis. One complete revolution, which is 360° or 2π radians, of the vector produces one *cycle* of the waveform; the time τ required to do so is the *period* of the waveform. The *frequency* of the waveform is the number of cycles per unit of time, normally per second:

$$f = 1/\tau \tag{1.5}$$

Frequency is usually measured in hertz (Hz), whose older name is cycles per second (cps).

The rotation rate of the vector ω covers 2π radians in one cycle:

$$\omega = 2\pi/\tau = 2\pi f \tag{1.6}$$

The instantaneous current $I(\omega t)$ or I_{AC} is simply the maximum value (amplitude) of the current multiplied by the sine function, and the same relationship exists between the instantaneous voltage $E(\omega t)$ or E_{AC} and the maximum value of the voltage:

$$I_{AC} = I_{max} \sin(\omega t) \tag{1.7}$$

$$E_{AC} = E_{max} \sin(\omega t) \tag{1.8}$$

When two sine waves of the same frequency must be considered at the same time, they may differ only in amplitude and/or in *phase*. A phase difference occurs whenever the two maxima and minima do not occur at the same time. It can most conveniently be measured as the *phase angle* θ, which can have any value from 0°(in phase) through 180°(out of phase) to 360° (also in phase, since it is identical to 0°). Thus, one sine wave voltage out of phase with another could be written as

$$E(\omega t) = E_{max} \sin(\omega t + \theta) \qquad (1.9)$$

and if it were 90° out of phase could be written as

$$E(\omega t) = E_{max} \sin(\omega t + (\pi/2)) \qquad (1.10)$$

Only sine wave signals that are in phase add their amplitudes directly, as scalar quantities. All others add vectorially, so that signals out of phase by exactly 180° interfere destructively, and thus their amplitudes are subtracted.

1.2.4 Capacitive Reactance

Capacitances offer an infinite resistance to DC current or voltage, but they pass and react against a time-varying voltage, thereby producing *capacitive reactance*. Capacitive reactance arises because a capacitance cannot change the voltage across itself except by a flow of charge, a current. To change the potential across a capacitance instantaneously would require the impossibility of an infinite current. A capacitance stores energy in the form of charge separated by distance, and this is the energy that reacts against changes in the potential across the capacitance. For a capacitance:

$$I = dQ/dt = C\,dE/dt \qquad (1.11)$$

$$I_{AC} = I_{max} \cos(\omega t) = \omega C E_{max} \cos(\omega t) \qquad (1.12)$$

The current therefore leads, or is out of phase with, the voltage by the difference between the sine and cosine functions, which is 90°. The *capacitive reactance* X_C is in the form of Ohm's law equal to the ratio between the maximum potential difference and the maximum current, or

$$X_C = E_{max}/\omega C E_{max} = 1/2\pi f C \qquad (1.13)$$

Reactance has the same units, ohms, as does resistance. The capacitive reactance is, as is seen from the formula above, frequency dependent. It decreases (from infinity) as the frequency increases. Capacitive reactance adds directly for capacitances in series and inversely for capacitances in parallel, since capacitance and capacitive reactance are inversely proportional to each other.

The energy stored in a capacitance is stored or discharged by addition or removal of charge. If a charge Q is stored in a capacitance at a voltage E, the energy dischargeable from that capacitance is $QE/2$ or $CE^2/2$ as can be shown by integration over the discharge curve. In a physical parallel-plate capacitance, real or ideal model, the capacitor is viewed as two parallel plates separated at some distance x by an insulating material called a *dielectric*. The energy that can be stored in such a capacitor is a function of the geometric properties of the capacitor. If the distance of separation x is doubled, the energy stored in the capacitor doubles also because the same charge is now separated by twice the distance. This cannot happen if the capacitor is isolated, however, nor can the charge Q be changed since there is no path by which charge can enter or leave. Since energy is conserved, the energy relations given above show that the voltage must double, and the capacitance must therefore be halved. The capacitance of a parallel-plate capacitor is thus inversely proportional to the distance x that separates the plates.

If the plate area A of an isolated charged capacitor is doubled, the same charge is distributed over twice the area and thereby produces only half the electric field it did before. The halving of E at constant Q forces doubling of the capacitance, hence

$$C = A\epsilon/x \qquad (1.14)$$

The proportionality constant ϵ is known as the *permittivity*. When the two parallel plates are separated only by a vacuum, the proportionality constant ϵ_0 is known as the *permittivity of free space* or the *permittivity of vacuum*, and has the value of 8.854×10^{-12} F/m. If the parallel plates are separated by a dielectric medium other than vacuum, the *permittivity* ϵ of that medium rather than the permittivity of free space ϵ_0 must be used in the capacitance equation. The permittivity of other media is greater than the permittivity of free space. A *relative permittivity*, or *dielectric constant*, ϵ_r characteristic of the insulating medium can be defined

$$\epsilon_r = \epsilon/\epsilon_0 \qquad (1.15)$$

in which case the capacitance equation becomes

$$C = \epsilon_r\epsilon_0 A/x \qquad (1.16)$$

Capacitances, or capacitors, are commercially available as electronic components in the form of plates separated by a dielectric film. Their values range from millifarads down to a few picofarads. Such real capacitors have very small leakages of current through them, and the dielectric film will break down if the maximum voltage it is designed to withstand is exceeded.

1.2.5   Inductive Reactance

Inductances react against time-varying currents to produce *inductive reactance*. An inductance stores energy in the magnetic field established by the flow of current. The energy so stored reacts against any change in the current since a change in the current must also alter the field. To change the field instantaneously would require the impossibility of an infinite applied potential difference.

Inductance is not as significant in electroanalytical chemistry as is capacitance. The effect of a pure inductance is the reverse of that of a pure capacitance in that the current lags the voltage by 90°, the difference between the sine and cosine functions. The *inductive reactance* is given by

$$X_L = 2\pi f L \tag{1.17}$$

where L is the inductance in henrys. A perfect inductance therefore offers zero resistance or zero reactance to a DC current.

Inductances, called *inductors* or *chokes*, are commercially available as electronic components in the form of coils of wire that may contain a ferromagnetic core. Their values range from hundreds of henrys down to a few microhenrys. Real inductors always have small capacitances and resistances in addition to their inductances.

1.2.6   Impedance

The *impedance* offered by an electrical circuit is the sum of its resistance, capacitive reactance, and inductive reactance. The resistive component of the impedance behaves identically with constant and time-varying electrical signals, but the capacitive and inductive components do not. Capacitive reactance decreases with frequency, while inductive reactance increases with frequency.

It should be noted that, unlike resistance, impedance is a complex quantity that for full expression requires specification of two values rather than one. The complex quantity may be considered vectorial, in which the components are the absolute magnitude and the phase angle, or it may be considered to be to lie in a complex plane whose perpendicular axes are the real (resistive) component and the imaginary or complex (capacitive–inductive) component. Thus the statement that the total impedance of a circuit is the sum of its resistive, inductive, and capacitive components is correct only if the quantities are added vectorially. Since the current–voltage phase shift in a capacitance is 90° and the current–voltage phase shift in an inductance is 90° in the opposite direction, they differ by 180° and are *out of phase*. The scalar magnitude of an impedance is then given by

$$Z^2 = R^2 + (X_C - X_L)^2 \tag{1.18}$$

This equation is not a full expression of the vectorial or complex impedance.

Similarly, use of Ohm's law in the impedance form

$$E = IZ \tag{1.19}$$

requires that the potential and the current also be considered as complex quantities.

The reciprocal of impedance, called the *admittance*, is also a complex quantity, and requires complex potential and current when Ohm's law is used in the form

$$E = I/Y. \tag{1.20}$$

1.2.7 Nonsinusoidal Waveforms

Any periodic waveform, however complex, can be considered to be made up as a sum of sinusoidal components differing in amplitude and in phase angle. When the sine waves summed to make up a nonsinusoidal waveform are all multiples of some single fundamental frequency, the series is known as a *Fourier series*, and the multiple frequencies are known as *harmonics*. Any single-valued periodic waveform can be represented as a *Fourier series expansion*. In a Fourier series expansion, the harmonics differ only in amplitude and in phase. A time-dependent voltage might be represented as

$$E = E_{DC} + E_1 \sin(\omega t + \theta_1) + E_2 \sin(2\omega t + \theta_2) + ... \tag{1.21}$$

In this equation, the *DC level* E_{DC} is the average value about which the signal varies; the other terms in E are the maximum values of the amplitude of the fundamental, second harmonic, ..., n-th harmonic; and the terms in θ are the corresponding phase angles.

In electroanalytical chemistry the most useful nonsinusoidal periodic waveform is the square wave. The construction of a square wave from its sinusoidal components is shown in Figure 1.2. The Fourier series expansion of a periodic square-wave voltage whose peak voltage is E_p includes only the odd harmonics:

$$E = 4E_p[(\sin \omega t)/\pi + (\sin 3\omega t)/3\pi + (\sin 5\omega t)/5\pi + ...] \tag{1.22}$$

The sawtooth waveform, which consists of a series of ramps, can be represented by a Fourier series in both odd and even harmonics:

$$E = 2E_p[(\sin \omega t)/\pi - (\sin 2\omega t)/2\pi + (\sin 3\omega t)/3\pi - ...] \tag{1.23}$$

The higher-frequency components are required to produce the sharp rises and falls of the leading and trailing edges of the pulses. Any circuit, electrical or electrochemical, that does not pass both higher and lower frequencies with equal ease will distort a square-wave or sawtooth signal. A sinusoidal signal will not be

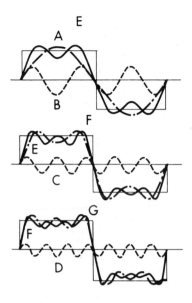

Figure 1.2   Construction of square wave from harmonic components. (*A*), fundamental; (*B*), third harmonic; (*C*), fifth harmonic; (*D*), seventh harmonic; (*E*), fundamental plus third; (*F*), fundamental plus third plus fifth; (*G*), fundamental plus third, fifth, and seventh. Square wave resultant is shown in lighter line.

distorted if its fundamental (and only) frequency, high or low, is passed by the circuit, although the amplitude may be reduced.

1.2.8   Transient Signals

An aperiodic or transient signal cannot be resolved into periodic waveforms in the manner described above. It can be resolved into frequency components, but the frequencies form a continuous spectrum rather than a series of discrete frequencies. This resolution is called *Fourier transformation*. Fourier transformation is normally carried out by calculation using a digitized form of the analog transient signal, and requires both analog-to-digital conversion and use of a digital computer. The accuracy of the transformation depends upon the number of points digitized, the accuracy of the analog-to-digital conversion, and the computing facilities used.

1.2.9   Circuit Response to Signals

A purely resistive circuit responds to signals that vary with time in exactly the same way it responds to signals that do not vary with time. A circuit that includes reactance does not. The general response of reactive circuits to time-variant signals is beyond the scope of this text, but the simple case of a voltage step function applied to a series *RC* circuit can be treated.

When a voltage step is applied to a series RC circuit, the voltage is apportioned between the resistance and the capacitance in the manner of a voltage divider. Initially none of the voltage appears across the capacitor and all of it across the resistor since for a capacitor $E = Q/C$ and the initial charge Q is zero. As charge flows into the capacitor, E_C increases exponentially toward the total applied voltage of the step and E_R therefore decreases exponentially to zero. Mathematically, since $E = E_C + E_R$,

$$E_C = \exp(-t/RC) \tag{1.24}$$

$$E_R = E - \exp(-t/RC) \tag{1.25}$$

The value of the product RC gives the time required for this to happen in terms of seconds ($\Omega \times F = V\,s/C \times C/V = s$). Hence the RC product is also known as the *time constant* τ of the circuit.

The potentials across the capacitive and resistive parts of the voltage divider are given in terms of time constants in Table 1.5. The use of parallel RC networks gives similar results, which are usually interpreted in terms of their electrically equivalent series circuits. Electrochemical cells are also often treated as series RC circuits.

TABLE 1.5   Response of Series RC Circuit to Voltage Step E

Time t	E_C	E_R
1.0τ	$0.632E$	$0.368E$
2.0τ	$0.865E$	$0.135E$
2.3τ	$0.900E$	$0.100E$
3.0τ	$0.950E$	$0.050E$
4.0τ	$0.982E$	$0.018E$
4.6τ	$0.990E$	$0.010E$

The response of RC circuits to more complex voltage or current functions can be calculated on the basis of the time constant of the RC circuit and the frequency or frequencies of the applied waveforms. Only for simple applied waveforms of low amplitude can such an approach be useful in electroanalytical work. As amplitude increases, simple series RC models of electrochemical cells are no longer able to adequately model the behavior of the actual cell, because actual cells are nonlinear devices that can be approximately treated as linear only when the excitation signals are small.

1.3   ELECTRONIC BUILDING BLOCKS

The generation, control, and modification of electrical signals is necessary in order to obtain electrochemical data in useful form. The brief introduction to electronic building-block circuits here is sufficient for understanding the apparatus and techniques; excellent texts are available for further study (see References G1-G3 at the end of Chapter 2).

1.3.1   Power Supplies

Electronic circuits require electrical power which can be provided either by batteries or from the AC mains. In most laboratory equipment the AC mains are the source of the power required for operation; the AC power is converted to DC power as required by transformer-filter circuits called *power supplies*. These supply the required power and any necessary constant voltages (*bias voltages*) for circuit operation. Both power supplies and batteries provide voltages relative to a *circuit ground*, usually the power supply negative return or the battery negative terminal.

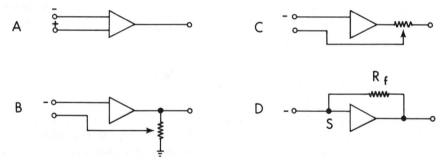

Figure 1.3   Amplifier circuits with negative feedback. (*A*) general symbol for amplifier; (*B*) voltage feedback; (*C*) current feedback; (*D*) operational feedback. The positive (noninverting) input is not used, and not shown, in circuits (*B*), (*C*), and (*D*).

Use of multiple power supplies with different circuit grounds in interconnected circuits may lead to *ground loops*, in which current flows between two ground points because there is a small potential difference between them. The effects of ground loops on sensitive measuring circuits are unpredictable, deleterious, and extremely difficult to track down or remove. Only *one* circuit ground should exist in an electronic measuring system, however complex it may be.

1.3.2   Amplifiers

An *amplifier* is an electronic circuit whose purpose is to increase the amplitude of an electrical signal using power from an external source. Amplifiers can be designed to amplify current, voltage, or both. For a voltage amplifier, the amplification factor, or *gain*, is the ratio of output voltage to input voltage; for a current amplifier, the gain is the ratio of output current to input current. An ideal voltage amplifier has infinite *input impedance* and thus zero input current, so there is no load on the signal (input) source and its output voltage is independent of output current and thus of output load. An ideal current amplifier has zero input impedance, and thus its input voltage is zero, and its output current is independent of output voltage. Real amplifiers lie between these two extremes.

The output of an amplifier can be of the same polarity as the input signal or of opposite polarity. Some amplifiers provide both polarities of outputs and inputs. The input signal may be a single signal (relative to circuit ground) or two signals. In the latter case the amplifier is called a *difference amplifier* because it amplifies only the difference between the two input signals. In an ideal difference amplifier, no signal common to both inputs should appear at the output. The deviation of real difference amplifiers from ideality in this respect is given by the *common mode rejection ratio*, or CMRR. The CMRR is the ratio of the gain for a difference signal to the gain for a common signal, and should be as large as possible.

The general symbol for an amplifier is given in Figure 1.3. Only two inputs and a single output are shown; ground lines, bias lines, and connections to the power supply are understood to be present. Such an amplifier may consist of a single integrated circuit containing several cascaded stages of amplification. The manufacturer's specifications include the supply voltage(s), the range of acceptable input signals, and the output into load, as well as the gain.

Amplifiers are often used in configurations in which all or part of the output is fed back to the input. If the output is fed back to the noninverting (+) input, it is called *positive feedback*. Positive feedback increases the gain of the amplifier, but its effect on virtually all other desirable characteristics of the amplifier is deleterious, and it is rarely used. *Negative feedback* occurs when the output is fed back to the inverting (−) input of the amplifier. This reduces the gain but improves the stability, frequency response, linearity, input resistance, noise, and virtually all other desirable characteristics of the amplifier. Feedback of voltage has the same effect for voltage amplifiers as feedback of current has for current amplifiers.

Negative feedback is possible in three modes: voltage feedback, current feedback, and operational feedback, as shown in Figure 1.3. In both the current feedback and voltage feedback circuits, a portion of the output is fed back in series with the input. In the operational feedback mode, the output voltage is connected through the *feedback resistor* R_f to the amplifier input in parallel with the input voltage. The effects of operational feedback are the same as those of voltage feedback except that the input impedance is low. The connection point designated S is called the *summing point*. An amplifier of high quality and high gain used in operational feedback mode is called an *operational amplifier*. Operational amplifiers together with resistors, capacitors, and other circuit elements are capable of mathematical operations and can be used as building blocks to construct *analog computers*. More important in electroanalytical chemistry is their key role in control and data-acquisition instrumentation.

1.3.3   Applications of Operational Amplifiers

A selection of basic operational amplifier circuits useful in electroanalytical instrumentation is shown in Figure 1.4. The basic measurement circuit, Figure 1.4*B*, is a current-measuring circuit in which the summing point S is maintained at a ground potential by the output signal fed back via R_f. Such a summing point is called a *virtual ground* because it is held at ground potential though it is not physically connected to ground. Any current drain or impedance placed between the output of circuit 1.4*B* and ground will not affect the input of circuit 1.4*B*, and thus the circuit serves to isolate the current source at its input from the meter or other current-measuring device used.

Circuit 1.4*A* accomplishes the same isolation for voltage measurement. The resistor R to virtual ground at point S converts the voltage input into a current input. This current input is measured in the manner just discussed. The voltage input, like the current input, will be unaffected by the actual measuring device used. The value of resistor R is selected on the basis of the type of voltage input and the currents that the operational amplifier can handle.

Both circuits 1.4*A* and 1.4*B* can be used for the measurement of the sum of several currents or voltages, since several simultaneous inputs to the summing point are possible. Voltages must be summed as currents through input resistors.

Circuits 1.4*C* and 1.4*D* are used for analog computation or, in electroanalytical work, integration of current to give charge and for differentiation of input and output signals. Circuit 1.4*C* is a differentiating circuit. As the input voltage changes, the summing point can be held at ground potential only by passage of current back through R_f. This current is proportional to the rate of change of the input voltage and also to the magnitude of the input capacitance. Any constant voltage is blocked by the capacitance, and no feedback current is produced by its presence. Since the output voltage is proportional to the output current that is fed back, it is proportional also to the rate of change of the input voltage:

$$E_{out} = -RC \, dE_{in}/dt \qquad\qquad (1.26)$$

This differentiating circuit, like any other, serves to amplify noise, because noise has components with a high rate of change of voltage with time. Some discrimination against this is possible by putting a small capacitance in parallel with R_f or a small resistance in parallel with the input capacitance. The differentiating effect will not be *seriously* affected unless the signal being differentiated is a high-speed signal.

The integrator circuit Figure 1.4*D* simply accumulates the feedback current on the capacitor C in the feedback loop. The potential across the capacitor is proportional to the charge upon it and can be monitored by any measuring device

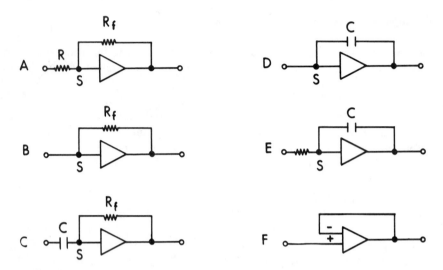

Figure 1.4 Useful operational amplifier circuits. (*A*) voltage measurement; (*B*) current measurement; (*C*) differentiation; (*D*) integration; (*E*) sweep generation; (*F*) voltage follower. Summing point S shown where present. Voltage follower circuit (*F*) uses noninverting input; this input is not used, and not shown, in circuits (*A*) to (*E*). Switching circuit used to discharge capacitor C prior to use is not shown in circuits (*C*) to (*E*).

that does not draw a significant input current. Provision must be made to remove charge from the capacitor between integrations by short-circuiting it. Use of this circuit over many seconds can give erroneous results because of current leakage across the capacitor or stray currents of very low level integrated over the measurement time.

Circuit 1.4*E*, a modification of the integrator, will provide a ramp voltage because of the integrating effect of capacitor C when a constant voltage is applied to its input.

The *voltage follower* circuit Figure 1.4*F*, differs from the other operational amplifier circuits discussed above in that there is no summing point and no virtual ground. The output voltage is fed back to the *inverting* input while the input voltage uses the *noninverting* input. The effect of this negative feedback is that the gain of the amplifier is essentially unity, the input impedance of the amplifier is very high, and the output impedance seen by the load is very low. The input signal is effectively isolated from any measurement circuit.

1.3.4 Digital Logic

A *digital* circuit, component, or signal is one in which only two values, or *logic levels*, are possible. These levels are designated 0 (false) and 1 (true). The voltage levels corresponding to these may vary with the circuit, but in digital circuits there either is an acceptable signal or there is not, whereas in analog circuits such as amplifiers a range of signal levels must be dealt with. Digital circuits thus operate in the binary number system. An array of logic levels organized in either space

TABLE 1.6   ASCII Code Representation

Lower Nybble	Hex 0	Hex 1	Hex 2	Hex 3	Hex 4	Hex 5	Hex 6	Hex 7
0	NUL	DLE	SP	0	@	P	`	p
1	SOH	DC1	!	1	A	Q	a	q
2	STX	DC2	"	2	B	R	b	r
3	ETX	DC3	#	3	C	S	c	s
4	EOT	DC4	$	4	D	T	d	t
5	ENQ	NAK	%	5	E	U	e	u
6	ACK	SYN	&	6	F	V	f	v
7	BEL	ETB	'	7	G	W	g	w
8	BS	CAN	(	8	H	X	h	x
9	HT	EM	)	9	I	Y	i	y
A	LF	SUB	*	:	J	Z	j	z
B	VT	ESC	+	;	K	[	k	{
C	FF	FS	'	<	L	BKS	l	|
D	CR	GS	−	=	M	]	m	}
E	SO	RS	.	>	N	↑	n	~
F	SI	US	/	?	O	←	o	DEL

(*parallel data*) or time (*sequential* or *serial data*) corresponds to one or a series of numbers in the binary number system. For example, the parallel array 00100011, read left to right, is $2^5 + 2^1 + 2^0 = 35$; the same serial array in the same order would have the same value. Any given place, corresponding to a power of 2 in the binary number system, is called a *bit*. The maximum decimal number that can correspond to a given array of n bits is $2^n - 1$. The maximum decimal number corresponding to an 8-bit array such as the example above is 255, but 256 unique states are possible since 00000000 corresponds to decimal zero.

Groups of binary bits frequently used in digital logic are given certain names. A group of 4 binary bits is a *nybble* and a group of 8 is a *byte*. One byte or more, used as a single entity, is called a *word*. The precision of data contained in a word depends on the number of bits in the word, the *word length*. One byte has a precision of 1/256 or 0.4%; a 12-bit word has a precision of 1/4096, or 0.02%; while a 16-bit word has a precision of 1/65,536, or 0.002%. Data in this form are said to be in *integer mode*.

Real mode, in which noninteger data are stored, consists of two integer sections, one containing the value and the other containing its exponent. The length of the sections depends upon the particular computer in which they are used. For example, in the Commodore PET microcomputer, numbers in integer mode are stored in two successive bytes, most significant first. The most significant bit of the most significant byte is taken as the sign bit, so the acceptable range of integers is from +32,767 to –32,767. Numbers stored in real mode, also called *floating-point mode*, are stored in 5 successive bytes of which the first contains the exponent (including its sign) and the remaining 4 contain the mantissa, in order of significance. The sign of the mantissa is carried in the most significant bit of the most significant byte of the mantissa, which is the second byte of the 5. This machine can store real numbers of either sign whose magnitude is between 10^{-39} and 10^{+38}. The precision of the mantissa is 31 bits, which is equivalent to about 9 significant decimal digits.

For human convenience, straight binary numbers are often given in modified form. One method, grouping binary bits into groups of 3 and assigning each group its decimal equivalent, is called *octal notation* because 8 numbers are possible—the integers 0 through 7. The more useful *hexadecimal*, or simply *hex*, *notation* groups the binary bits into groups of 4 and assigns each an equivalent symbol, using the numbers 0 through 9 followed by the letters A through F. Representation of 1 byte requires 2 hexadecimal characters. A third representation, *binary-coded decimal*, or BCD, notation, uses a group of 4 bits to represent each digit of an ordinary decimal number. Instruments often have output jacks that present the data in BCD form because it can easily be obtained in that form from the circuits driving a digital display unit. These forms must all be converted into straight binary form before they can be used in digital calculations.

Alternatively, a group of bits may be taken to represent a single alphabetic or numeric character. The almost universal code used is the American Standard Code for Information Interchange, referred to as *ASCII notation*, adopted in 1968. Older IBM equipment uses a similar code called Extended Binary Coded Decimal Interchange Code (*EBCDIC notation*), although this is apparently being slowly phased out in favor of ASCII notation. In ASCII notation, seven bits are used and thus there are 128 possible characters as shown in Table 1.6. The eighth (and most significant) bit of a byte is taken as the *parity bit*, which often is not implemented (*no parity*). Parity is simply redundant coding which acts as an error check. In the seven bits of an ASCII code there are either an even or odd number of logic ones. An eighth bit is added so as to make the number of ones always even (*even parity*) or always odd (*odd parity*). The parity bit can then be checked against what it should be on the basis of the other seven bits every time the character is recorded or stored. A mismatch (*parity error*) indicates an error of at least one bit.

Some of the characters of the ASCII code are used for control purposes by various devices (Table 1.7). These are often produced on a terminal by striking a control key and an alphabetic key in sequence. For this reason SOH is also known as control-A, STX is known as control-B, and so on. Characters which are not included in the ASCII set are coded differently by different manufacturers, and the eighth (and most significant) bit of a byte may then be used to distinguish the ASCII set from the additional character codes.

1.3.5   Gates and Digital Signals

The logic of digital signals is *Boolean logic*, in which the symbol 1 stands for true (signal present), and the symbol 0 stands for false (signal not present). Electronic components capable of carrying out digital logic operations are called *gates*. A gate corresponds to a particular logic operation. Thus an *AND gate*, corresponding to the "and" logical operation, has a 1 (true) output if and only if *each* of its two (or more) inputs is true. An *OR gate*, corresponding to the inclusive "or" logical operation, has a true output if *any* of its two (or more) inputs is true. Gates that invert their output perform the logical "not" operation, and with inversion the two

TABLE 1.7   ASCII Control Characters

Code	Meaning	Code	Meaning
NUL	Null	DLE	Data Link Escape
SOH	Start of Heading	DC1	Device Control 1
STX	Start of Text	DC2	Device Control 2
ETX	End of Text	DC3	Device Control 3
EOT	End of Transmission	DC4	Device Control 4
ENQ	Enquiry	NAK	Negative Acknowledge
ACK	Acknowledge	SYN	Synchronous Idle
BEL	Bell	ETB	End of Transmitted Block
BS	Backspace	CAN	Cancel
HT	Horizontal Tab	EM	End of Medium
LF	Line Feed	SUB	Substitute
VT	Vertical Tab	ESC	Escape
FF	Form Feed	FS	File Separator
CR	Carriage Return	GS	Group Separator
SO	Shift Out	RS	Record Separator
SI	Shift In	US	Unit Separator
SP	Space	DEL	Delete

gates described above become *NAND gates* and *NOR gates*. Gates that only invert, *NOT gates*, are uncommon; this logical operation is often performed by NAND gates with one input held true. Any Boolean logical operation, however complex, can be preformed exclusively with NAND gates or with NOR gates. Advantage is taken of this in the design of more complex digital circuits. The symbols for gates are shown in Figure 1.5.

Modern logic circuits are designed as integrated circuit chips that usually operate from +5-V DC power supplies. They consider 0 V (ground) as logic 0, or false, and +5 V (supply power) as logic 1, or true. The exact ranges of cutoff depend upon the type of integrated circuit and are given in the manufacturer's specifications. Most logic circuits are designed to operate with *transistor-to-transistor logic*, or *TTL logic*. Gates using TTL logic have a *propagation delay*, or time between change at the input and change at the output, of about 10 ns. They have a *fanout* (number of outputs which can be fed from a single input, or the reverse) as high as 15, although half that is a more reasonable figure. The difference between the maximum effective input signal for logic state 0 and the minimum effective input signal for logic state 1, the *noise immunity* of the gate, can exceed 1 V. These figures are superior to those of most competing logic circuits. Logic circuits are usually designed to drive one *TTL load*, or load equivalent to one TTL logic gate, at the output. Additional circuits may be required if more than one input line is connected to a single output line.

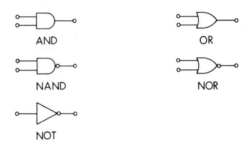

Figure 1.5 Symbols for logic gates.

1.3.6   Digital-to-Analog Converters

Digital data are not directly usable to produce many useful analog control signals required in electroanalytical work, such as the input control voltage to a potentiostat. Digital data output is often less convenient for recording or human interpretation than are the same data in analog form. For these two reasons, *digital-to-analog converters* are used. A digital-to-analog converter, or DAC, is a single integrated circuit (or package of circuits) that accepts parallel digital data inputs and gives an analog voltage output.

Most modern DAC devices can be set up, or *configured*, in different ways. They all have a *resolution* that is determined by the number of parallel inputs (bits) the device can accept. The precision of the output is inversely proportional to the word length the DAC can accept as discussed earlier. The most common DAC devices have a resolution of 8, 10, 12, 14, or 16 bits. Cost increases sharply with resolution. An 8-bit DAC can be used directly with a standard 8-bit output port; latching of the digital signal can be done either by the port or by the DAC circuit. Use of DAC devices of greater precision requires either time-multiplexing of the data through a single 8-bit port or use of two ports, one of which carries the most significant byte and the other the least significant byte. Either alternative requires additional programming of the computer.

A DAC may be configured in *unipolar* mode, in which the output voltage ranges (generally positively) upward from zero (at, for an 8-bit DAC, 00000000) to maximum (at 11111111), or in *bipolar mode*, in which the output voltage ranges from negative maximum (at 00000000) through zero (at 10000000) to positive maximum (at 11111111). The magnitude of the maximum output voltage depends upon the supply voltage and the configuration used. Common ranges of output voltages are ±1 V and ±10 V. It is common practice to attach a power amplifier to the output of the DAC, since the DAC itself usually is not designed to be capable of providing significant power to external circuits.

The basic circuit of a DAC is an operational amplifier used to sum the input voltages from the parallel digital input lines through the appropriate array of binary-weighted summing resistors, as shown for a four-bit DAC in Figure 1.6.

Since the input lines are all driven from the same supply, they all have the same (or zero) input voltage.

Certain DAC devices can be configured to accept data in BCD form as well as in straight binary code. These are usually designed to accept three or three and one-half BCD decimal digits. These devices are useful in providing output to, say, a recorder from a circuit designed to give digital readout on a BCD instrument display.

Some BCD devices have a zero or offset analog voltage input that can be used to establish 0.0 V output for the appropriate (00000000 or 10000000) digital input. Alternatively, zero offset can be provided by additional bias circuits at the output of the following power amplifier. If the DAC output is being used to drive a potentiometric x–y or x–t recorder, the zero controls on the recorder may suffice for this purpose, but for other applications a zero-offset bias circuit will almost certainly be needed.

1.3.7   Analog-to-Digital Converters

An *analog-to-digital converter* is necessary in order to transform signals from the real analog world into digital form that can be manipulated by digital computers. Any analog signal can be used if it is first converted to an appropriate analog *voltage* signal, which is then applied to the input of an analog-to-digital converter (ADC). An ADC is a single integrated circuit (or package of circuits) that accepts an analog voltage input and gives parallel digital data outputs. These outputs may be in straight binary or in BCD form.

The use of ADC circuits is not as straightforward as is the use of DAC circuits. This is because the DAC circuit is a completely analog circuit and will respond to the data on the digital inputs with no fixed timing delays. The use of latched inputs is really necessary only to free the data lines for other purposes while the output analog signal is being used. If the inputs change more rapidly than the amplifier can respond, which is unlikely, the analog voltage output of a DAC will not reach stability. This is easily resolved by an input latching circuit that holds the digital data until the analog output is no longer needed. In ADC circuits, however, it is necessary not only to latch or hold the analog input signal but also to wait for the digital output, because there is a finite *conversion time* between the appearance of a stable analog voltage at the input of the ADC and the appearance of a stable digital output on its output lines.

The DAC circuit *may* have a single control line that activates the latching input at some point whose time is controllable. The ADC circuit *must* have control lines that latch the input and start the conversion process. Most ADC circuits also give an output signal when the conversion process is complete and stable digital data are available at the output. These signals are used for communication with the microprocessor CPU to ensure that the ADC receives the input at the desired time and puts the correct data out on the data lines.

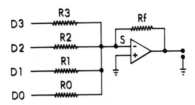

Figure 1.6   Four-bit unipolar DAC circuit. If the value of R3 is R, that of R2 is $2R$, that of R1 is $4R$, and that of R0 is $8R$; the feedback resistor R_f is made equal to $(8/15)R$. Analogous binary weighting is used for DAC devices of greater precision. The value of the output voltage depends on the input voltage as well as on the digital logic.

The precision of an ADC is specified in the same way as is that of a DAC, as a *resolution* in terms of bits; it is ±1 bit in the least significant bit (LSB), or ±½ LSB. The *conversion rate*, another name for the conversion time, is an important parameter, which may range from a few milliseconds to 100 ms and more. At the time of writing, ADC circuits with resolution of 8 bits and conversion times below 40 microseconds are common. The cost of an ADC increases if better resolution or shorter conversion time is required and increases very markedly if both are required.

Whereas DAC circuits require input in parallel form and most ADC circuits output data in parallel form, some ADC circuits can output data in serial form. The two types of ADC circuits used are the *successive-approximation* type, which is the most popular, and the *dual-slope*, or *integrating*, type. The successive-approximation type compares the analog input voltage to the output voltage of an internal DAC, successively setting the bits of an internal register from most significant bit to least significant bit as 1 or 0, depending on whether the output of the DAC whose input is the register is greater or less than the input voltage. The *convert pulse* from the CPU starts the process, and a *status pulse* is put out by the ADC when the conversion is complete. This communication occurs on the control lines of the CPU. The same control arrangement is used with the dual-slope type of ADC, which is basically an integrating counter. The input voltage is gated by a switch into pulses at a preset frequency, and these pulses are integrated over some fixed period of time. An internal reference voltage is then treated in the same way, but with opposite polarity so that the integrator returns to zero. The number of pulses of the reference voltage required to return the integrator to zero is directly proportional to the input voltage and is in digital form. The magnitude of the reference voltage serves as a scaling mechanism.

Although it is not, strictly speaking, an ADC circuit, a voltage-to-frequency converter will serve as one with some assistance from the CPU. The voltage-to-frequency converter produces an output train of square-wave pulses whose frequency is directly proportional to the analog voltage input. The proportionality constant is a property of the unit design and is usually between 10 and 100 kHz/V. The pulses can be counted for a fixed period of time, usually under CPU control, and the output count is directly proportional to the input voltage. This method of analog-to-digital conversion has a high resolution, but is slower than the other types of ADC circuits and imposes an additional burden on the CPU.

1.4 COMPUTERS

Electrochemical data are analog data, which by appropriate use of analog-to-digital converters may be transformed into digital data suitable for computer processing. This processing can be done by general-purpose large computers such as the IBM mainframes, by general-purpose minicomputers such as the DEC PDP series, or by microcomputers. The computer can be either a general computer used also for other purposes or a specifically dedicated machine. Direct acquisition of experimental data by general-purpose large computers, even in time-sharing systems, is wasteful of their time and is not cost-effective. The author has used file transfer of data acquired on a DEC PDP-11 to an IBM System/370 operating the time-sharing MTS system, thus securing the power of a large, central mainframe computer for data treatment while using a minicomputer system for data acquisition.

Use of minicomputers or microcomputers for data treatment and/or control of the electrochemical experiment is now a frequent practice. No one system or group of systems appears to be uniquely preferable. With modern integrated circuit technology, the boundary between minicomputer and microcomputer is no longer distinguishable. The discussion of microcomputers that follows applies, with minor extension, to minicomputer systems as well.

A microcomputer consists of a *central processing unit* (CPU), which can carry out instructions coded in binary numbers, *input–output* units (I/O), which connect the microcomputer to the outside world, and *memory blocks*, which contain data or instructions upon which the CPU will operate. These units, which consist of one or more integrated circuit chips, are linked together by groups of parallel wires called *buses*, which carry signals between them. Laboratory use of microcomputers is basically a matter of getting the data signals on, and the control signals off, the right bus at the right time. Buses are found internal to integrated circuit chips, external to the chips, and at the level of the boards upon which the chips are mounted. Those buses internal to integrated circuit chips are inaccessible to the user, but the buses at the external chip level and board level are both accessible and useful.

The design of microcomputers is dominated by the design of their CPUs. A CPU is almost always a single integrated circuit chip. These CPU chips are available with data buses 4 bits wide, 8 bits wide, and 16 bits wide; the latter are just coming into use. Four-bit CPU chips are used in game and control applications, but are not useful for serious scientific work, and 16-bit microprocessors are not yet widely available, so only 8-bit chips need be considered.

There are four families of chips (CPU plus related units) that dominate the 8-bit microprocessor market. These are the 8080 family (Intel), which includes the generally compatible Z80 group (Zilog); the 6500 family (Commodore); the 6800 family (Motorola); and the 1800 family (RCA). The 1800 family is useful for control applications, is low in cost, and requires less electric power to operate because of its CMOS construction, but it is more limited in computing power. The 6500 family has less computing power than the 8080 family or 6800 family, but is quite respectable. The 8080 family and 6800 family, which contain the greatest number of members, also contain the CPU chips of highest computing power. Units belonging to one family can be used, with greater or lesser difficulty, with units belonging to another family.

1.4.1 Buses: IEEE-696

Design of a complete data acquisition or control system based on one of these families from the chip level, while perfectly feasible, is extremely time-intensive and not cost-effective, because reasonably well-designed complete one-board microcomputer systems are available for less than $500 from commercial sources. Use of a commercial system means that the basic bus design at the external chip level is complete, and the user must either connect to the system as it stands or make connections only at the board level. The ease of the latter will depend upon the external bus connections (if any) provided by the board designer. Unfortunately, virtually all board manufacturers use their own individual buses. The sole exception to this is the hobbyist S100 bus, which has been standardized (IEEE-696). This 100-connector bus, originally designed for the 8080 family, is usable with other families as well. Most other buses have, and need, fewer connectors and lines. The bus structure of a system is more important than the CPU it contains for most data-acquisition/control applications.

Microcomputers using more than a single board are, or should be, bus based. In such machines, separate boards are usually found for the CPU and its ancillary units, for I/O, and for memory. Eight-bit CPU chips can generally access 64K (65,536 bytes) of memory; the 6809 and 16-bit CPU chips can access more. There are four functional types of memory: *random access memory* (RAM), also but less commonly known as *read–write memory* (RWM); *read-only memory* (ROM); *programmable read-only memory* (PROM); and *erasable programmable read-only memory* (EPROM). Usually only RAM and EPROM are the concern of the scientific user, with PROM and ROM being used by the designer to contain operating instructions for the CPU, interpreters such as BASIC, or other unchanging code. Memory boards may contain any combination of these types of memory, although most are designed for either RAM or ROM rather than both.

The external chip buses, which connect to the bus at board level, are of three types. In 8-bit CPU systems the *data bus* consists of 8 lines, to carry 1 byte in parallel form. The *address bus* consists of 16 lines to address 65,536 memory locations. The *control bus* consists of several lines whose number and function depend upon the particular CPU. Some CPU chips time-multiplex together all or

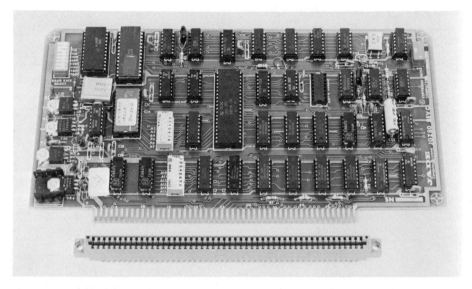

Figure 1.7 Typical design of S100 (IEEE-696) bus board. Board (above) is inserted into connectors (below) on backplane or base support. Board shown is Z80 CPU board (Jade Computer Products) used in author's laboratory. Note on-board regulators on far left, monitor ROM and on-board USART (left), and blank DIP socket (right) used for USART output from board. Pins on one side are 1-50, on opposite side 51-100.

parts of these different buses in order to reduce the number of physical connections to the chips, so that the same lines carry different information at different times. Latching circuits must be used to demultiplex these to permit reasonable access to the separate buses if they are to be used with external data-acquisition/control circuitry.

The S100 bus as standardized by the IEEE is a physical and electronic standard connector and signals which can operate as either an 8-bit or 16-bit data bus and can thus work with either an 8-bit or 16-bit CPU. A typical physical standard S100 bus circuit board is shown in Figure 1.7. The components of the S100 bus are shown in Table 1.8. There are several subbuses with noncontiguous pins. The *power supply bus* of 9 lines includes +8 V DC (1 and 51), +16 V DC (2), −16 V DC (52), and ground (20, 50, 53, 70, 100). The DC power on the bus need not be regulated. Regulators on each board convert the +8 V DC to the regulated +5 V DC required by most of the integrated circuit chips, while other less-needed voltages are obtained using regulators on each board from the +16-V DC and −16-V DC supplies. The use of on-board regulation reduces board-to-board noise transfer through the power supply lines. The *address bus* actually includes the 24 address lines A0 through A23, which permits *extended addressing* of 256 × 65,536 bytes (16.8 megabytes), though most 8-bit systems use only the first 16 of these. The *data bus* consists of 16 lines, usable as 8 input lines (DI0–DI7) and 8 output lines (DO0–DO7) on older systems or more recently as 16 input/output bits using DO0 through DO7 for the least significant 8 bits

TABLE 1.8   Pin Connections for S100 Bus

Pin	Signal	Pin	Signal
1	+8 V DC	51	+8 V DC
2	+16 V DC	52	–16 V DC
3	XRDY	53	GROUND
4	VI0	54	SLAVE CLR
5	VI1	55	DMA0
6	VI2	56	DMA1
7	VI3	57	DMA2
8	VI4	58	SXTRQ
9	VI5	59	A19
10	VI6	60	SIXTN
11	VI7	61	A20
12	NMI	62	A21
13	PWRFAIL	63	A22
14	DMA3	64	A23
15	A18	65	NDEF
16	A16	66	NDEF
17	A17	67	PHANTOM
18	STATDSB	68	MWRITE
19	CDSB	69	RFU
20	GND	70	GND
21	NDEF	71	NDEF
22	ADDSB	72	RDY
23	DODSB	73	INT
24	O(B)	74	pHOLD
25	pSTVAL	75	pRESET
26	pHLDA	76	pSYNC
27	RFU	77	pWR
28	RFU	78	pDBIN
29	A5	79	A0
30	A4	80	A1
31	A3	81	A2
32	A15	82	A6
33	A12	83	A7
34	A9	84	A8
35	DO1	85	A13
36	DO0	86	A14
37	A10	87	A11
38	DO4	88	DO2
39	DO5	89	DO3
40	DO6	90	DO7
41	DI2	91	DI4
42	DI3	92	DI5

43	DI7	93	DI6
44	sMI	94	DI1
45	sOUT	95	DI0
46	sINP	96	sINTA
47	sMEMR	97	sWO
48	sHLTA	98	ERROR
49	CLOCK	99	POC
50	GND	100	GND

and DI0 through DI7 for the most significant 8 bits (in order: 36, 35, 88, 89, 38, 39, 40, 90, 95, 94, 41, 93, 43). There are also 4 undefined lines, NDEF (21, 65, 66, 71), used by different users for different purposes; and 3 lines reserved for future use, RFU (27, 28, 69). The remaining 44 lines constitute the *control bus*; some of the control bus may not be used in simpler S100 systems.

Specification of the control bus signals in complete detail is beyond the scope of this text. The 8 signals designated VI0 through VI7 are vectored interrupt input lines to the CPU. Clock signals are provided on CLOCK and 0(B). The 8 designations preceded by s are status signal lines and can be disabled by bringing STATDSB low. The 5 designations preceded by p are control signals, which can be disabled by bringing CDSB low. The 24 address lines can be disabled by bringing ADDSB low, and the 8 data output lines can be disabled by bringing DODSB low.

Since there are no time-multiplexed signals on the S100 bus, all multiplexing and demultiplexing must take place on the board on which it is required.

1.4.2 Serial Digital Data Transfer

Serial digital data transfer is used to convey information to and from devices such as Teletypes, keyboards, terminals, printers, and display devices. Many such devices, and many computing devices, come equipped for serial data transfer. By far the most common standard for transfer of serial data are the EIA-RS232C interface originally developed by the Electronic Industries Association for communication of serial digital data between data terminal equipment (DTE) and data communications equipment such as an acoustic modem (DCE). The standard defines physical 25-pin connectors (Figure 1.8), with the male OB-25 connector attached to the DTE and the female DB-25 attached to the DCE. In present systems, the full standard (some 30 pages!) is usually not implemented; the computer itself may be configured as either DTE or DCE. Only a few of the 25 lines available are usually used, as shown in Table 1.9, although all are specified in the standard. As a minimum, lines 2 (TXD), 3 (RXD), and 7 (GND) are

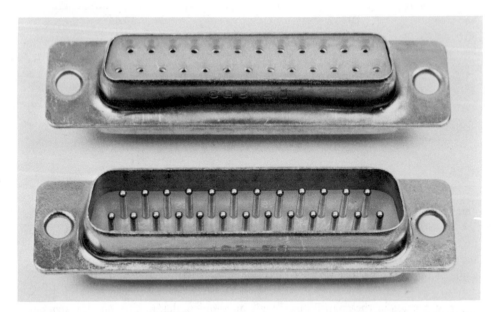

Figure 1.8 Connectors for RS232C standard. Connectors shown are chassis type; cable types are also available for this standard.

TABLE 1.9 Serial RS232C Signals Generally Used

Pin	Originator	Code	Name/Function
1	either	none	protective ground
2	DTE	TXD	Transmit Data
3	DCE	RXD	Receive Data
4	DTE	RTS	Request to Send
5	DCE	CTS	Clear to Send
6	DCE	DSR	Data Set Ready
7	both	GND	Ground
8	DCE	DCD	Data Carrier Detect
20	DTE	DTR	Data Terminal Ready
22	DCE	RIN	Ring Indicator

needed, with the GND lines on each device tied together and TXD of each device tied to RXD of the other. It may be necessary to jumper lines 4, 5, 8, and 20 together at each device in order to make either the DCE or DTE believe that the full standard has been implemented.

The RS232C standard will support both asynchronous (no clock signal) and synchronous modes of serial data transmission but laboratory equipment is almost invariably operated in asynchronous mode. In asynchronous mode a single logic zero is added to the front of each serial character as a *start bit*, and at least one logic one is added to the end of each serial character as a *stop bit*. The stop bit is maintained until the next character is ready for transmission. Logic zero is a voltage between $+1.5$ V and $+36$ V, often $+12$ V, while logic one is a voltage between -1.5 V and -36 V, often -12 V. Since RS232C circuits employ voltage levels, they should not be *directly* connected to devices requiring the presence or absence of current (such as the 60 mA or 20 mA current loops employed by Teletypes). Unless a suitable interface circuit is used, direct connection will probably destroy something.

The RS232C signal voltage levels are often converted to TTL logic levels suitable for direct connection to microprocessor circuits by means of the MC1488 (quadruple TTL-input RS232C-output) and MC1489 (quadruple RS232C-input TTL-output) integrated circuits, which require both positive and negative 12 V DC power. Implementation of an RS232C serial port from a serial port available on a microprocessor or an associated universal asynchronous receiver and transmitted (UART) integrated circuit is straightforward. The resulting lines will transmit data at rates up to 9600 baud over a distance of 30 meters. For longer distances, the RS232C connector can be connected to an acoustic or direct modem so that the data can be transmitted over ordinary telephone lines at data rates of 300 baud. Transmission at higher rates is risky over ordinary telephone lines, and rates above 4800 baud may not be feasible even on dedicated circuits unless precautions against extraneous electrical noise are taken.

1.4.3   Parallel Digital Data Transfer

Parallel digital data transmission can involve simply connecting the 8 or 16 input lines of the destination to those of the source, and can be done with a simple ribbon cable over short distances within a laboratory. Addition of buffering driver circuits may be necessary. This is all that is required to connect one parallel port with another. It does not, however, deal with the problem of connecting one instrument to another or to a computer, since control signals are needed in addition to the data signals.

In 1975, the IEEE established Standard 488, the standard digital interface for programmable instrumentation. The standard was actually developed by the Hewlett-Packard Company, who refer to it as the Hewlett-Packard Instrument Bus (HPIB). Tektronix refers to it as the General Purpose Instrument Bus (GPIB), and other manufacturers may use other names. Certain computers are equipped with this interface, including the Commodore PET and some S100 bus systems. The bus permits asynchronous data transfer in 8-bit parallel form between up to 15 devices at laboratory distances (2 meters).

The standard connectors for the IEEE-488 are 24-pin connectors, often with stacking capability, as shown in Figure 1.9. Of these 24 lines, eight are ground lines, eight are the eight parallel data lines, three are data control lines (Not

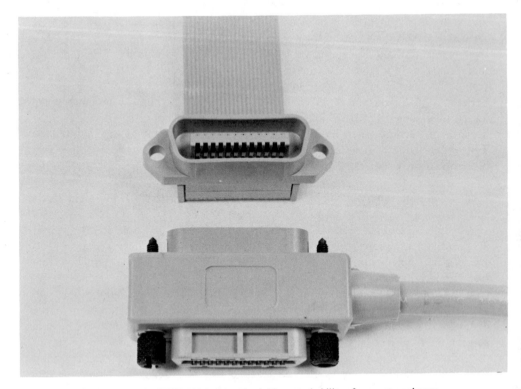

Figure 1.9 Connectors for IEEE-488 bus standard. Note stackability of connectors shown; attachment to both ribbon cable and other types is possible.

Ready for Data, NRFD; Data Valid, DAV; Not Data Accepted, NDAC), and five are instrument bus control lines. Of the attached devices, one and only one must be a *controller* (the Commodore PET assumes *it* is the controller and can't be programmed to do otherwise). The remainder can be *talkers, listeners,* or both. If the computer is equipped with this interface, and the instrument to which it is to be attached is also, only programming is necessary to completely interface them. If, however, either or both is not so equipped, interfacing by means of direct parallel ports used as control lines may be considerably easier than implementing this standard on an instrument or computer not designed with it in mind.

In operation, the controller first asserts one of the instrument bus control lines (attention, ATN). All devices on the bus then become temporary listeners to the data lines, upon which the controller asserts a data byte which is an address. If the data byte has values between 32 and 62, the device whose address is 32 *less* than the data byte (0-30) becomes a listener. If the data byte has values between 64 and 94, the device whose address is 64 *less* than the data byte (0–30) becomes a talker. A value of 63 (unlisten) stops all listeners, and a value of 95 (untalk) stops all talkers if it is placed on the data lines, when ATN is asserted. Other values less than 32 or greater than 95 serve other control functions.

After the controller has designated the talker and the listener or listeners, the talker waits until all listeners have released NRFD (NRFD is then high and any listener asserting NRFD pulls it low; NDAC is already asserted as low). The talker then puts the data on the data lines and asserts DAV by pulling it low. Each listener, when DAV is asserted, first asserts NRFD, then takes the data, and then releases NDAC. When the last or slowest listener has released NDAC, the talker knows that the data has been received. It then releases DAV. The listeners then assert NDAC and, when ready for the next data byte, release NRFD. When the last or slowest listener has released NRFD, the talker puts the next data byte on the data lines and so the cycle continues.

The maximum signal rate of one million bits/second is rarely approached even by the fastest of talkers in the laboratory environment. The actual rate of data transfer depends upon the rate at which the slowest destination can accept it.

1.5 REFERENCES

G1 H. V. Malmstadt, C. G. Enke, and S. R. Crouch, *Electronic Measurements for Scientists*. W. A. Benjamin, Menlo Park, 1974, 906 pp.

G2 H. V. Malmstadt and C. G. Enke, *Digital Electronics for Scientists*, W. A. Benjamin, New York, 1969, 542 pp.

G3 R. Zaks, *Microprocessors, Second Edition*, Sybex, Berkeley, 1977, 420 pp., paperback.

G4 A. Lesea and R. Zaks, *Microprocessor Interfacing Techniques, Second Edition*, Sybex, Berkeley, 1978, 416 pp., paperback.

1.6 STUDY PROBLEMS

1.1 Three resistances of 10 ohms, 100 ohms, and 1000 ohms are connected in parallel. Compute the resistance of this array as seen by an external signal source.

1.2 A certain power supply capacitor is rated at 95 mF and 15 VDC. What is the maximum energy which can be stored in this capacitor? How does it compare to one rated at 43 mF and 20 VDC?

1.3 Express the 256 decimal numbers which could be the values of a single byte in octal notation, in hexadecimal notation, and in binary representation.

1.4 The table which gives the logical output of a digital circuit as a function of all possible logical input patterns is called a truth table. Construct a truth table for a circuit consisting of (a) two parallel NAND gates, (b) two parallel NOR gates, (c) two serial NAND gates.

CHAPTER
2

MEASUREMENT OF
ELECTRICAL QUANTITIES

Electroanalytical measurements all involve measurements of electrical quantities that are relatable to chemical parameters. The methods of measuring these quantities are vital to the electroanalytical chemist, for they determine the accuracy, and indeed the possibility, of measurement of the chemically significant parameters.

2.1 CURRENT

Current is probably the single most important quantity measured in electroanalytical chemistry. The unit of current, the ampere, is a fundamental unit of the SI. Electroanalytically useful currents range downward from amperes to picoamperes. All methods of current measurement are based on either the magnetic field induced by passage of the current or the voltage difference arising from flow of the current through an impedance.

2.1.1   Indication of Current: The Galvanometer

The basic instrument used for indication of current is the moving-coil galvanometer. The galvanometer consists of two magnetic pole pieces and a fixed iron core that provide a radial magnetic field. The current is passed through a rotatable coil of wire to which is attached a pointer or a mirror to indicate current passage. Normally the pivots of the coil serve also as the current leads to it and a restoring force is provided by a coil spring. This arrangement is shown in Figure 2.1.

When a flat coil of wire is placed in a fixed radial magnetic field and current is passed through the coil, the magnetic field produced about the coil by passage of the current through it interacts with the fixed magnetic field already present to produce a force that causes the coil to rotate. If the rotational force is opposed by the force of a spring, the coil will rotate until the deflecting and restoring forces are equal; the angle of net rotation, or angular displacement, of the coil is then directly proportional to the current flowing through the coil. Attachment of the coil to a mirror, or even the use of a multiple-reflection optical system, can provide

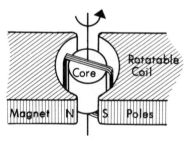

Figure 2.1   The galvanometer.

a path that is long enough that currents of less than 1 nA produce visible angular displacements.

2.1.2   Electromechanical Ammeters

A galvanometer is capable of measurement of current, but the currents that can flow through the coil are limited, and a direct current measurement using a galvanometer is rarely practical. An *ammeter* consists of a fixed resistor, or a series of switch-selectable fixed resistors, in parallel with the resistance of the moving coil of the galvanometer as shown in Figure 2.2. The resistors R1, R2, or R3 are selected by switch S. They are generally lower than the resistance of the galvanometer coil and hence shunt most of the current through themselves instead of through the coil, giving rise to their name of *shunt resistors* or "shunts." The fraction shunted through the coil is constant, and hence the mechanical ammeter gives an angular displacement directly proportional to the current flowing through it. Unlike the reading of its counterpart, the mechanical voltmeter, the reading of the ammeter will always be correct for the current flowing through it. It must be borne in mind, however, that insertion of an ammeter (or even a galvanometer) into the circuit must add some resistance, even though it may be a comparatively low resistance, and that the current may be different in the absence of the ammeter.

Use of a mechanical rather than an electronic current meter can avoid grounding and shielding problems related to the AC line in sensitive measuring circuits. The precision with which an ammeter of the galvanometer type can be read depends upon the scale and the quality of the meter movement. Precision of better than 1% is possible with research-grade meters, but not with smaller or cheaper instruments. The use of a galvanometer in an ammeter configuration poses one other problem in that the resistance of the ammeter as seen by the external circuitry does not remain constant as the range is changed by use of different shunt resistors. If the load resistance imposed by the ammeter is always only a negligibly small part of the total load, it will not adversely affect the readings on any range. If, however, on any range the resistance of the ammeter is a significant part of the load resistance, errors will be introduced whose magnitude will change drastically with range setting.

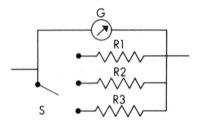

Figure 2.2   Galvanometer in ammeter configuration.

An ammeter of the galvanometer type can give a precise current measurement provided that the current does not change more rapidly than the coil can physically move in response to those changes. The period of practical physical galvanometer or meter movements is some 3 to 5 s, and currents changing significantly over a 1-s time interval are not faithfully reproduced by motion of the coil. Mechanical ammeters are not suitable for measurement of rapidly changing currents such as are often encountered in electroanalytical techniques.

Currents measured by a galvanometer or an ammeter of the galvanometer type can be recorded manually or, using a light-spot galvanometer, directly on photographic film. These techniques, while precise and accurate for the measurement of slowly changing currents, are inconvenient; film recording is now obsolete. Since these are the only ways to easily record galvanometer output, virtually all methods used to record instantaneous current are methods of recording the instantaneous voltage drop across a standard resistor.

2.1.3   Electronic Ammeters

Although solid-state circuits do permit the construction of true electronic ammeters, most devices purporting to be such are actually electronic voltmeters reading the voltage drop across a standard resistor. Any convenient scale can be obtained by proper selection of the resistor value. The speed of response of an electronic voltmeter depends on its circuitry. The volt-ohm meter, which uses as its output device a mechanical meter, can be no more rapid in response than an electromechanical ammeter. Devices with digital displays are also limited in response speed by the readout circuits as well as by the speed of analog-to-digital conversion, although the response appears rapid in human terms. When digital recording or automatic data acquisition is attempted, the limitation of response speed becomes apparent. The effect of an electronic ammeter on a circuit being studied depends upon its input impedance as seen by that circuit. Modern electronic ammeters of the voltmeter type have an input impedance essentially equal to the impedance of the standard resistor through which the current is passed, since the input impedance of the voltmeter can be very high.

Electronic ammeters have been built that require no physical or electrical contact with the circuit being studied. These devices, often called *power meters*, measure the magnetic field induced in a wire by passage of current through it. The meter circles the wire and intercepts the field. Power meters are not sufficiently precise at the milliampere and lower currents usually encountered in electroanalytical work and are rarely used. The magnetic field decreases with decreasing current, and pickup of stray fields from other equipment becomes significant.

2.1.4   Recording of Currents That Vary with Time

Currents which vary with time are virtually never recorded as currents; instead, they are passed through a precision fixed resistor, and the voltage across the resistor is then recorded. There are no adverse effects inherent in this measurement procedure other than the effect of the inserted resistance on the circuit being studied.

2.1.5   Standards of Current

The ultimate standard of current is the definition of the SI unit of current, the *ampere*. The ampere is a fundamental unit of the SI, and is defined as the constant current that, if maintained in two straight parallel conductors of infinite length, of negligible circular cross section, and placed one meter apart in vacuum, would produce between these conductors a force equal to 2×10^{-7} Newtons per meter of length. The ampere is thus relatable to the more fundamental units of length and mass by experimental means. The experimental apparatus, the *current balance*, consists of a pair of coils so arranged that the force between them when they are carrying current can be accurately measured. Modern current balances are capable of making such measurements to a precision of better than 4 ppm. Current balances are rarely seen outside of national or international metrology laboratories and are not practical for use as electrochemical laboratory standards.

No practical standard of current suitable for laboratory use exists. Measurement of current in the laboratory is therefore standardized using Ohm's law together with laboratory standards of voltage and of resistance, and current-measuring devices are calibrated using this procedure.

2.1.6   Sources of Current

Constant current can be supplied by DC power sources from commercial suppliers or by locally built units. Locally built or commercial units are of two types: those in which the constant current is supplied by batteries and those in which the constant current is supplied from the AC mains. Constant direct current supplies operating from the AC mains are inherently inferior owing to rectification ripple and possible line variation, although excellent commercial design has significantly reduced this problem, and the convenience of using commercial instrumentation is superior to that of locally built units. Battery-based circuits, however, are unsuitable for long-duration or high-current use because of their limited capacity and internal polarization. A useful compromise is a source powered by batteries rechargeable from the AC mains, *provided* that the recharging circuit is disconnected when the power supply is in use. The writer has used commercial nickel–cadmium and lead–acid batteries successfully in this configuration.

Regardless of the source of the current, its constancy must be maintained despite changes in the resistance of the cell through which it is being passed. This is done using a large, constant, series resistor, an electromechanically adjustable large series resistor, or an electronic control. To permit large series resistors, use of high-voltage 45- or 90-V batteries, even several in series, is recommended. The current is more constant even though the batteries are of lower capacity. The constancy of the current produced using a high voltage and a large series dropping resistor will depend upon the relative magnitude and constancy of the supply voltage. This method is not useful where large currents are required, because battery polarization and Joule heating of the dropping resistor will cause changes in the resulting current. Practical devices are not normally suitable for currents greater than a few milliamperes, and even then only for short periods of time.

In the 1940s and 1950s constant current supplies were designed using electromechanical control circuitry such as the servosystems of potentiometric strip-chart recorders. This approach is now obsolete owing to the slow response time and inherent complexity of the electromechanical unit as compared with an electronic system. The most common electronic system is known as a *galvanostat* or an *amperostat*.

When a source of current that can vary with time in a controllable way is required, electronic control of a programmable power supply is required. Programmable power supplies useful in electroanalytical chemistry are usually galvanostat circuits based upon operational amplifiers. These are programmed using a generated voltage waveform that the current output of the galvanostat follows.

2.1.7   Control: The Galvanostat

A *galvanostat* is an electronic system, normally based on operational amplifiers, that controls the current through a cell regardless of such changes in the cell resistance as may occur. A galvanostat is actually a programmable power supply, but if it is programmed with a constant input voltage it will provide a constant current output. The constancy of the output current then depends wholly on the galvanostat circuitry and the input voltage.

The operation of a simple galvanostat circuit is shown in Figure 2.3. The amplifier is used in the operational feedback mode in which the output is returned to the input, thereby establishing the summing point of the input as a virtual ground held at ground potential by the output of the amplifier. The counterelectrode and the indicating electrode are in the operational feedback loop. The current that flows through the cell must therefore equal, and oppose, the current produced by the input voltage applied across the fixed input resistor to the

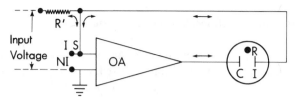

Figure 2.3 Operational-amplifier galvanostat. Input voltage applied through fixed resistor R' to summing point S produces constant current that is equaled and opposed by current in operational feedback loop, which is also the current flowing between the counterelectrode C and the indicating electrode I. The reference electrode R is not used for control purposes in a simple galvanostat circuit.

same virtual ground. The cell *current* will therefore follow the changes in the programming input *voltage*.

2.2   CHARGE

Charge is an important quantity in electroanalytical work because it is directly related, through the Faraday constant, to the number of moles of electrons that have passed through the circuit. Measurement of charge, in the Faradaic rather than the electrostatic sense, is actually measurement of current passage over time since $Q = It$. The SI unit of charge, the *coulomb,* is the ampere-second.

2.2.1   Measurement: Constant Current

The charge passed through a circuit by a *constant current* is very easily measured since the integral of a constant current over time is then simply the product of the magnitude of the current and the interval of time, both of which are measurable with a high degree of precision. The precision of the measurement usually depends significantly on only the constancy with which the current is maintained.

If the current is not constant, but can vary over time, an integrating method must be used to obtain the charge passed. There are four methods of integrating the current to obtain charge: mathematical, electromechanical integrators, electronic integrators, and chemical coulometers.

2.2.2   Measurement: Mathematical Integration

In the *mathematical* or *graphical* method, the current is recorded as a function of time and then later integrated either using calculation or by graphical instruments such as a planimeter to obtain the area under the recorded curve, which is the charge. While useful in qualitative studies and in chronoamperometric investigations, the planimeter and other similar methods such as cut-and-weigh have a precision not better than 1% and in many cases worse. They are, therefore, not suitable for quantitative analytical coulometry. The calculation method, while offering better precision, normally requires digital data-acquisition systems of reasonable speed in addition to access to the necessary digital computing facilities.

2.2.3   Measurement: Electromechanical Integration

Electromechanical integrators were used in the 1950s and 1960s, but must now be considered obsolete. The best precision reported was 0.1%, which is acceptable. These devices have now been replaced by electronic integrators, and further electromechanical development is unlikely.

2.2.4   Measurement: Electronic Integration

The most common method of *electronic integration*, an analog method, is to let the instantaneous charge (which is proportional to current) be built up as actual charge on a capacitor over a period of time. The voltage across the capacitor, which is a measure of the charge held, is then determined. If the quality of the capacitor is such that no significant charge leaks off during the period of charge cumulation, this method is capable of results with good precision. A useful integrator circuit consists of an operational amplifier with an integrating capacitor in the feedback loop.

A digital method can be used if the current being measured is applied as a series of identical small current pulses. If the number of coulombs in a single pulse is accurately known, multiplication by the number of pulses (which can easily be counted by a digital counter) gives the charge passed with the same relative error as the relative error with which the charge of a single pulse is known, if the number of pulses is sufficiently large. This method is highly practical for coulometric titrations, since often electrogeneration conditions can be established that are not adversely affected by the pulsating nature of the current. This method should not, however, be used without care both as to the apparatus and as to the electrochemical reactions being employed.

2.2.5   Measurement: Chemical Coulometers

Very high precision of charge measurement is possible with a *chemical coulometer*. The silver coulometer (Richards and Heimrod, S1; Rosa and Vinal, S2) and iodine coulometer (Washburn and Bates, S3), while superior to all others in precision, are inconvenient to use in analytical determinations. The hydrogen–oxygen gas coulometer is capable of an acceptable 0.1% or better if proper precautions are taken (Lehfeldt, S4; Lingane, S5) and is considerably more convenient. For current densities below 0.05 A/cm^2 the hydrogen-oxygen gas coulometer reads significantly low (Page and Lingane, S6) and the hydrogen–nitrogen coulometer is recommended instead. For further details consult the references cited in Lingane (S7).

2.3   VOLTAGE

Voltage, as used generally and in this text, is a synonym for the more descriptive term *potential difference*. The potential difference between two metallic conductors is the difference in electron pressure or driving force on electrons between them. For this reason, the measurement of any single absolute electron pressure is impossible, however useful a theoretical construct it may be. A complete circuit, involving a return path for electrons, is necessary in measurements of voltage. If a complete circuit is not present, the small current required by the measuring apparatus will alter significantly the charge balance across that apparatus and destroy any significance the reading would have. If this circuit involves one electrode it must also involve a second electrode in the same electrolyte in order to be complete. The writer was once consulted when difficulties were encountered in pH measurement; one terminal of the pH meter was connected to a glass electrode in the solution and the other terminal was connected to an external earth ground. No measuring device can give reasonable information in such a situation, but more sophisticated high-impedance electronic voltmeters may give misleadingly stable, and totally meaningless, readings.

2.3.1   Electromechanical Voltmeters

The *electromechanical voltmeter* consists of a galvanometer in series with a resistance much larger than that of the meter as shown in Figure 2.4. The switch S selects appropriate values of R_S to obtain the different desired voltage ranges of the voltmeter. Electrically, this can be considered as a series combination of two resistances, the galvanometer resistance R_m and an additional series resistance R_S. The values of these are chosen with R_S much larger than R_m so that the total resistance is approximately equal to R_S. Then, since $I = E/R$, the current I is proportional to an applied constant voltage E, and the meter, which actually has an angular deflection proportional to *current*, can be calibrated in terms of voltage.

Electromechanical meters place a *load* upon the circuit being studied. This load is their *input impedance* whose value is $R_S + R_m$. This resistance is not greater than the *figure of merit* of the meter, usually 20 kΩ/V, at the voltage of the measurement. The figure of merit multiplied by the voltage gives the input impedance of the meter. More correctly, since R_S is much larger than R_m, the actual impedance at the input is given by the figure of merit multiplied by the maximum reading of the particular voltage scale in use. This load must cause the measured potential difference, which is the real potential difference with the meter *in position*, to be lower than the desired potential difference, which is the real potential difference with the meter *out of the circuit*, because the meter provides an alternative path for electrons. The significance of this difference between the measured potential and the desired potential will depend on the nature of the electrodes being studied, and should not be assumed to be an insignificant difference.

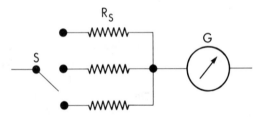

Figure 2.4   Galvanometer in voltmeter configuration.

2.3.2   Electronic Voltmeters

An *electronic voltmeter* is one in which the load of the output device is carried by the output of an amplifier circuit. A circuit employing a voltage follower amplifier, shown in Figure 2.5, is a typical configuration. The electronic meter may use either an electromechanical voltmeter or an electronic device such as a digital readout to display its output. Differences in output devices are irrelevant to the effect of the electronic voltmeter on the circuit being studied, although the type of output device may affect the precision with which measurements can be made. The significant difference between an electronic voltmeter and an electromechanical voltmeter is that the power necessary to drive the output device is provided by the amplifier rather than by the circuit being studied. Hence the load *on the circuit being studied*, which is the input impedance of the amplifier, can be much less. The amplifier input impedance can be as high as $10^{13}\ \Omega$ (Orion Model 801A digital pH meter, FET input), although lower levels of $10^7\ \Omega$ would be found with a cheaper vacuum-tube voltmeter.

Design and local construction of electronic voltmeters are neither trivial nor recommended, since many excellent devices are available from commercial sources.

2.3.3   The Potentiometer

The *potentiometer*, invented by Poggendorf in 1841, is the oldest and probably still the most precise method of measuring voltage. The basic circuits of a potentiometer are shown in Figure 2.6.

Operation of the potentiometer depends on the use of galvanometer G to detect the passage of current. The cursor or wiper C on the resistive slidewire AB picks off a fraction of the voltage *AB*. The voltage across AB is adjusted to a standard known value before measurement by adjustment of R such that the voltage *AB*, or some specified fraction of it, is exactly equal to that of the standard cell E_2. This voltage is provided by E_1, which need not be a standard cell but whose voltage should not be subject to short-term (5–10-min) drift; long-term stability is of lesser importance. With the voltage *AB* now known, switch S connects the unknown emf into the circuit, and cursor C is moved along AB until no current flows when tapping key K is momentarily closed. At this point the voltage *AC* is

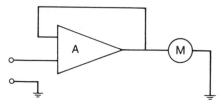

Figure 2.5 Basic electronic voltmeter. A, amplifier; M, readout device.

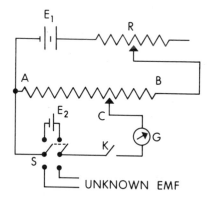

Figure 2.6 Principles of operation of a potentiometer. E_1, battery, usually 6 to 12 V DC; E_2, standard cell, usually a Weston cell; R, variable resistor for standardization; slidewise A–B with sliding contact C; G, galvanometer; D, tapping key; S, DPDT switch. Standard cell E_2, switch S, and resistance R can be eliminated if the voltage drop across AB is measured by an external calibrated voltmeter.

exactly equal in magnitude and opposite in sign to the unknown voltage. The value of *AC*, and hence also that of the unknown voltage, is read from the physical position of cursor C. It is highly desirable, although not absolutely necessary, that the resistance of slidewire AB be linear, in which case measurement of fractional distance AC/AB gives immediately the fractional voltage *AC/AB*.

The advantage of the potentiometer over other methods of voltage measurement is that at null balance it draws no current from the circuit being studied and hence does not alter the voltage being measured. When the potentiometer is off balance, there will be a significant current flow, albeit only momentary. Thus electrochemical systems that are significantly disturbed by momentary current flows should not be studied with a potentiometer.

Comparison between a modern potentiometer and a modern electronic voltmeter is instructive in view of the trend toward electronic, and frequently digital-readout, voltmeters in preference to manual potentiometers. While the voltmeter is more rapid, more convenient, and draws no significant current from the system being studied at any time, it has a disadvantage in that it will read a potential difference that may well not be that of any reversible electrochemical system and may be due instead to traces of impurities or other artifacts. A potential can even be measured by these devices in an incomplete circuit, as would be the case in potential measurements using a combination electrode in which the electrolyte has dropped below the level of the reference junction. The potentiometer, which when off balance does draw significant current, can be used to make micropolarization measurements by potential settings positive and negative of the balance point by a few millivolts, which is a useful test for the reversibility of a measured potential. Calibration of the linearity of an electronic voltmeter basically is the province of the manufacturer, although tests against external standards usually are provided for. Calibration of a potentiometer depends upon use of an external standard cell; linearity is again the province of the manufacturer. Nonlinear devices, if calibrated at many points, are precise but sufficiently inconvenient to be unknown.

Construction of a precise potentiometer in the laboratory is comparatively simple, and a research-grade instrument can be produced locally from resistance decade boxes and a galvanometer if both are of high quality. Research-grade potentiometers available commercially are superior to locally manufactured devices in both precision and convenience. Sources in North America include the Leeds and Northrup Company and the Rubicon Company, both of Philadelphia, Pennsylvania.

Commercial research potentiometers often come in three ranges: 0 to 1.6 V, 0 to 0.16 V, and 0 to 0.016 V. The precision available depends upon the range used. For example, the Leeds and Northrup K-5 research potentiometer has error limits of 0.003% of reading $+$ 3 μV, 0.005% of reading $+$ 0.3 μV, and 0.005% of reading $+$ 0.1 μV on these three ranges. Calibration of the instrument can improve these error limits by about a factor of 2. The upper limit of 1.6 V is established by the source of opposing electromotive force, normally a battery. Measurement of voltages above 1.6 V is possible in principle by using series connection of batteries, but this practice can overload and burn out the potentiometer resistors and hence should never be attempted. Measurement of voltages above 1.6 V with a potentiometer requires a "volt box," or precision voltage divider, such that the voltage read by the potentiometer is below 1.6 V. Volt boxes are available from commercial manufacturers, but can also be easily assembled in the laboratory from resistance decade boxes. Use of a volt box permits potentiometric measurement of voltages as high as 1500 V.

2.3.4   Recording of Voltages That Vary with Time

None of the three devices mentioned above is effectively capable of measuring voltages that change at any significant rate with time. Manual voltage measurement with any of these devices requires that the voltage be effectively constant for at least a few seconds. Measurement of a voltage that changes significantly with time therefore means recording that voltage in some form so that it can be studied later at leisure. It is impossible to employ an electromechanical voltmeter for this purpose, because the inertia of the movement causes it to lag behind the voltage change significantly.

The electronic voltmeter can be used for measurement of a voltage that varies with time if the actual output device is also electronic rather than electromechanical, as is the case in digital electronic voltmeters. These devices can often be operated in a sample-and-hold mode of controllable sampling rate as well as operated as continuous monitors. In such a measurement, the voltmeter is clocked by an adjustable oscillator, so that the instantaneous voltage is sampled, and displayed, every preset interval of time. The display is held until the next sampling interval. The display can be noted manually or fed to a recording device such as magnetic computer tape or a card or tape punch. The effect of the recording of data using an electronic voltmeter upon the system being studied is identical to that of the electronic voltmeter used manually and is almost always insignificant.

Potentiometric Recorders

The potentiometer, like the electronic voltmeter, can be modified for recording. Such a device, called a *potentiometric recorder*, is shown in Figure 2.7. Most strip–chart or *x-y* chart recorders are potentiometric recorders. In potentiometric recorders, the unknown voltage is balanced with an internally adjusted potential difference obtained from an adjustable slidewire resistance just as it is in manual potentiometry. The slidewire is attached to servomotors operated by the amplified difference between the unknown voltage and the internal adjustable voltage. An imbalance between the two causes the servomotors to drive the slidewire cursor, and a pen attached to it, in the direction of restoring balance. The speed of operation of a potentiometric recorder depends on the gain of its internal amplifiers and the design of its servomotors. Instruments currently available usually have a 1-s response time, where response time here means the speed with which the pen can move completely across the chart; a few are even faster. Attempts to follow changes in voltage that are of the same magnitude as, or faster than, the response speed of the recorder will lead to errors in the recorded trace, and use of potentiometric recorders for voltages changing significantly in less than 10 s is ill-advised. When the potentiometer is balanced, it draws no current from the circuit being studied; when the potentiometer is off-balance, however, significant current can be drawn with possible adverse effects on the electrochemical system being studied. Since the signal driving the servomotors is the amplified difference between the unknown and internally adjusted voltages, such a difference must necessarily exist during recording of any time-varying voltage. Therefore, use of a potentiometric recorder will affect the voltage being measured; the significance of this will depend on the design of the measurement circuit and the recorder, since use of amplifiers of high input impedance can reduce the error to insignificance. Local construction of potentiometric recorders is ill-advised. The quality and price of commercial units are variable, and care must be taken in selection of a commercial unit.

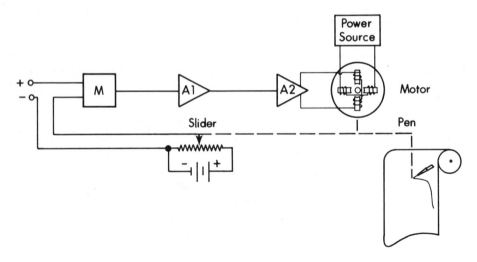

Figure 2.7   Operation of potentiometric recorder. M, signal modulator; A1, voltage amplifier; A2, power amplifier. Motor is phase-sensitive. Chart drive motor not shown. Electrical linkages are solid lines; mechanical linkages are broken lines.

When a potentiometric recorder has both of its perpendicular axes designed for potentiometric voltage inputs, it is called an x–y recorder. Such recorders are particularly convenient in electroanalytical techniques such as cyclic voltammetry, in which the direction of a voltage sweep is reversed in the course of an experiment. In other experiments, such as those involving rotating ring–disk electrodes, recorders with two separate voltage inputs on the same axis may be convenient. In either case, their use is a matter of convenience rather than necessity. If an x–y recorder is purchased for electroanalytical use, it should be capable of operation in an x–t mode as well.

The Oscilloscope

When voltages change more rapidly with time than a potentiometric recorder can track them, an *oscilloscope* or a transient recorder must be used for accurate measurements. An oscilloscope is essentially an x–y display device in which two electrical signals are independently amplified and fed as voltages to two pairs of control plates that control the vertical and horizontal displacement of the electron beam of a cathode ray tube (CRT). The horizontal displacement plates can be, and usually remain, connected not to an external signal but to an internally generated sawtooth wave or *sweep* whose frequency is established by the use of calibrated front–panel controls. Thus the oscilloscope becomes an x-t device. The start of a sweep can be triggered, by proper selection of control settings, manually, at the frequency of the 60-Hz AC power line, or by the external signal appearing on the y axis. If the external signal is used, controls are provided that enable selection of the *trigger direction* (positive-going signal or negative-going signal) and the *trigger level* (magnitude of signal or signal change required to effect triggering of the sweep). An additional alternative, use of a second external signal to trigger the sweep, is often provided with similar trigger-direction and trigger-level controls. Virtually all modern oscilloscopes have triggered sweep circuits, which are essential for most electroanalytical applications.

Modern oscilloscopes generally have input impedances of 1 MΩ, high enough that essentially no current is drawn from the circuit or system being studied. If, however, the electrochemical system being studied has extremely high resistance, such as is found in the glass pH electrode or in certain nonaqueous electrolytes, this input impedance may not be high enough to avoid significant polarization, and special measuring devices may be required.

The response speed of the modern research oscilloscope is sufficiently rapid for any electrochemical measurement and will introduce insignificant distortion. Typical signals on modern research oscilloscopes range from DC to 100 MHz (bandwidth, Hewlett-Packard 1740). Even oscilloscopes of lower cost can handle signals to 100 kHz, and response to even higher frequencies can be obtained (at higher cost). By these standards most electroanalytical signals are slow.

Oscilloscopic tracings can be recorded photographically; Polaroid film with an ASA rating of 3000 is recommended. Films of lower ASA rating may be incapable of recording extremely rapid voltage transients. Measurement of voltage from oscilloscope photographs is significantly less accurate than similar measurement from potentiometric recorder charts, and oscilloscopic recording is inherently inferior in precision to potentiometric recording. A triggered sweep, with which all modern laboratory oscilloscopes are equipped, is necessary in photographic recording. Most modern oscilloscopes are adequate for electrochemical work of low precision, say ±3%; high-precision electrochemical work should employ other methods of readout. Oscilloscopes are capable of time-interval measurements that are more precise than their measurements of electrical quantities.

Some oscilloscopes, known as *storage oscilloscopes*, are equipped with special cathode ray tubes and circuits that enable them to store a transient waveform on the screen for many minutes. This is both convenient and, in recording, economical of film, since only the more useful curves are photographed. They are, however, considerably more expensive than oscilloscopes of the same quality without the storage feature.

Transient Recorders

A *transient recorder* is a digital device introduced by Biomation, Inc., as a separate unit for laboratory purposes. The same type of device is often found as the data input section of general-purpose computerized data-acquisition and handling systems, electroanalytical and otherwise. It consists of a fast ADC unit and a block of RAM together with a controllable precise time-base oscillator, a DAC, and appropriate trigger and timing controls. Upon receipt of the trigger signal, the ADC samples the input voltage waveform at the rate set by the time-base oscillator, placing each successive data point into a successive location in RAM until the block of RAM is full. The contents of the RAM, which is the digitized transient waveform, can then be output at any desired rate in digital form or, through the DAC, in analog form for display on an oscilloscope or potentiometric recorder.

The trigger signal can be supplied as a separate external input or may be taken from the normal input signal in the usual manner of a triggered oscilloscope. Alternatively, the recorder may be used in a continuous acquisition mode in which the memory is being continually refilled with new data on a first-in first-out basis *until* a trigger signal is received. The memory RAM then contains the input signal for the time interval *preceding* the trigger signal. Use of this mode permits acquisition of data when the only really distinctive trigger signal follows, rather than precedes, the data of interest.

2.3.5   Control: The Potentiostat

The first automatic *potentiostat*, or device for maintaining a constant potential difference between electrodes, was developed in 1942 by Hickling (S8). Many types of potentiostats were developed through the 1950s. The most common modern types are based upon operational-amplifier (OA) circuits. A typical basic OA design is shown in Figure 2.8.

The operational amplifier potentiostat circuit is quite straightforward. The amplifier output is connected to the counterelectrode, whose potential does not matter; the output is also fed back to the inverting input of the amplifier. This feedback takes place through the reference electrode circuit, since this is connected in the operational feedback loop. The effect of this feedback is to decrease the potential difference between the inverting and noninverting inputs of the amplifier. Since the noninverting input is connected to ground, the junction, or *summing point*, S of the circuit is driven to ground potential also by the operationally fedback output of the amplifier and thus is called a *virtual ground*. The potential difference between the working electrode, connected to ground, and the reference electrode, connected to virtual ground, is therefore zero, and the working electrode is held at the potential of the reference electrode by the amplifier output. The potential of the working electrode can be set at any value (relative to the reference electrode) by insertion of a voltage difference in the reference electrode circuit. Information on potentiostats and galvanostats is available in articles in *Analytical Chemistry* (G4), in manufacturers' literature (Wenking, Taccussel, Princeton Applied Research), and in the book by Von Fraunhofer and Banks (G3). A typical modern laboratory potentiostat/galvanostat is shown in Figure 2.9.

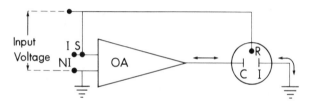

Figure 2.8   Operational-amplifier potentiostat. Input voltage is applied between inverting (I) and grounded noninverting (NI) inputs. Output to counterelectrode C while working electrode W is grounded. Operational feedback loop is through reference electrode R to summing point S.

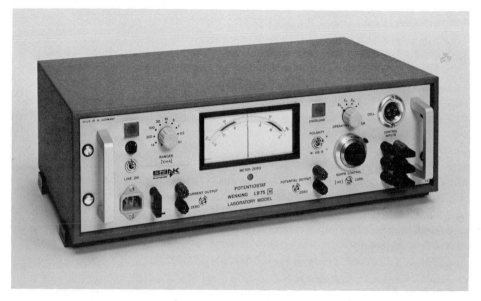

Figure 2.9   Commercial potentiostat/galvanostat. Wenking Model LB75, courtesy of Brinkmann Instruments (Canada), Ltd.

2.3.6   Standards of Voltage

The SI unit of voltage or potential difference, the *volt*, is a derived unit, obtained from the ampere and nonelectrical basic units. The volt is defined as the voltage between two points of a conducting wire carrying a constant current of one ampere when the power dissipated between these points is one watt. A watt is defined as one joule per second, and a joule of work is done when a force of one newton acts through a distance of one meter; the newton is defined in terms of the kilogram, meter, and second. Standards of voltage for laboratory use cannot be constructed from the SI definition; physical standard cells must be employed.

Standards for voltage were originally based, more or less by default, upon the *Daniell* copper–zinc cells developed in 1836, which were used as sources of electricity for telegraphy:

$$Zn/ZnSO_4(aq)//CuSO_4(aq)/Cu$$

In these cells various types of separators and different concentrations were used, but in all of them the potential changed with time through interdiffusion of the solutions. Potentials of fresh cells ranged between 1.07 and 1.14 V. The Daniell cell was replaced by the *Clark cell* (S9) in 1872, later modified by Rayleigh (1884). The modified Clark cell uses a saturated zinc amalgam as the negative electrode. The electrolyte is an aqueous solution saturated with both zinc sulfate and mercury(I) sulfate and the positive electrode is mercury:

$$Zn(sat.,Hg)/ZnSO_4(sat.\ aq.),\ Hg_2SO_4(sat.\ aq.)/Hg(l)$$

The potential of a Clark cell is 1.434 V (15°C), as accepted internationally following 1893.

Saturated Weston Cells

Modern physical standards of voltage are based on the *saturated Weston cell* (S10) developed in 1891, upon which international agreement was reached at the London Conference on Electrical Units and Standards in 1908. The construction of a saturated Weston cell, sometimes called a *normal Weston cell*, is

$$Cd(Hg)/CdSO_4 \cdot 8/3H_2O(s),\ Hg_2SO_4(s),\ H_2O/Hg(l)$$

The negative electrode is a two-phase saturated cadmium amalgam, originally 12 w % cadmium but now generally 10 w % cadmium. The electrolyte is an aqueous solution saturated with both mercury(I) sulfate and hydrated cadmium sulfate; an excess of both solid phases is necessary to ensure saturation at all temperatures.

The electrolyte may contain 0.015 to 0.025 mol/dm^3 sulfuric acid, which serves to prevent hydrolysis of the mercury sulfate and to reduce the attack of the electrolyte upon the glass. The presence of the acid does decrease the potential difference of the cell slightly; the potential of a cell containing 0.025 mol/dm^3 sulfuric acid is 30 μV lower than that of a neutral cell. The positive mercury electrode is usually covered with the excess solid mercury(I) sulfate. The cell is often assembled in an H-shaped sealed glass vessel as shown in Figure 2.10, although other configurations have been used. Platinum wires sealed through the glass are used for contacts.

Since at any temperature the activities of all of the participants in the electrode reactions are constant, the voltage of this cell, which is the potential difference between its electrodes, is constant also. This constancy of activity will be achieved only by saturation of the electrolyte with both mercury sulfate and cadmium sulfate and of the cadmium amalgam with sufficient cadmium to form the

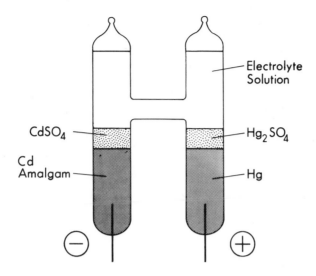

Figure 2.10   Saturated Weston cell, H-type.

two-phase system. The cell voltage is a function then only of temperature and pressure.

Groups of saturated Weston cells are the practical primary standard of voltage, and are stable virtually indefinitely. They are, however, quite sensitive to temperature. Thermal damage that is permanent will occur below the freezing point of the electrolyte (−20°C) and above the transition temperature of hydrated cadmium sulfate (43.6°C), and thermal hysteresis is severe even within this range. For neutral or acid cells using 12% cadmium amalgam, the potential difference between 0°C and 40°C is given to the nearest one microvolt by the accepted formula of 1908:

$$E = 1.01865 - 40.6 \times 10^{-6}(t{-}20) - 0.95 \times 10^{-6}(t{-}20)^2 + 0.01 \times 10^{-6}(t{-}20)^3 \tag{2.1}$$

while for cells using 10% cadmium amalgam the equation of Vigoureaux and Watts (S11) is

$$E = 1.01865 - 39.39 \times 10^{-6}(t{-}20) - 0.903 \times 10^{-6}(t{-}20)^2 + 0.0066 \times 10^{-6}(t{-}20)^3 - 0.00015 \times 10^{-6}(t{-}20)^4 \tag{2.2}$$

With temperature coefficients this large, constant temperature conditions are essential, and control to within 0.02 K is required for potentials constant to within 0.1 μV. The group of 44 primary standard cells maintained by the U.S. National Bureau of Standards (USNBS) is held within 0.01 K of 28.00°C. Some of these cells have been in use since 1906; this group differs by no more than 1.3 μV from the group kept at the international center in Sevres, France, for about the same period of time, and its potential varies by no more than 0.6 μV/yr.

Passage of a small amount of current through a saturated Weston cell will not permanently change any of the activities and thus will not permanently alter the cell voltage. Passage of even a small current will, however, temporarily alter the activities at the surfaces of the electrodes, and hence the cell voltage as well, until the solubility equilibria are reestablished by diffusion. This process is slow, and hence as little current as possible should be drawn from a saturated Weston cell or, for that matter, any standard cell.

Weston cells of the saturated variety can be calibrated to the nearest 0.1 μV by national standardizing organizations such as the U.S. National Bureau of Standards. Such cells, shipped while held at 28°C in special containers, are used as standards for interlaboratory comparisons of voltage.

Unsaturated Weston Cells

It is perfectly practical to use a saturated Weston cell, or a properly thermostated group of them, as a laboratory standard. Such cells and groups are commercially available with oil or air constant temperature baths. However, these cells are both temperature- and shock-sensitive. The *unsaturated* Weston cell is therefore preferred as a practical secondary or laboratory standard of voltage. These cells are identical to the saturated Weston cell except that the electrolyte is saturated with cadmium sulfate at 4°C; since no excess of cadmium sulfate is present, the cell is unsaturated at all other temperatures. These cells have a much smaller temperature coefficient, about -0.5 μV/K, and can withstand shipment if proper precautions are taken. Their potentials range between 1.0188 and 1.0196 V and are valid to $\pm 0.005\%$. However, these potentials decrease with time by 50 to 100 μV/yr, and are no longer reliable when their potentials are less than 1.0183 V; their lifetime is therefore about 10 years. Unsaturated cells should be checked at least annually against a saturated cell, which is stable for decades with proper treatment.

The internal resistance of Weston cells ranges from 100 Ω to 750 Ω; the higher values are found for saturated cells. The current drawn should never exceed 100 μA and the cells are destroyed by even brief short-circuits. Even unsaturated Weston cells require a week to settle after shipment or handling before a stable potential can be counted upon.

Unsaturated Weston cells of high quality are sold by Weston Electronic Instrument Corporation, Newark, New Jersey, and by Eppley Laboratory, Inc., Newport, Rhode Island; these are normally certified by the U.S. National Bureau of Standards.

Ruben Cells

Although Weston cells are superior to all others known (at the time of writing) in stability of voltage when used as reference standards, they are not designed for rugged use. Many commercial electronic devices operated from the AC mains are equipped with internal voltage standards. These voltage standards are commercial "mercury cells", or Ruben cells (Friedman and McCauley, S12), which come in standard commercial battery sizes. They are manufactured by Mallory Battery Company, Tarrytown, New York; Union Carbide (Eveready); and others.

Rechargeable Ruben cells can be made, but commercial cells used as voltage standards are not of the rechargeable variety. The cell, which uses an alkaline electrolyte, can be represented as either

$$Zn(s)/KOH, ZnO(s), H_2O/HgO(s), Hg$$

or

$$Zn(s)/KOH, K_2ZnO_2, H_2O/HgO(s), Hg$$

since the zinc oxide used in cell construction actually exists in the form of the zincate ion in the cell. The cell discharge reaction forms zinc oxide and mercury. The standard potential of the Ruben cell at 25°C is 1.344 V, which is in close agreement with the potentials of commercial cells. This potential is comparatively insensitive to both temperature, varying only 3 to 4 mV from $+50$°C to -50°C, and to the state of discharge of the cell. Unlike that of lead–acid or carbon–zinc cells, the discharge curve of the Ruben cell is virtually flat until the cell is exhausted. For this reason, and also because the nominal potential of a mercury cell, 1.35 V, is less than that of a carbon–zinc cell, 1.5 V, mercury standard commercial cells *must not* be replaced with other types. Such replacement should be done on a regular basis. The life of internal standard mercury cells will depend upon instrument use and design; they *may* serve for a year or more in some equipment.

Other Standards of Voltage

There is no need to employ standards of voltage other than the Weston cell in laboratory work, although other cells of somewhat inferior quality can be made. Any precision potentiometer, properly standardized against a Weston cell, is itself a standard of voltage; electronic voltmeters are less trustworthy in this regard.

Voltage *sources*, either powered from the AC mains or battery operated, should not be used as voltage *standards*, nor should standard cells be used as *sources* to power measurement circuits.

2.3.7   Sources of Voltage

Laboratory sources of voltage are of two types: batteries and sources operated from the AC mains.

Lead–acid batteries, which are nominally 2 V/cell so that a standard six-cell automotive battery will deliver 12 V, will on discharge have cell voltages ranging from 1.85 down to 1.00 V, with about 1.75 V being typical; normal full charge may reach 2.4 V, which is safe and a normal automotive float voltage. The carbon–zinc Leclanche cells are nominally 1.5 V, but actually range downward from 1.5 or 1.6 V, depending upon their state of discharge. Rechargeable nickel–cadmium batteries may also be used. Batteries have the advantage of being true DC sources of voltage with no inherent AC ripple component. For most electroanalytical purposes, however, modern regulated power supplies available from many commercial sources have sufficiently low ripple that they can be used, and are much preferable from the standpoint of convenience.

2.4   RESISTANCE

The SI unit of resistance, the *ohm* (Ω), is simply the quotient E/I of the volt and the ampere obtained from Ohm's law. The quotient I/E of the ampere and the volt gives the SI unit of conductance, the *siemens*. The siemens (S) is the reciprocal of the ohm as conductance is the reciprocal of resistance, and is identical to the now obsolete *mho*. Electrical resistance measurements are rarely used in electroanalytical work except as related to conductance measurements. Conductance and its analytical uses are the subject of Chapter 4, in which further details of these methods will be found. Direct or indirect resistance measurements are not possible in electrolytic systems using DC voltages or currents and Ohm's law. This is because of the capacitive reactance present in all electrode systems, as discussed further in Chapter 5, and also to the Faradaic effects of direct current passage, as discussed further in Chapter 6.

Ohm's law can be used to measure electrolytic resistance, but in a rarely useful way. If in the course of other studies the instantaneous potential difference and current flow are measured, their quotient correctly gives a resistance. This resistance, however, will be a function of the time-dependent values of the current and the potential difference.

2.4.1   Measurement: The Ohmmeter

Electrical resistance of the ohmic type—that is, impedance containing neither a capacitive nor an inductive reactance—can be measured in the laboratory in two ways. The first, the so-called *ohmmeter* or ohmmeter function on multifunction voltmeters, is simply a voltmeter reading the *IR* drop across a resistor through which a known current is being passed, or an ammeter reading the current through the resistance when a preset voltage drives a small current through it. The ohmmeter *cannot* give correct resistance measurements when a nonohmic component of resistance is present, as in electrochemical cells.

2.4.2   Measurement: Bridges

Resistance or its reciprocal, *conductance*, is most precisely measured using various balancing bridge circuits. Perhaps the most common resistance bridge circuit is the Wheatstone bridge shown in Figure 2.11. The operation of a Wheatstone bridge is simple; no current flows at balance, and at balance $R_c = R_3(R_1/R_2)$.

A resistance bridge such as the Wheatstone bridge is adequate for measurement of ohmic resistance or conductance but inadequate for measurement of electrolytic resistance or conductance. This is because electrolytic resistance contains nonresistive components, and, when an AC signal is applied as it necessarily must be to prevent electrode polarization, the cell exhibits an *impedance* rather than a resistance. The resistive component of the impedance imposes no phase shift, but the inductive reactance and the capacitive reactance do, in opposite directions. There is virtually no inductive component of impedance present in an electrochemical cell, but there is a capacitive component whose magnitude is quite significant, the *double-layer capacitance*. Thus, rather than balance a cell resistance as in the Wheatstone bridge circuit, it is necessary to balance a cell impedance. An impedance bridge such as the Wien bridge shown in Figure 2.12 is used for this purpose. The cell here is taken as a series combination of the electrolytic resistance, which is the reciprocal of the conductance, and the double-layer capacitance. The condition of balance is that the capacitive and resistive components must *both* be balanced, and their separate determination is therefore possible. Such bridges are generally fed with audiofrequency AC signals, nominally 1 kHz. Different types of bridges with similar functions are also used. The precision of bridge measurements can be as high as 0.001%, but for titration use 0.01% or 0.05% precision is sufficient.

If the amplitude of the AC signal is larger than a few millivolts, or if the frequency of the AC signal is greater than about 1 kHz, the simple series circuit model of the electrolytic cell breaks down. The stray capacitance of the cell and of the electrode leads, which appears in parallel with the series circuit, now becomes significant. More important, the double-layer capacitance is paralleled by a significant leakage path for Faradaic current, which appears as a capacitive and resistive network. Analysis of these more complex circuits is beyond the normal scope of electroanalytical chemistry, and is taken up in the general references of Chapters 4 and 5.

2.4.3   Standards of Resistance

The SI unit of resistance, the ohm, is defined in terms of the ampere and nonelectrical units of the SI; this definition does not lead to practical standards of

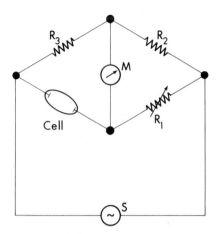

Figure 2.11   Basic Wheatstone bridge circuit. M, current detector such as galvanometer; R_1, R_2, R_3, resistors; R_c is the resistance of the electrochemical cell.

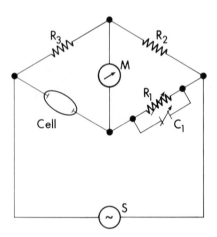

Figure 2.12   Basic Wien bridge circuit. Measuring device M must respond to AC.

resistance or conductance. Standards of resistance are based upon the standards of voltage and current and upon Ohm's law. The use and fabrication of such standards is a practical and reasonably simple laboratory procedure.

Resistance standards invariably consist of a length of alloy wire of appropriate cross-sectional area packaged with appropriate leads. Resistance standards or precision resistors constructed of wire such as manganin have a very low temperature coefficient, about 0.001%/K at room temperature, and are stable to within 1 ppm over 25 years if properly handled. Precision resistors are constructed by trimming of resistance wire to the appropriate length and balancing them in a bridge against a known standard that was calibrated using Ohm's law in conjunction with voltage and current standards.

The most precise standards of resistance commercially available are certified fixed resistors (sold by the Leeds and Northrup Company and the Rubicon Company, Philadelphia, Pennsylvania), which are reliable to a few thousandths of a percent. More common laboratory standards are fixed precision resistors, usually with a precision of 0.05% or 0.1%, sold by electronic supply houses such as General Radio Company, Cambridge, Massachusetts. Decade boxes containing fixed precision resistors are convenient but expensive.

Excessive current will alter the resistance of standard resistors by Joule heating, since heat increases the resistance of wire-wound resistors. This is not normally a problem unless the maximum current specified by the manufacturer is exceeded, because the temperature coefficient of the wire is low. Excessive current can also *destroy* such resistors by Joule heating, and must not occur, even momentarily. Immersion in oil or water baths is rarely helpful; such high-wattage resistors as require this are uncommon in electroanalytical laboratories.

Standards of resistance are designed to function properly only in DC circuits. Although they may be used safely within their power-dissipation limits in AC circuits, their impedance should not be assumed to equal their resistance. Inductive reactance can be significant in high-frequency applications or when the signals, such as square waves, contain high-frequency components. Standards of impedance are not normally used in electroanalytical chemistry, although standards of inductance and of capacitance suitable for laboratory use are commercially available from Leeds and Northrup Company and other manufacturers.

2.4.4   Sources of Resistance

Ordinary commercial electrical resistors serve adequately as sources of resistance for electroanalytical purposes and last indefinitely as long as their wattage rating is not exceeded. Dummy cells for testing electroanalytical equipment can be constructed from ordinary resistors and capacitors, with occasional addition of diodes in a back-to-back configuration. An ordinary commercial resistor calibrated using a resistance bridge and standard resistors can serve as a secondary standard of resistance.

2.5   TIME

The SI unit of time, the *second* (s), is a fundamental unit of the SI. Time is an important parameter in modern electroanalytical measurements, since the trend has been to shorter measurement times known with greater accuracy. It is therefore necessary to consider here devices and standards for measurement of time, although most of these have been developed and marketed for other purposes. Measurement of absolute or chronological time has no place in electroanalytical chemistry; it is the measurement of intervals of time or relative time that is significant.

2.5.1 Mechanical Devices

Measurements of precise time intervals longer than a few minutes are rare in electroanalytical studies. For these occasional measurements of long time intervals, ordinary precise clocks or hand-operated stopwatches are often of sufficient precision. For shorter periods of time, these devices are inadequate since they cannot be of greater precision than the response time of their human operators or recorders.

Electromechanical devices of greater precision, and certainly greater reliability, eliminate the human response time error. Such devices consist of a clock operated by a synchronous motor. The motor is turned on and off by one pole of a double-pole switch or relay; the second pole connects the electrochemical cell to the desired source of current or voltage. Since the poles operate essentially simultaneously, the times of motor operation and of cell operation are identical, and the time shown on the clock will be the time during which the electrochemical cell was connected into the circuit.

Such devices have been widely used in coulometry, and precision of 0.1% is easily achieved by simple commercial circuits for coulometric times of a few minutes or more; the 0.1% precision applies *per operation*, however, and the error in coulometric measurements using intermittent currents may be many times as great as that using a single measurement. The measurement of the time elapsed is limited in precision by the precision of the device itself as well as by the precision of the frequency of the AC mains , which determines the rotation speed of the motor.

Further improvements in the measurement of time by electromechanical devices is possible, although for precision better than 0.1% the writer prefers all-electronic units. Electromechanical devices precise to 0.01% are commercially available, designed to minimize lag and coasting of the motor by a solenoid-operated brake. The error per operation of the device should not be taken at face value, however, since the motor rotation is locked to the 60-Hz frequency of the commercial AC mains, which may vary by 0.1% or more over intervals of a few minutes even though the generators are adjusted so that over a period of 24 h or more the frequency variation is less. Studies of the significance of this factor are available (Craig, Satterthwaite, and Wallace, S13; Fry and Baldeschweiler, S14; Gerhardt, Lawrence, and Parsons, S15); institutional power supplies may be even less reliable than larger generating systems. The solution to this problem of the AC mains, a precise local power supply, is not generally practical and, even if practical, would not resolve the inherent mechanical factors of the clock that limit precision. An all-electronic unit is to be preferred.

2.5.2   Electronic Counters

Electronic time-measuring devices of high precision are digital devices and are counters rather than clocks, although their output may appear in clock form. They count either the oscillations of the 60-Hz AC mains or the oscillations of a precise-interval electronic oscillator. Electronic timers that use the AC mains as oscillator cannot be more precise than the frequency of the AC mains themselves as discussed above. Moreover, the precision of the counter is always plus or minus one count, or 1/60 s, and so any counter operated from the AC mains is only satisfactory for periods of a few minutes or more.

Electronic timers that contain precise internal oscillators surmount the limitations of those based upon the AC mains. The precision of the timer is then determined by the precision and frequency of the internal oscillator. The frequency is usually at least several kilohertz so that the one-count limitation is negligible. The precision of the oscillator frequency is determined by its design. The best oscillators employ crystal-controlled designs, for which the oscillator frequency is determined by the physical dimensions of the crystal itself. The crystal dimensions, and therefore the frequency, change slightly with temperature; thus, precise thermostatic control of the environment of the crystal is used to obtain extreme precision.

2.5.3   Recording Devices

Time intervals in electroanalytical studies are often measured as distances between two events registered during continuous recording of some output signal. These signals are normally varying voltages (or varying currents that by passage through a fixed resistor are converted to varying voltages). The recording device is either a potentiometric recorder of the x–t type or an oscilloscope with a time-base horizontal sweep. The latter requires photographic recording or a storage oscilloscope.

The precision of either recorder or oscillographic time measurement depends upon the electroanalytical technique used as well as upon the characteristics of the recording device, and no generally useful quantitative comments can be made here. However, certain useful comments on the measurement procedures can be made.

Potentiometric strip-chart recorders operate with a synchronous motor geared down to drive the chart at some selected constant rate. These gear trains and motors have in them a certain amount of slack, and when the chart drive is turned on it may take a second or so for the drive to reach correct speed. It is therefore best to start the chart drive motor just before, rather than simultaneously with, the signal imposed upon the cell. If the signal is triggered manually, as most are, it is useful to start the chart drive and then start the signal as the pen crosses a convenient vertical line. A similar problem exists with triggered-sweep oscilloscopes in that the sweep starts after, rather than immediately upon, the oscilloscope trigger signal. If the oscilloscope trigger signal is given just before the signal is imposed upon the cell, the initial-condition information is not lost. The magnitude of the difference desired depends upon the technique used. Some modern research oscilloscopes are equipped with delay circuitry that can be used to accomplish this. The writer has used a mercury-wetted relay to close the circuit to the cell, while the activation voltage used to close the relay also serves as the oscilloscope trigger signal; the slight mechanical delay in the relay is sufficient for comparatively slow sweep rates.

Measurement of time depends upon the use of a grid inherent in the paper or the photograph that has been used to make the recording, or alternatively upon ruler measurement of distance. Suitable grids are found on commercial strip-chart paper or upon the flat sheets used for x–y recorders (which have no chart drive slack but are subject to error in paper placement). A grid may be imposed upon oscilloscopic photographs by a double exposure of each film, once to obtain the grid and once to obtain the output, if the grid can be illuminated with the camera fixed in position. It is not normally possible to obtain a clear grid and a clear signal in the same exposure, because different exposure and lighting conditions are required.

Oscilloscopic sweeps are not inherently less precise than recorder chart drives, but the inferior precision obtained in photographic recording from an oscilloscope face makes chart recording the preferred method; photographic recording is used only for output signals that change too rapidly to be followed by potentiometric recorders.

2.5.4   Standards of Time

The fundamental standard of time is the operational definition of the SI base unit of time, the second, as the duration of 9,192,631,770 periods of the radiation corresponding to the transition between the two hyperfine levels of the ground state of the cesium-133 atom. Frequency–time standards based on this definition are commercially available at a cost of about $25,000 (Hewlett-Packard); deviations between units and from the USNBS standard are 1 part in 10^{12}. Rubidium-based standards are less precise, about 1 part in 10^{11}, at about half the cost of a cesium standard. Neither is commonly found in electroanalytical laboratories.

Quartz oscillators of high quality have a frequency stability of at best 5 parts in 10^{12} over a short term. Over longer periods of time, such as days, even temperature-controlled quartz crystal oscillators can be relied upon to only perhaps 1 part in 10^{9}. These are the normal secondary and laboratory time standards. Some electroanalytical instruments incorporate crystal-controlled oscillators as internal time standards, but it is not common for the crystals to be thermostated. Such oscillators are often of the same 1- to 100-MHz types as are used for microprocessor clocks. Their typical precision is ±0.0025% of nominal frequency at 25°C; a further variation of ±0.0035% (of the 25°C value) occurs over the operating temperature range of –20° to +70°C. This precision is adequate for virtually all electroanalytical work.

2.6 REFERENCES

General

G1   H. V. Malmstadt, C. G. Enke, and S. R. Crouch, *Electronic Measurements for Scientists*. W. A. Benjamin, Menlo Park, 1974, 906 pp.

G2   H. A. Strobel, *Chemical Instrumentation: A Systematic Approach, 2nd ed.*, pp. 160–185, and references cited therein.

G3   J. A. Von Fraunhofer and C. H. Banks, *Potentiostat and Its Applications*, Butterworths, London, 1972, 254 pp.

G4   W. M. Schwarz and I. Shain, *Anal. Chem.* **35**, 1770 (1963), and following articles in this issue.

Specific

S1   T. W. Richards and G. W. Heimrod, *Proc. Amer. Acad. Arts Sci.* **37**, 415 (1902); **44**, 91 (1908); *J. Amer. Chem. Soc.* **37**, 692 (1915).

S2   E. B. Rosa and G. W. Vinal, *Bull. U.S. Nat. Bur. Stand.* **13**, 447, 479 (1916).

S3   E. W. Washburn and S. J. Bates, *J. Amer. Chem. Soc.* **34**, 1341 (1912).

S4   R. A. Lehfeldt, *Phil. Mag.* **15**, 614 (1908).

S5   J. J. Lingane, *J. Amer. Chem. Soc.* **67**, 1916 (1945).

S6   J.A. Page and J.J. Lingane, Anal. Chim. Acta *16*, 175 (1957).

S7   J. J. Lingane, *Electroanalytical Chemistry, 2nd ed.*, Wiley-Interscience, New York, 1958, pp. 452–459.

S8   A. Hickling, *Trans. Faraday Soc.* **38**, 27 (1942).

S9   L. Clark, *Proc. Roy. Soc. (London)* **20**, 444 (1872).

S10   E. Weston, German Patent 75,194 (Jan. 5, 1892).

S11   P. Vigoureaux and S. Watts, *Proc. Phys. Soc. (London)* **45**, 172 (1933).

S12   M. Friedman and C. E. McCauley, *Trans. Electrochem. Soc.* **92**, 195 (1947).

S13   R. S. Craig, C. B. Satterthwaite, and W. E. Wallace, *Anal. Chem.* **20**, 555 (1948).

S14   E. M. Fry and E. L. Baldeschweiler, *Ind. Eng. Chem., Anal. Ed.* **12**, 472 (1940).

S15   G. E. Gerhardt, H. C. Lawrence, and J. S. Parsons, *Anal. Chem.* **27**, 1752 (1955).

2.7 STUDY PROBLEMS

2.1 Compute the charge that must be passed in an electrolytic cell used for the purification of blister copper through the $Cu(II)/Cu$ redox process if exactly 1 kg of copper metal is desired. The current through the cell is 100 A. Compute how long this electrolysis would take.

2.2 Given that the density of metallic thallium is 11.85 kg/dm^3, compute the total charge necessary to deposit one monolayer of thallium metal on a hanging mercury-drop electrode (HMDE) of area 0.0176 cm^2 from a solution of thallium(I) ions.

2.3 A student measured a potential difference of 1.10 V in a cell whose internal resistance was known to be 500 Ω on the 1.5-V scale of an electromechanical voltmeter whose figure of merit is known to be 20 $k\Omega/V$. By how much does the resultant reading differ from what the potential difference would have been with no voltmeter in the circuit? Repeat this calculation for cell resistances of 50, 5000, and 50,000 Ω.

2.4 A time-measuring circuit employs a 5-kHz oscillator in conjunction with a digital counter. Calculate the absolute precision possible with this time-measurement circuit and the relative precision possible in the measurement of a 1-min time interval. Repeat the calculation for a similar circuit employing the 60-Hz line frequency.

2.5 A strip-chart recorder is used to record a plot of current against potential. It is necessary to measure the potential at which the current rises sharply. If the horizontal scale of the recorder is 50 mV/cm, estimate the best precision of the voltage you could expect to measure on the chart. (Use ordinary 10 × 10 cm graph paper, and estimate the length error by repetitive measurement of the distance between two nonadjacent grid lines.)

2.6 In a certain electroanalytical technique, the output of the sensors is in terms of current that must be recorded accurately as a function of time. The current is significant only *after* a conditioning sequence whose time is about 3 min and *before* the effects of voltage changes affect the measurements, which will begin to happen within 100 ms after the end of the conditioning sequence. The conditioning sequence is controllable either manually or automatically, and an electrical trigger signal can be obtained at the end of it. Explain how you would measure this current and what precision the current measurement would then have.

CHAPTER
3

STRUCTURE OF
ELECTROANALYTICAL CELLS

Electroanalytical chemistry is that branch of chemical analysis that employs electrochemical methods to obtain information related to the amounts, properties, and environments of chemical species. The overlap between electroanalytical chemistry and chemical analysis, on one hand, and electroanalytical chemistry and physical electrochemistry on the other, is considerable. Electroanalytical chemistry has made contributions to and received contributions from the two fields it spans. While electroanalytical chemistry can reasonably be considered a subdivision of analytical chemistry generally, and hence the distinction is one of subclassification only, there is a real philosophical difference between electroanalytical chemistry and physical electrochemistry, which is often called simply "electrochemistry." Physical electrochemistry is concerned with the theory of electrode processes and with the application of those theories. Hence, physical electrochemistry concentrates primarily on the electrode process itself and secondarily upon the ionic phase, generally a liquid that is a solvent for ions, in which the electrode is placed. Electroanalytical chemistry is concerned with the theory of electrode processes insofar as, and only insofar as, that theory is concerned with a property of some material *other than those of the electrode and the solvent*. This material may be a sought-for constituent in a sample, a material that is to be studied, or an interference whose properties or presence is not desired. The purpose of the study may not involve the development or improvement of method of analysis, but it will be related to the measurement of the properties of substances, for that is the purpose of chemical analysis.

The philosophical difference between physical electrochemistry and electroanalytical chemistry is reflected in works on the subject. Many textbooks on physical electrochemistry exist and are listed at the end of Chapter 5. Those books and monographs dealing with the different specific areas of electroanalytical chemistry are listed at the end of their respective chapters. The comparatively small number of works covering the full scope of electroanalytical chemistry are listed at the end of this chapter.

3.1   THE ELECTRODE AS A PROBE

Isaac Maurits Kolthoff (S1) once defined electroanalytical chemistry as the application of electrochemistry to analytical chemistry. Such a definition is partially accurate, but if analytical chemistry is thought of as the art of determining how much of what is where, it may be misleading. It is preferable to consider electroanalytical chemistry as that area of analytical chemistry and electrochemistry in which the electrode is used as a *probe*, to measure something that directly or indirectly involves the electrode, but in which the desired information is not the fundamental operation of the electrode process. The fundamental operation of the electrode process is rightly the concern of physical electrochemistry. Thus in the main areas of present electroanalytical development, such as voltammetric techniques, microcoulometry, and ion-selective electrodes, the basic analytical thrust is the development, study, and improvement of probes for studies whose fundamental interest lies beyond that of the electrode process.

It is then fitting that electroanalytical study be centered on the electrode, considered as a probe in the electrolyte. Fundamentally, an *electrode* is a junction between an ionic conductor and an electronic conductor; it is the *interphase*, or more correctly the *interphase region*, in which current changes from being carried by the movement of ions in the electrolyte to being carried by the movement of electrons in the electrode proper. In practice, since a complete circuit is always needed for the continuous flow of electric current carried either by ions or by electrons, every such circuit must involve at least two electrodes—two junctions, or junction regions, where the type of conductance changes. These two regions are necessarily heterogeneous, not homogeneous, because an interphase must be present. An interphase is regional, not a planar phase boundary as an *interface* might be. It is actually two adjacent regions—one extending out from the electrode into the electrolyte, which is *not* identical to the bulk solution, and another extending from the surface of the metal electrode inward, which is *not* identical to the bulk of the electrode. The interphase region must be looked at somewhat like a ham sandwich in which two slices of ham are the two parts of the actual interphase, and the two slices of bread are the bulk electrode and the bulk solution. It is clearly a *region*, and an anomalous region, rather than a plane of demarcation.

Analytical chemists usually create, insert, or probe with electrodes in order to find out something about the solution, or electrolyte, in which the electrodes are inserted. Thus we tend to think too much about the bulk solution and not enough about the electrode or the interphase region that surrounds it. This can be unfortunate, because that which is measured by an electrode is that which occurs in the interphase region, not that which occurs in the bulk solution. For example, if the electrode reaction results in the production or loss of hydrogen ions, the pH in the interphase region may differ drastically from that in the bulk solution. This is often a problem in the study of organic compounds, most of whose electrochemical reactions involve gain or loss of protons, and thus buffers must be used in these studies. But even then the kinetics of the buffer reactions must be rapid relative to any electrode processes involved, and the concentration of buffer *in the interphase region* must be high enough to supply or absorb all of the protons involved, so that the pH in the interphase region remains constant. Any additional millimoles of buffer outside the interphase region are of no value until they can enter the interphase region by diffusion or convection.

3.2   ELECTROANALYTICAL CHEMISTRY: ADVANTAGES AND LIMITATIONS

At an electrode surface, a transfer between electronic and ionic conduction takes place. Thus electroanalytical techniques can deal only with material in solution that can reach the electrode surface. Some relationship, direct or indirect, must exist between ions and the substance or phenomenon being studied. Thus electroanalytical chemistry must be done in some medium, or electrolyte solution, in which ions can exist. These solutions are often aqueous, although any medium in which ions can exist and the substances are soluble can be used.

The metallic conduction side of the interphase is rarely of itself a limiting factor since almost any metal will serve as a location for electronic conduction. Noble metals are usually used to avoid reaction between the electrode and components of the electrolyte. The usual choices are platinum, palladium, gold, silver, and mercury, of which mercury is the most common since it is the only liquid noble metal.

Electroanalytical techniques can be used to study anything that directly or indirectly undergoes a reaction involving electron transfer. Although the inorganic tradition is particularly strong in electroanalytical chemistry and many determinations of inorganic metal ions have been made, many organic compounds can be determined directly and even more indirectly. The electroanalytic controllable parameter of potential is similar to the spectroscopic controllable parameter of wavelength in that it permits adjustable selectivity in analysis, but the selectivity of electroanalytical methods is normally inferior to that of spectroscopic methods.

Another significant limitation of electroanalytical techniques is the fact that there must always be at least two electrodes, and hence one must consider not only what is going on at a single probe electrode but also what is going on at the second electrode, even if the second electrode is not of direct interest and is present only to complete the circuit. In practice, we usually want to consider only one electrode at a time; in principle, both must always be considered at the same time.

Suppose, for example, that we want to study hydrogen evolution from acid solution at a single electrode such as platinum. Two physical electrodes are needed. These can be separated into two half-cells, the only connections between them being one ionic conductor and one electronic conductor. We can then speak of half-reactions, each occurring in a half-cell, with charge flowing between the half-cells through these conductors. It might be thought that such physical separation of these two electrodes would now enable concentration on only one of them. Unfortunately, that is not the case. By such physical separation the *products* of one electrode reaction can be separated from the other electrode, because with either a porous separator or a salt bridge of proper design migration will not move them fast enough to interfere with normal electrochemical measurements. There are some problems introduced by the ionic conductor, such as possible contamination by the salt bridge ions and increased cell resistance, but these can be overcome. But separation of the *products* does not thereby produce separation of the *observations*. Suppose we impose upon such separated half-cells a potential difference, or make potential difference the independent variable, while observing current as the dependent variable. If the two electrodes and the solutions in which they are immersed are identical, the reaction at one of them will be the hydrogen evolution we wish to observe, while at the other the reverse reaction will take place. The reaction in which we are not interested, however, occurs with greater difficulty. Hydrogen dissociation is very slow compared to the reverse reaction of hydrogen evolution because the breakup of molecular hydrogen to individual atoms is slower (by orders of magnitude) than the recombination step. Thus the actual current in the circuit will be controlled, or limited, by the reaction that is not of interest, rather than the reaction of interest. It is therefore necessary, although it may not be simple, to make certain that the electrode it is intended to study is the only one actually being studied, by proper attention to experimental design.

3.3   CELL ELECTRODES

An electrochemical cell or device must consist of at least two electrodes and one electrolyte. By *electrode* is meant an interface at which the mechanism of charge transfer changes between electronic (movement of electrons) and ionic (movement of ions), and by *electrolyte* is meant a medium through which charge transfer can take place by the movement of ions. In a less formal sense, we use *electrode* to indicate the electronic conductor and *electrolyte* to indicate the ionic conductor in an electrochemical cell. In either event, the simplest possible electrochemical cell can be visualized as in Figure 3.1.

In any electrochemical cell, there must be a minimum of two physical electrodes used. In a cell used for electroanalytical measurements, however, there are always three electrode functions. This is more obvious in nonequilibrium measurements when significant current is drawn from the cell, because the three electrode functions are then normally carried out by three different physical electrodes, but it remains true for equilibrium measurements even if these employ only two physical electrodes.

The first of the three electrodes with their associated functions is the *indicating electrode*. This is also known as the *test electrode*, and, undesirably, as the *working electrode*. This is the electrode at which the electrochemical phenomena being investigated are taking place. Its function is to serve as a location for

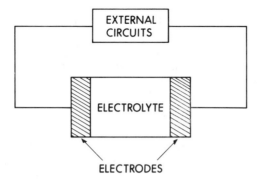

Figure 3.1   Electrochemical cell.

electrochemical measurements. It may or may not be constructed of inert material.

The second functional electrode is the *reference electrode*, which is also known as the unpolarized electrode or unpolarizable electrode. This is the electrode whose potential is constant enough that it can be taken as the reference standard against which the potentials of the other electrodes in the cell can be measured. By "constant enough" we mean that the change in its potential with current, time, or other variables need not be zero, but must be sufficiently small that the measurements are not adversely affected. It should not be, and normally is not, constructed of inert material.

The third functional electrode is the *counterelectrode*, which is also known as the auxiliary electrode. This is the electrode that serves as a source or sink for electrons so that current can be passed through the cell. In general, neither its current nor potential is measured or known. It is usually constructed of inert material.

The indicating electrode cannot be combined in function with either of the others, and therefore always exists as a separate physical electrode, but the other two electrode functions can sometimes be combined so that only two physical electrodes are present in the system. The only advantage of this is greater convenience, particularly in cell geometry and in external cell instrumentation. The disadvantage is that the functions of the reference electrode and counterelectrode are basically incompatible, in that the stability of the potential desired of the reference electrode will of necessity be adversely affected by the passage of current required of the counterelectrode. Only if the magnitude of the current in the cell is known to be small enough that this adverse effect, called polarization of the reference electrode, is negligible can a two-electrode cell be used in analytical electrochemistry. Such is the case in classical aqueous polarography using a saturated calomel reference electrode, for example. The use of a two-electrode cell in preference to a three-electrode cell will never constitute a real improvement; it may produce a physically simpler cell but never one that is better characterized.

3.3.1   Electrode Current Density, Area, and Function

The important parameter in relation to variation in potential caused by flow of current through an electrode is not the current but the *current density*, or current per unit area. Current density is usually symbolized by j, while current is usually symbolized by I. In electroanalytical work, the area of the electrode is taken as the geometric area unless otherwise specified. Thus the area of a square foil electrode is taken as its measured square area counting on both sides of the foil plus the minor contributions from the edges and the supporting wire, and the area of a wire electrode is taken as that of a cylinder of the length and diameter of wire, which is the product of π, the diameter, and the length. The real area, or area on a molecular scale, is proportional but not equal to the geometric area. The ratio of real area to geometric area is always greater than or equal to unity; for liquid metals such as mercury it is approximately unity, but for solid electrodes it is several to many times greater than unity. This ratio, the *roughness factor* of the electrode, is only approximately constant from one electrode material to another even when the surface preparation procedures are identical. It differs radically between liquid and solid electrodes, and may be different for the same electrode used in different or successive experiments. This factor must be kept in mind when comparing measurements made in different systems or cells. Polishing, or even electropolishing, the surface of a solid electrode can reduce the roughness factor, but it will still be much larger than unity.

For the indicating electrode the current drawn depends on the measurement being made, so that the current density at this electrode may vary considerably from one type of measurement to another. For the reference electrode, since all net current flow works against the stability of the reference potential, or polarizes the reference electrode, the current density should be as close to zero as possible. This can be achieved by making the reference electrode current small, the reference electrode area large, or preferably both. Where possible the area of the reference electrode is made much larger than that of the indicating electrode. For the counterelectrode, it is impossible to make the current small because the current is necessarily equal to the current flowing through the indicating electrode. It is nevertheless desirable to make the-counterelectrode current *density* much less than that of the indicating electrode so that limitations on the current through the cell imposed by cell reaction will arise at the indicating electrode, where they will be observed, rather than at the counterelectrode, where they will not. Hence it is necessary to make the area of the counterelectrode much greater than that of the indicating electrode whenever possible, or provide in some other way that significant polarization of the counterelectrode does not occur. It is also necessary, of course, to ensure that products of reactions at the counterelectrode and the reference electrode do not reach the indicating electrode, where they will be observed. This normally requires the use of porous separators or salt bridges to compartmentalize the electrolyte.

The potentials of some reference electrodes will not change when currents of a few microamperes, the normal magnitude of polarographic currents, pass through them. Such an electrode can serve simultaneously as reference electrode and auxiliary electrode. The saturated calomel electrode (SCE), which is the usual reference electrode used in polarography, is unpolarizable (has virtually no change in potential) under polarographic conditions so that a two-electrode system is common in aqueous polarography. This simplification will not work properly for nonaqueous systems, because resistance and hence ohmic drop is orders of magnitude greater than in aqueous work, or for reference electrodes that are more polarizable than the SCE, and is fundamentally inferior to a three-electrode system.

3.4   ELECTRODE CONSTRUCTION

Some types of electrodes are specific to certain electroanalytical techniques and are discussed with them in the appropriate later chapter. The more generally useful electrodes are *inert electrodes* and *reference electrodes*. Inert electrodes are electrodes whose materials of construction do not become involved in chemical or electrochemical reactions taking place at their surfaces. Inert electrodes are useful as counterelectrodes and, for many nonequilibrium methods, as indicating electrodes. No truly inert electrode material usable in all cell electrolytes at all potentials exists, but the more noble metals are preferred; normal choices are, in the order given, mercury, platinum, gold, and, less often, silver and palladium. Noble metals have the advantage that they are generally not attacked during passage of current and therefore the available potential range is limited only by the electrolyte.

One must occasionally consider the absorption of a reactant or product into the basic structure of the electrode. This behavior is most notable with gases, although the formation of metal alloys is a very similar phenomenon widespread in fused-salt electrochemistry, as in the deposition of lithium on platinum. It is especially important in the case of hydrogen in Pd–H systems, but the solubility of hydrogen in other metals is also significant.

3.4.1   Liquid Inert Electrodes: Mercury

Electrochemical processes are heterogenous processes that take place at electrodes, that is, at *interphases* between an electronically conducting medium (the electrode) and an ionically conducting medium (the electrolyte). The phase junctions can be liquid–liquid, solid–liquid, or solid–solid. Some of the characteristics of the interphase region are generally ascribed to the electrode proper, as constrasted with, say, the Nernst diffusion layer, which is normally ascribed to the electrolyte, or with phenomena such as adsorption that are related to the electrode reaction.

At the molecular level, the surface of even a highly polished solid electrode resembles the Himalaya mountains more than it does a flat plain. The surface area of such an electrode is much larger than its geometric area, and the various sites on it are by no means equivalent. Qualitatively, the difference between real and geometric area can be described by a "roughness factor" that is the ratio between real and geometric area, but quantitatively this breaks down because the surface of a solid contains nonequivalent sites. For liquid electrodes, the surface area is comparable to the geometric area, and the sites can be considered equivalent. The number of electronically conducting liquids, however, is very small. In practice, the only liquid electrode used is mercury, whose liquid range of -39 to $+357°C$ includes almost all electroanalytical measurements. Gallium has been tried in high-temperature applications, but is oxidized so easily it is virtually useless.

For routine electroanalytical work, triply distilled mercury available from chemical supply houses is of sufficient purity. The quantity of mercury used in an electroanalytical laboratory and the ease of recycling mercury, however, make its purification useful. In some cases the impurity level in available commercial mercury is too high; this frequently becomes a problem in anodic-stripping voltammetry.

There are three classes of impurities that are normally found in mercury used in the laboratory: surface oxide and scum, dissolved base metals, and dissolved noble metals. Surface oxide and scum is removed by filtration through perforated filter paper ("pinholing") until a clean surface is obtained. Base metals such as zinc and cadmium are oxidized out of the mercury by drawing air through the mercury under a 2-mol/dm^3 nitric acid solution with a vacuum aspirator. Compressed air should *not* be used since it often contains traces of oil from the compressor. This process should be carried out in a hood (for ventilation), in a large basin (to catch spills), and with a trap between the vacuum flask and the aspirator (to catch mercury bumping over). For mercury that is not grossly contaminated, the process is completed in 2 or 3 days; when bubbles of mercury 1 to 3 cm in diameter persist in the solution for a few seconds, the base metals have been removed. The mercury can then be washed, dried, pinholed, and used. The mercury so purified has a concentration of base metals lower than that in commercial distilled mercury, but the noble metals (platinum, gold, silver) have not been removed. These can be removed only by distillation, which is usually done under vacuum or in a stream of air. Further details of mercury purification are given by Coetzee (S2).

Mercury is toxic, and spills should be contained (by trays under apparatus) and removed (by suction or a spiral of amalgamated copper wire); vapor pressure of mercury is reduced by surface dirt or powdered sulfur. Although there is no significant hazard in reasonable work with mercury in an electroanalytical laboratory *that is well-ventilated*, distillation or high-temperature work is not without hazard owing to the much higher vapor pressure of mercury at elevated temperatures.

Mercury can be used for reductions over a wide range of potentials, but is fairly easily oxidized, especially in chloride media. Mercury is more useful than solid noble metals for reduction because reduction of aqueous hydrogen ion on mercury has a much higher overpotential than it does on the noble solid metals; thus reductions that are inaccessible because of hydrogen evolution on other metals can be studied on mercury. Solid noble metals such as platinum are used at more positive potentials, but the replacement of mercury by such a metal involves the addition of irregular surface, and a uniform surface behavior can no longer be expected. For example, voltammograms run on a rotating platinum electrode are always less well-defined than similar polarograms run on a dropping-mercury electrode, though the currents obtained are greater.

3.4.2   Construction of Mercury Electrodes

Many types of mercury electrodes have been constructed. Probably the most widely used is the *dropping-mercury electrode* (DME) used in polarography, which is taken up in Chapter 15. The simplest form of mercury electrode, the *mercury-pool electrode*, is simply a pool of mercury placed in the bottom of a cell. This electrode is useful for controlled-potential coulometry, chronopotentiometry, and other purposes. In a Pyrex cell, it is most conveniently constructed by leaving a depression of appropriate size in the bottom of the cell and sealing a short length of platinum wire through the bottom or side of the depression to establish electrical contact with the external circuit. A stopcock (preferably Teflon) at the bottom of the cell to serve as a drain is sometimes convenient. Alternatively, a vacuum aspirator inserted through the top of the cell can be used to remove the mercury as well as the electrolyte. When more than one mercury-pool electrode is necessary in a cell or when the bottom of the cell is not available, small pool electrodes can be constructed into the ends of J-shaped Pyrex tubing and dipped into the cell. Platinum wires and contacts in the long arm of the J-tube, which extends above the cell electrolyte, serve for electrical contact.

The *hanging-mercury-drop electrode* (HMDE) is probably second only to the DME in popularity for nonequilibrium electroanalytical measurements. Such an electrode can be prepared by sealing a platinum wire into the end of a soft glass tube, polishing the end flat, etching the platinum back slightly with aqua regia, and amalgamating the platinum. Drops falling from an ordinary DME can be caught in the electrolyte by a small Pyrex spoon and hung on the etched-back end of the wire while still under the electrolyte (Ross, DeMars, and Shain, S3). It is far more convenient to extrude individual drops through a vertical capillary from a mercury reservoir by means of a micrometer-driven syringe. Apparatus for doing so is commercially available (Metrohm, Princeton Applied Research). More elaborate devices for extrusion of mercury drops from a static reservoir under electronic control are also available (Princeton Applied Research).

The *mercury thin-film electrode* (MTFE) consists of a thin (1- to 100-μm) film of mercury laid down on some solid electrode support so that the electrode exposes only mercury to the electrolyte. Gold is not a suitable solid electrode substrate, because gold amalgam is formed, but platinum may be used. Carbon is preferable, however, because it does not interact with mercury. The preferable forms of carbon are the wax-impregnated graphite electrode (WIGE) or glassy carbon polished flat because a smoother and less porous surface can be obtained for deposition. The film is produced by cathodic deposition from an acid aqueous solution of Hg(II) in a cell with a mercury pool anode. In stripping voltammetry an MTFE can sometimes be prepared in the course of the procedure; this is discussed in Chapter 16.

3.4.3   Solid Inert Electrodes: Platinum, Gold

A considerably wider choice of solid noble metals is available. In order of frequency of use they are platinum, gold, and silver, followed by the occasionally used palladium, rhodium, and iridium. Various polycrystalline forms including sheets, rods, and wires are commercially available in high purity, and the materials are readily machined to useful shapes. Single crystals of the noble metals can be prepared and have been used in electrode studies beyond the scope of this work. All of the noble metals have an overpotential for hydrogen evolution much less than the overpotential for hydrogen evolution on mercury, and in water and other protic solvents they are usable to potentials about 1 V less negative than is mercury. On the anodic side they are less easily oxidized than is mercury, and the potential range extends about 1 V more negative than it does on mercury in most solvents. Like mercury, the noble metals are more easily oxidized and therefore have a shorter potential range when the electrolyte contains complexing anions such as halides or cyanide. The effect of these is most significant for gold, which forms stable tetrahalo and tetracyano complexes of Au(III).

Solid noble metal electrodes are not as chemically inert as might be assumed, and the preparation of a reproducible solid electrode surface is not a trivial problem. All of the noble metals adsorb hydrogen on their surfaces, although gold does so to a lesser degree than platinum. Palladium absorbs hydrogen into the bulk metal in appreciable quantities and should not be used as a cathode in protic solvents.

3.4.4   Construction of Platinum and Gold Electrodes

Fabrication of large platinum metal electrodes from platinum foil is possible with ordinary scissors; contact to a platinum electrode is easily achieved by welding the foil to a platinum wire. Heating of the two pieces to red heat (glassblowing torch) and a light hammer blow on a flat surface forms a proper weld. The wire can then be sealed in soft glass or, preferably through a graded seal, into Pyrex. Platinum microelectrodes can be prepared by sealing a wire through the closed end of a soft glass tube, followed either by cutting off the excess and polishing (the usual method) or by melting back the platinum to a small ball with a glassblowing torch. The geometric areas of these microelectrodes are similar to the area of an HMDE, but the roughness factor is greater, and the real surface area is considerably larger.

Surfaces of solid electrodes are much more difficult to prepare reproducibly than are mercury surfaces, especially if the measurements made are sensitive to electrode area, as almost all nonequilibrium electroanalytical measurements are. Only planar electrodes can generally be given a thorough pretreatment. The planar electrode is first polished until it is smooth and bright in the manner of a metallographic specimen using successively finer grades of abrasive such as alumina, silicon carbide, or jeweler's rouge. The electrode is then rinsed, dried with tissue, and used.

Further treatment is often used to enhance reproducibility. This consists of heavy surface oxidation, either chemical (chromic–sulfuric acid, hot nitric acid) or electrochemical (>10 mA/cm^2 for perhaps 1 s), rinsing, and then careful electrochemical reduction of the electrode surface at a potential less negative than hydrogen ion reduction or solvent reduction. Aqueous work usually requires this further treatment.

3.4.5 Solid Inert Electrodes: Carbon

As an inert electrode material, carbon is useful for both oxidation and reduction in both aqueous and nonaqueous solutions. Only graphitic forms of carbon conduct and are useful as electrodes, but there are several of these. Ordinary spectroscopic-grade graphite rods can be used for work in which the area of the electrode need not be well defined, but the porosity of these rods does not permit their use with most nonequilibrium electroanalytical techniques. This problem can be overcome by impregnating the graphite with hot paraffin wax to give *wax-impregnated graphite electrodes* (WIGE). The active area of such electrodes, though reduced, is much more reproducible. A similar material is obtained by mixing graphite powder with Nujol, forming a carbon paste. Carbon paste packed into an insulating tube with an internal metallic contact gives a *carbon-paste electrode* whose surface can be smoothed with a knife. The paste can be extruded and the excess removed by knife to give a renewable electrode surface, though much less conveniently than a DME.

More recently electrodes made of glassy carbon or pyrolytic graphite have become common (Tokai and other manufacturers). They are much superior because of their lower-porosity surfaces.

3.4.6 Construction of Carbon Electrodes

Carbon electrodes can be constructed by normal machining techniques in any desired form. Care must be taken when glassy carbon or pyrolytic graphite is used because these materials are inherently anisotropic as they are formed by pyrolysis of hydrocarbons. Machining must always be done relative to the inherent planar structure of the material because planes that cut across the planes of the graphitic sheet structure are much more porous than the plane that parallels the graphitic sheet structure. The edges of the machined form can be sealed, by force-fit into Teflon or by paint or lacquer, to prevent access of electrolyte to them. Glassy carbon or pyrolytic graphite can be polished to flat surfaces. The materials are anisotropic, so cut edges must not be exposed to electrolyte solutions. These materials are usually employed as machined flat surfaces in holders machined from an insulator such as Teflon.

3.5 CELL ELECTROLYTE: THE SOLVENT

The cell electrolyte may be a simple aqueous solution, a complex mixture of salts in solution, a paste, or a solid; in principle, a gas could serve as electrolyte, though such devices are normally considered to lie within the domain of physics rather than that of chemistry. The usual cell electrolyte consists of several components, each of which influences the observations made using the cell. These components are the solvent, the supporting electrolyte, the buffer, and sometimes other components. Only the solvent is essential, but other components are almost invariably present also.

No single cell electrolyte can be universally recommended even for a specific technique such as classical polarography, but a few suggestions from polarographic practice can be given. A common chloride medium for classical polarography is $0.1M$ KCl, LiCl, NH_4Cl, or $(CH_3)_4NCl$, but this medium is not buffered. A common buffered system is $0.5M$ acetic acid and $0.5M$ sodium acetate, which has a pH of 5; addition of supporting electrolyte is unnecessary with a buffer concentration this high. A comparable system is $1.0M$ ammonia and $1.0M$ ammonium chloride, whose pH is about 9.5, but many metals precipitate as hydroxides when the pH is this basic even though formation of complex ions with ammonia occurs. Other buffers used include citrate, malonate, and phosphate, which also have complexing effects on metal ions. Addition of EDTA, generally to buffered electrolytes, is used when stronger complexation is necessary. Noncomplexing media are usually based on the perchlorate anion. The cell electrolytes used in other nonequilibrium electroanalytical techniques are often based on these. Organic compounds soluble in aqueous media are studied in aqueous media. When the solubility in water is not sufficiently high, mixed solvent electrolytes can be prepared by addition of 50% by volume of a nonaqueous *solvent* to one of these *electrolytes*. Nonaqueous solvents frequently used in this way include methanol, ethanol, dioxane, dimethylformamide, and acetonitrile.

3.5.1 Water as a Solvent

The *solvent* is the liquid that makes up the vast majority of the electrolyte solution. By far, the majority of work is done in aqueous media, in which the solvent is water. Water has many advantages as an electrochemical solvent. It is obtainable with comparative ease at a very high degree of purity. It will dissolve to at least some extent more different compounds, and more different types of compounds, than any other known solvent. It is not adversely affected by exposure to air or atmospheric moisture, does not degrade with time either in storage or in use, and has a reasonable range of potential over which it is stable. More studies have been described using water as a solvent than any other. It is, then, the most well-behaved solvent available.

Several methods of water purification are available, and their use is generally necessary. Some electroanalytical measurements must be run in aqueous solutions as they naturally occur (natural waters, seawater, biological fluids). Others must or can be made on naturally occurring solutions conditioned by the addition of supporting electrolyte, buffer, or perhaps acid, so that purification procedures cannot be used. Ordinary doubly distilled laboratory water is adequate for most routine work. Demineralization with ion-exchange columns is not as trustworthy as distillation, because the columns will not remove nonionic impurities. Indeed, organic materials may be leached from the resin. A column of activated charcoal will often remove many of these organic impurities by adsorption. When a sensitive method such as anodic-stripping voltammetry is to be performed, additional distillation, preferably from alkaline permanganate to decompose organic impurities, is recommended. If organic impurities are not a problem, exhaustive electrolysis of the solvent containing the supporting electrolyte may be useful.

There are two disadvantages to the use of water as an electrochemical solvent. The first is that certain materials, especially large-molecule organic compounds, are not soluble, or are not soluble enough, in water to permit its use for their study. This disadvantage has been partially, but not completely, overcome by the development of more sensitive electroanalytical techniques such as differential pulse polarography. The second disadvantage of water is that water is a protic solvent. As a consequence of the inevitable presence of hydrogen ion in aqueous solutions, it is mandatory that the pH of the solution studied be controlled, generally by a buffer system. Any electrochemical reaction that can or does involve protons, as most organic reductions do, will be different in protic and aprotic solvents and will also be pH-dependent in protic solvents. For example, in protic solvents the reduction of quinones generally proceeds as a two-electron, two-proton reversible reduction to the corresponding hydroquinones. In aprotic solvents the reduction of quinones is generally to the radical dianion, since protons are not available from the solvent.

3.5.2 Mixed Solvents

Use of a *mixed solvent*, which means a mixture of two miscible liquids of some fixed composition, for all practical purposes in electroanalytical chemistry means a mixture of an organic solvent and water. Use of a mixed solvent can effect improvement of solubility, and that is the only reason for its use. The best approach to mixed solvents is to remain as close to water as possible and as far away as necessary. The usual organic solvents used are dioxane, methanol, and ethanol; the proportions vary with the solubility desired. Dioxane has the advantage that solutions up to 75% dioxane and only 25% water are still essentially aqueous, but peroxide formation and slow degradation of dioxane is a serious problem. In general, organic solvents for precise work must be freshly distilled, preferably from a reducing agent, to reduce the impurity level.

There are several disadvantages other than inconvenience inherent in the use of a mixed solvent in preference to water. The first of these is the choice of reference electrode. The usual and best general choice is the aqueous saturated calomel electrode (SCE). Using the aqueous SCE in a solvent that is even partially nonaqueous will introduce a liquid junction potential whose magnitude is often significant, but this is preferable to going to a nonaqueous reference electrode. This liquid junction potential means that there will be a systematic potential shift as compared to potentials measured in a similar fully aqueous solvent. A second disadvantage arises from the fact that mixed solvents, like water, are protic solvents and therefore their pH must be controlled. This can be achieved with buffers, usually the same buffers as in fully aqueous electrolytes. However, the change of solvent shifts also the entire pH scale within that solvent, and while the internal scale will be self-consistent, it will not be consistent with the fully aqueous pH scale. It is adequate to measure a pH in a mixed solvent as is done in water, the usual method being an aqueous glass electrode standardized in fully aqueous buffers. These measurements are operationally valid and internally consistent, but are not valid for comparison with pH values measured in aqueous solution or in other mixed solvents.

3.5.3 Nonaqueous Protic Solvents

Nonaqueous protic solvents are varied, ranging from alcohols through both organic and inorganic acids. In terms of solubility there is little advantage to the use of a completely nonaqueous protic solvent over a mixture that is partially aqueous, and the usual reason for the use of nonaqueous protic solvents is that the reaction mechanism is desirably different from that which is observed in the presence of water. Investigations in nonaqueous protic solvents center on the solvent as well as the solutes. Each solvent constitutes a separate solvent system with its own pH scale, scale of potentials, and general electrochemistry. Many aqueous techniques (glass electrode, DME) can be used effectively in one or more of them. Common nonaqueous protic solvents include methanol, ethanol, glacial acetic acid, liquid ammonia, and amines.

The convenience of nonaqueous protic solvents is less, generally markedly less, than the convenience of mixed solvents because water is now an impurity that must be excluded. References to nonaqueous solvent work in general are available in several sources (Lingane, G1; Bard, G2; Delahay, G3). Little of this work is electroanalytical in nature.

3.5.4 Aprotic Solvents

Aprotic solvents, which are necessarily nonaqueous, are even more varied than aqueous protic solvents and range from molten salts through organic compounds such as acetonitrile. Since protons need not be present, aprotic solvents permit the study and use of reactions that are not accessible in protic media. While pH in the

Bronsted sense is meaningless in these media, pH in the Lewis sense remains meaningful; for example, electrochemistry in $NaCl-AlCl_3$ mixtures requires understanding of the acidity of $AlCl_3$ relative to $AlCl_4^-$. The reactions are sometimes useful for analytical procedures, but interest in nonaqueous aprotic solvents is for fundamental reasons that prescribe the solvent chosen more often than for analytical purposes. Thus while the disadvantages of nonaqueous aprotic solvents relative to water listed below are real, water is rarely a reasonable alternative.

The purification of an aprotic solvent is generally more difficult than is purification of water, as is true for both of the examples mentioned above. Not only are the procedures generally more difficult, but the starting materials available commercially are generally less pure, and, of course, water is now a significant impurity. Second, like some components of mixed solvents, some aprotic solvents react with atmospheric oxygen or water; some are subject to autodegradation.

Another disadvantage is that the literature available (G1, G2, G3) for aprotic solvents, like that available for mixed solvents, is significantly less than that available for aqueous solutions. This is of greater importance in aprotic solvents because aqueous data and analogies cannot be used. While the amount of literature available varies somewhat from one solvent to another, it is always very much less than for aqueous solutions. The inadequacy of literature can seriously hamper the design of experiments, especially regarding the reference electrode and the solubility of compounds. The reference electrodes are necessarily totally nonaqueous, since use of an aqueous reference would introduce protons, but it is not obvious how far proposed reference electrodes can be trusted. Often they are not well-characterized or have faults. For example, use of the common Pt(II)/Pt reference electrode in fused LiCl–KCl is suspect because temperature-dependent equilibria exist between Pt(IV) and Pt(II) at temperatures above 450°C, the normal temperature of measurement; the Ag(I)/Ag couple is not subject to this complication and is preferable, but much of the available literature has employed Pt(II)/Pt for studies over a considerable temperature range. The solubility of compounds in many aprotic solvents is unknown, and unsuspected reactions between materials being studied and the solvent may occur.

A final disadvantage of aprotic solvents may arise from technological problems, of which there are two general classes. The first is limitations imposed by available materials. For example, a liquid-metal electrode for electroanalytical studies in fused LiCl–KCl at 450°C is not available; at this temperature mercury cannot be used effectively, because it has a very high vapor pressure, while the other metals that are liquid, such as gallium, lead, and bismuth, are less noble than mercury, and surface oxide formation or lack of potential range makes them virtually useless. The second, which is significant in organic aprotic solvents but not in molten salts, is that the cell resistance will usually be considerably higher, perhaps $20 - 100$ kΩ rather than the $20 - 500$ Ω usually found in practical aqueous cells. This higher resistance forces the abandonment of two-electrode systems in favor of three-electrode systems in which IR drop does not appear between the indicating and reference electrodes. The higher resistance often cannot be reduced by use of more supporting electrolyte since the concentrations required exceed its solubility.

3.6   CELL ELECTROLYTE: OTHER COMPONENTS

Cell electrolytes always include a solvent and any ions or molecules of interest. Those used for electroanalytical purposes generally include at least a supporting electrolyte and a buffer system as well, and may include additional species deliberately added to the electrolyte such as complexing agents or surface-active materials.

3.6.1   Supporting Electrolyte

The *supporting electrolyte* is an inert soluble ionic salt added to the solvent, generally in 10-fold or 100-fold excess over the concentration of the species being studied. The inertness meant here is the ability to avoid oxidation or reduction at the indicating or reference electrode during the course of the electrochemical measurements.

There are three functions of the supporting electrolyte. First, it carries most of the ionic current of the cell since its concentration is much larger than that of the other species in solution. Thus it serves to complete the circuit of the electrochemical cell and keep the cell resistance at a low value. This is necessary only for nonequilibrium electroanalytical measurements, in which the current that flows is not negligible. Second, it maintains a constant ionic strength. This is necessary because the structure of the interphase region should not change significantly if a reaction occurs there. A stable structure is created on the electrolyte side by adding a high concentration of an inert salt. The maintenance of a constant or reproducible structure of the interphase region is necessary for both equilibrium and nonequilibrium measurements alike, since the structure of the interphase region affects all of the electrical properties measurable by an external electrical circuit. Third, it suppresses the effect of the *migration current*. Migration current is the current that arises as a result of the movement of ions caused by an electric field—here, the electric field produced by the potential difference between the indicating and counterelectrodes of the cell. This motion of charge can be detected as a current, the migration current. The *net* migration current observed is reduced by the presence of a large excess of ions that are not electrochemically active at the potentials in use, because they can carry an ionic current without permitting its conversion into electronic, and hence net or measured, current at the electrodes. Migration current is of consequence only when current is a measured parameter, which can happen only in nonequilibrium electroanalytical measurements.

Potassium chloride, KCl, is a commonly used supporting electrolyte salt for two reasons. First, and least important, KCl is easily available in high-purity form. It is far more important that the mobility of the potassium ion and of the chloride ion are almost exactly equal; about 49% of the ionic current is carried by motion of potassium ion and 51% by motion of chloride ion. Potassium nitrate, KNO_3, is an alternative when chloride cannot be used; the mobilities of potassium ion and of nitrate ion are also similar. Potassium ion can be discharged, although only at quite negative potentials. However, when a high concentration of potassium ions is present, the foot of their polarographic reduction wave may extend sufficiently positive to cause unwanted interference. This can be avoided by use of tetraalkylammonium salts, since tetraalkylammonium ions are even more difficult to discharge. Interference of chloride ion often arises when mercury electrodes are used owing to chloride facilitation of mercury oxidation by formation of insoluble calomel, Hg_2Cl_2. Chloride interference can be avoided by use of nitrate or perchlorate salts. Since nitrate ion is not as electrochemically inert as is perchlorate, perchlorate is often preferred even though potassium ion and perchlorate ion differ more in mobility than do potassium and nitrate ion.

3.6.2 Buffer System

A buffer is necessary to maintain the constancy of pH in protic solvents. It is absolutely mandatory if protons are or can be involved in the electrode reaction, because it is necessary that a constant pH be maintained *in the interphase region of the indicating electrode.* Such a buffer, which is the normal chemical buffer made up of an acid–base conjugate pair, must have a high buffering capacity *at the interphase.* This can be achieved only if, in the interphase region, the concentration of the buffer system is high and the buffer system has a high intrinsic speed of protonation and deprotonation. Not all acid–base conjugate pairs are equally rapid in intrinsic protonation speed; carbonate buffers are slower than acetate buffers.

A high buffering capacity in the interphase region is especially relevant when nonequilibrium electroanalytical methods are employed. Buffers adequate for potentiometric measurements may be inadequate for measurements of polarographic half-wave potentials; buffers adequate for measurements of polarographic half-wave potentials may be inadequate for chronopotentiometric measurements, because the strain on the buffer increases with increasing current density. Any one buffer will have, of course, only a limited pH range over which it is adequate, but the size of this range decreases with increasing current density.

Like the supporting electrolyte, the concentration of the buffer may be 50 or more times the concentration of the sought-for or test species. The buffer concentration is thus large enough to significantly affect the ionic strength of the solution. It also in part serves as a supporting electrolyte, and advantage can be taken of this in studies where the concentration of actual supporting electrolyte salt that can be added is limited by considerations such as solubility.

3.7   CELL ATMOSPHERE

Some electroanalytical measurements can be run in cells under an atmosphere of ordinary air. Oxygen, however, is electrochemically active, and the solubility of oxygen in water is sufficiently great that oxygen reduction interferes with most electroanalytical techniques in aqueous media, mixed solvents, or even in nonaqueous media. Additional complications such as reoxidation of electroreduced species can also be expected if oxygen is present. As a consequence, most electroanalytical measurements are carried out under an atmosphere of inert gas. Normally this is nitrogen, although argon is occasionally used instead. Use of other gases or gas mixtures is rare, and only occurs when mandated by the specific electroanalytical measurements being made.

Nitrogen is preferred to argon only for reasons of cost. Most commercial nitrogen is manufactured by fractional distillation of liquid air and contains enough oxygen that a solution deaerated with it still contains enough residual oxygen to seriously affect classical polarography and any more sensitive technique. This oxygen interference can be prevented by use of nitrogen without oxygen, such as high-purity commercial nitrogen, or, preferably, nitrogen run through a deoxygenation system. Commercial argon generally contains much less oxygen than does commercial nitrogen, but deoxygenation is still a worthwhile precaution.

Deoxygenation can be carried out using a chemical reductor. The most common types are hot, finely divided copper over which the gas is passed, aqueous chromium(II) sulfate solutions, and aqueous acid vanadyl sulfate solutions (Meites and Meites, S4). Alkaline pyrogallol or reduced anthraquinones are also effective. For aqueous or mixed solvent studies, aqueous reduction is preferable, while dry reduction is preferable in totally nonaqueous media.

Passage of gas through a reductor of either variety may carry material from the reductor into the cell as fine particles or as a fine spray. An intermediate wash bubbler, or presaturator, should therefore be used between the deoxygenator and the cell. In aqueous work the presaturator can be filled with water. In work with mixed solvents the presaturator must be filled with the same solvent mixture as the cell. The presaturator fill must be replaced frequently, because it will preferentially lose the organic phase, which normally has a much higher vapor pressure than does water. In nonaqueous work, a drying tube filled with dessicant serves the same function as the presaturator; alternatively, a bubbler may be filled with the nonaqueous solvent.

3.8   CELL DESIGN

It is impossible to produce a single cell design for electroanalytical studies because of the wide differences between different techniques. Cells designed for specific uses are available through laboratory supply houses.

A few brief general comments on cell design are useful here. The normal material of cell construction is Pyrex glass, for reasons both of visibility and chemical inertness. Teflon or other plastic materials may be used instead, but a greater content of organic components normally leads to greater contamination by organic materials and sensitive measuring techniques may be affected. The size of the cell is variable and depends on the size and degree of dilution of the material being studied. If material availability poses no problem, 25 or 50-cm^3 cells, or larger, are used for the sake of convenience, but when material availability is limited reduction to a sample solution volume of 5 cm^3 is reasonable and to a sample volume of 0.5 cm^3 is possible, the latter with considerable difficulty and adverse effects upon the results obtained.

The most fundamental problem in cell design is the placement of the electrodes in the cell. In electroanalytical measurements one usually measures either current or potential as a function of time. This measurement is of the *overall* current or potential difference as seen by an external measuring device. Since electrons do not appear labeled as to their origin, and electrode potentials reflect the levels of all electrochemically active species in the cell, erroneous electrode placement can give rise to error in interpretation. This is perhaps obvious in a two-electrode cell configuration, but is not so obvious in a three-electrode cell configuration.

Consider a simple two-electrode configuration for an aqueous polarographic cell in which a dropping-mercury electrode (DME) is symmetrically centered. The DME serves as the indicating electrode, and a pool of mercury at the bottom of the cell serves as the second electrode. Since the area of the mercury pool is large relative to that of the DME, the pool will be comparatively unpolarized and will serve as a reference electrode, though of unknown potential. Conversion of a two-electrode system to a three-electrode system requires, in this case, insertion of a reference electrode at some point in the cell. The reference electrode will measure the potential between the indicating electrode and itself. More correctly, if the reference electrode as a unit is considered to include the required salt bridge and if the potential of the reference electrode itself is considered to include all potentials within this unit including the liquid junction potentials of both ends of the salt bridge, as is generally the case, the potential difference it measures will include, in addition to the actual potential of the indicating electrode, any potential difference between the point of insertion of the reference electrode and the surface of the indicating electrode. Such potential differences, if any, will be measured and assigned to the indicating electrode. In the present cell, in which low currents flow and electrolyte is of low resistance, the ohmic drop anywhere in the cell is low and placement of the reference electrode is not critical.

If, however, a cell is used in which either high current is drawn or the cell resistance is high, let alone both, the *IR* or *ohmic* drop through the solution will be significant and, moreover, will vary with time as does the current. The portion of this ohmic drop seen by the external measuring circuit connected between the indicating and reference electrodes will depend upon the position of the reference electrode relative to the lines of current flow between the counterelectrode and the indicating electrode. It is therefore desirable to keep the reference electrode as close as reasonable to the indicating electrode and as far as possible from the direct lines of current flow between the indicating and counterelectrodes. In extreme, and in electroanalytical studies unusual, cases one can use a fine-pointed capillary salt bridge, known as a *Luggin–Haber* (Haber, S5) *capillary*, to maintain electrolytic contact with the reference electrode. The potential sensed by the end of the capillary is the potential measured by the external circuit, and since ionic current does not flow from the solution in it through its walls to the indicating or counterelectrodes, it effectively shields the reference electrode path. Studies of isopotential lines in electroanalytical cells have been reported by Harrar and Shain (S6).

3.9   REFERENCES

General

G1   J. J. Lingane, *Electroanalytical Chemistry, 2nd ed.*, Interscience, New York, 1958. Obsolete now, but still a good general text.

G2   A. J. Bard, Ed., *Electroanalytical Chemistry: A Series of Advances*, Marcel Dekker. New York, 1966 to date, Vols. 1–11, 1979. Best source of electroanalytical technique reviews.

G3   P. Delahay, *New Instrumental Methods in Electrochemistry*, Interscience, New York, 1954, 437 pp. Mathematical but excellent.

G4   B. B. Damaskin, *The Principles of Current Methods for the Study of Electrochemical Reactions*, G. Mamantov, Trans., McGraw-Hill, New York, 1967.

G5   G. Charlot, J. Badoz-Lambling, and B. Tremillon, *Electrochemical Reactions*, Elsevier, New York, 1962. Inferior in clarity of expression and in analytical scope to Lingane, but in general scope superior.

G6   R. W. Murray and C. N. Reilley, *Electroanalytical Principles*, Wiley-Interscience, New York, 1963. A reprint of Part I, Vol. 4, pp. 2109–2232 of *Treatise on Analytical Chemistry*, I. M. Kolthoff and P. J. Elving, Eds. A useful but brief discussion of some of the principles of electroanalytical methods.

G7   W. C. Purdy, *Electroanalytical Methods in Biochemistry*, McGraw-Hill, New York, 1965, 354 pp. While the applications included in this text are biochemically oriented, the treatment of electroanalytical methods is general and well-organized.

G8   B. Tremillon, *Chemistry in Non-Aqueous Solvents*, Reidel, Dordrecht/Boston, 1974, 275 pp. A good, basic, and brief introduction to the topic.

G9   J. J. Lagowski, Ed., *The Chemistry of Non-Aqueous Solvents*, Academic Press, New York, 1966–1978. A collection of six volumes of authoritative review articles on various topics in the field.

G10   G. J. Janz and R. P. T. Tomkins, *Nonaqueous Electrolytes Handbook*, Academic Press, New York. Volume I, 1972, 1108 pp.; Volume II, 1973, 933 pp. A collection of tabular data.

G11   M. R. Rifi and F. H. Covitz, *Introduction to Organic Electrochemistry*, Marcel Dekker, New York, 1974, 432 pp. An introduction with emphasis on synthesis and industrial applications.

Specific

S1   I. M. Kolthoff, *J. Electrochem. Soc.* **118**, 5C (1971).

S2   J. F. Coetzee, "Mercury", in *Treatise on Analytical Chemistry*, I. M. Kolthoff and P. J. Elving, Eds., Part II, Vol. 3, Interscience, New York, 1967, pp. 235–249.

S3   J. W. Ross, R. D. DeMars, and I. Shain, *Anal. Chem.* **28**, 1768 (1956).

S4   L. Meites and T. Meites, *Anal. Chem.* **20**, 984 (1948).

S5   F. Haber, *Z. Phys. Chem.* **32**, 193 (1900).

S6   J. E. Harrar and I. Shain, *Anal. Chem.* **38**, 1148 (1966).

CHAPTER
4
IONS IN SOLUTION: CONDUCTANCE

Many substances, when dissolved in some liquid solvent, dissociate to some extent into charged particles called *ions*. The dissociation is an equilibrium whose position depends on the substance and the solvent in which it dissolves. Since a compound or mixture is electrically neutral, a solution made by dissolving it in any solvent must be neutral also, which means that the total charge on all positive ions (*cations*) must equal the total charge on all negative ions (*anions*). This statement is known as the *law of electroneutrality*.

When a substance dissociates into ions and the dissociation is essentially complete, the substance is called a *strong electrolyte*; incomplete dissociation is found for *weak electrolytes*. In either case, the resulting ions interact with surrounding solvent molecules or ions, a process known as *solvation*, to form charged clusters known as *solvated ions*. These solvated ions can move through the solution under the influence of an externally applied electric field; such motion of charge is known as *ionic conduction*, and the resulting current is *ionic current*. The ionic current is determined by the nature of the ions, their concentrations, the solvent, and the electric field imposed.

An electrode is an anomalous region of transition between ionic and electronic conductors. It is the purpose of the present chapter to discuss ionic conduction of current in that part of the electrolyte sufficiently removed from the electrodes that it does not influence or be influenced by the interphase regions surrounding the electrodes. This region, called the *bulk solution* or *bulk electrolyte*, is uniform in concentration and is assumed to be uniform in electric field as well.

The simplest useful electrical analogue model of an electrochemical cell is a simple resistor, conveniently known as the *cell resistance*, whose origin lies solely in the transport of ions through the bulk of the solution. In the bulk electrolyte current is carried only by means of ions. If a direct current is imposed upon a chemical cell, chemical reactions will occur at the electrodes in accordance with Faraday's laws. If an alternating rather than a direct current is used, that Faradaic reaction that takes place on one half-cycle is reversed on the following half-cycle. If, in addition, no products can escape from the interphase region, no *net* Faradaic current can flow. There are still flows of current, however, and such currents, which do not produce chemical changes in materials, are called *nonFaradaic* current. One of these is the current due to the current-carrying ability, or *conductance*, of ions. Since the bulk electrolyte is unaffected by electrochemical reactions at the electrodes, it is not affected by alternating current flow in a different way than by direct current flow if due allowance is made for the periodic reversal of the field. Thus measurements of ionic conduction are normally made by AC techniques to avoid complications due to Faradaic processes.

4.1   CONDUCTANCE OF ELECTROLYTES

Conductance, whether ionic or electronic, is the reciprocal of resistance. Ionic conductance, which for the bulk solution is the only resistance present, is the reciprocal of ionic resistance. Conductance, as such, is rarely of analytical use because the conductance of any conductor, solution or metal, depends on the size and shape of the conductor. To remove the dependence on the size and shape of the conductor requires use of *conductivity* κ. Conductivity is defined by

$$\kappa = l/AR = Gl/A \tag{4.1}$$

where l is the length of the conductor and A is its cross-sectional area. The value of the conductivity is characteristic only of the conducting medium, since the dependence on the geometry of the conductor has now been removed. The SI unit for conductivity is siemens per meter S/m (cgs: Ω^{-1}/cm).

The conductivity of a solution of a strong electrolyte such as KCl decreases as the solution concentration decreases. For dilute solutions, or solutions sufficiently dilute that the ionic environment does not change significantly upon further dilution, the conductivity should decrease as it does with concentration only because the number of charge carriers per unit volume decreases. It is therefore convenient to define the *molar conductivity* or *specific molar conductance* Λ of an electrolyte:

$$\kappa = \Lambda c \tag{4.2}$$

which factors out the dependence upon concentration. If the concentration is expressed in moles per cubic meter (mol/m^3), the appropriate SI unit for Λ is siemens meter cubed per meter per mole (S m^3/m mol), which is S m^2/mol. Other concentration units include the mol/dm^3, or *molarity*, M, and the mol/kg solvent or *molality*, m. These units are used in conductance measurements, as well as generally, for expression of concentration. Experimentally the value of Λ is found to be independent of concentration for any electrolyte whenever the solution is sufficiently dilute. For interpretation of processes in solution this constant value of Λ is of fundamental importance because Λ^0, the *molar conductivity* or *specific molar conductance*, is independent of concentration, provided the solution is sufficiently dilute.

The cgs metric formula for Λ, called the *equivalent conductance*, is $\Lambda = 1000$ κ/c, where c is the concentration in equivalents per cubic centimeter. This unit corresponds to the physical system in which one equivalent of electrolyte is enclosed by two electrodes placed exactly 1 cm apart, the area of the electrodes being adjusted depending upon the concentration of the solution. Such a value, in Ω^{-1} $cm^2/equiv$, is in SI the S cm^2/mol for a univalent ion, and hence equivalent conductance values are 10,000 times greater than molar conductivity values. Molar conductivities can be measured for any salt, acid, or base in aqueous or nonaqueous solution (see Table 4.1). For interpretation of processes in solution, however, the value of Λ^0 is of more fundamental importance than that of Λ because Λ^0, the molar conductivity *extrapolated to zero concentration* or at infinite dilution, is characteristic only of the ions and the solvent and is independent of any ionic interactions; at infinite dilution ions can have no interactions with each other.

Kohlrausch found that for dilute solutions of strong electrolytes, extrapolation of measured values of Λ to infinite dilution was approximately linear when done against the square root of concentration. This means that the data suggest the empirical equation:

$$\Lambda = \Lambda^0 - Bc^{1/2} \tag{4.3}$$

in which B is an empirical coefficient. This is the form the extrapolation should take, in fact, on the basis of the Debye–Huckel approach discussed in the following section.

At infinite dilution, where ions must necessarily conduct independently of each other, *Kohlrausch's law* of the independent migration of ions takes the status of an axiom rather than that of an experimental fact. It is, nevertheless, empirically true that each species of ion present at infinite dilution contributes a fixed amount to the total ionic conductivity regardless of the nature of any other ions present. This means that the total conductivity of a sufficiently dilute solution is given by the sum of the individual ionic conductivities of the i ions:

$$\kappa = \Sigma_i \kappa_i \tag{4.4}$$

It is convenient to define the *molar ionic conductivities* λ of individual ions in the same manner as molar conductivities of electrolytes; for the ith ion:

$$\kappa_i = \lambda_i c_i \tag{4.5}$$

which expresses, at sufficient dilution, Kohlrausch's axiom as

$$\kappa = \Sigma_i \lambda^0{}_i c_i \tag{4.6}$$

The molar conductivity of an electrolyte at sufficient dilution is then simply the sum of the molar ionic conductivities of the ions produced by dissociation of the

TABLE 4.1   Conductivities of Solutions at 25°C

Material	Conductance (Ω^{-1} cm^{-1})	Conductance (S/m)
Sulfuric acid[a]	9.65×10^{-3}	0.965
Nitric acid[a]	3.72×10^{-3}	0.372
Sodium chloride[c]	3.66	366
Water[b]	0.85×10^{-6}	0.85×10^{-4}
NaCl (sat. aq)	0.2513	25.13
0.1M HCl	0.0391	3.91
0.1M CH$_3$COOH	0.52×10^{-3}	0.052
0.1M KCl	12.88×10^{-3}	1.288
1.0M KCl	0.1118	11.18

[a]Pure acid.
[b]Water, in equilibrium with the carbon dioxide of the air.
[c]Pure fused salt, 850°C.

salt:

$$\Lambda^0 = \Sigma_i n_i z_i \lambda^0_i \tag{4.7}$$

where Λ^0 is the *molar conductivity*, z_i is the absolute value of the charge on ion i, and n_i the number of moles of the ith ion produced by the dissociation of one mole of the salt.

Values of λ^0_i, the *molar ionic conductivities* or *equivalent conductances of individual ions*, can be obtained from measured values of Λ extrapolated to give Λ^0. A partial table of such values of 25°C is given in Table 4.2 and a more extensive list in Appendix 2. Since λ^0_i is a constant characteristic only of the ions, measurement of κ permits following the variation of c_i with time, as in the course of a titration.

4.1.1   Transport Number

When more than one ion is present in a solution it is useful to describe the fraction of the ionic conductance due to each of the ions present. The *transport number* of the ion t_i, sometimes called the *migration number* or *transference number*, is therefore defined as the fraction of the conductance due to the ith ion:

$$t_i = K_i / \Sigma_i j_i \tag{4.8}$$

or, in sufficiently dilute solutions,

$$t_i = \lambda^0{}_i c_i / \Sigma_i \lambda^0{}_i c_i \tag{4.9}$$

When the solution contains only a single electrolyte, the above equation simplifies to

$$t_i = \lambda^0{}_i / \Lambda^0 \tag{4.10}$$

Transport numbers vary with the nature of the salt and of the solution as well as concentration; experimental methods exist for obtaining them (Spiro, G5). The transport numbers of many ions in an aqueous solution of a single equi-equivalent salt are approximately 0.5. Transport numbers do change with concentration in solution of a single salt, but only slightly. However, since the transport number of an ion is the fraction of the total ionic conductance due to that ion, the transport number of any particular ion or ions can be reduced to virtually zero by addition of a large concentration of some salt that does not contain them to the solution.

Note that the units for molar ionic conductivity, S m²/mol (siemens square meter per mole), are units of velocity under a uniform potential gradient. Hence these values are also referred to as *limiting ionic mobilities*. For aqueous hydrogen ion a value of 0.35 S m²/mol is converted as follows, using the Faraday constant as 10^5 C/mol:

3.5×10^{-2} S m²/mol

3.5×10^{-2} A m² mol⁻¹ V⁻¹

3.5×10^{-2} A m mol⁻¹ per V/m

3.5×10^{-2} C m mol⁻¹ s⁻¹ per V/m

3.5×10^{3} m/s per V/m

The absolute mobility of an uncharged particle, u_i, is simply its diffusion coefficient D_i divided by the product of the Boltzmann constant and the absolute temperature; in molar terms,

$$u_i = D_i N / RT \tag{4.11}$$

since the Boltzmann constant is simply the gas constant R divided by the Avogadro number N. For a particle of macroscopic dimensions moving through an ideal hydrodynamic continuum with velocity u, this force is opposed by the viscous drag of the medium until these forces are in balance. The ideal case of a spherical particle moving through an ideal hydrodynamic medium was treated by Stokes (S1). Under these conditions the mobility is given by *Stokes' law*:

$$u_i = 1/6\pi\nu r_i \tag{4.12}$$

where ν is the viscosity of the medium and r_i is the radius of the particle. Equating these gives the *Stokes–Einstein equation*:

TABLE 4.2   Equivalent Ionic Conductances[a]

Cations	λ^0 (Ω^{-1} cm² equiv⁻¹)	Anions	λ^0 (Ω^{-1} cm² equiv⁻¹)
H⁺	349.99(8)	OH⁻	198.4
K⁺	73.54	Cl⁻	76.39
Na⁺	50.12	Br⁻	78.18
Li⁺	38.7(0)	I⁻	76.88
½ Ba²⁺	63.6(6)	NO_3^-	71.50
½ Ca²⁺	59.53	HCO_3^-	44.5(2)
½ Mg²⁺	53.0(8)	CH_3COO^-	40.9(2)
½ Pb²⁺	69.5	$C_6H_5COO^-$	32.4(0)
½ Cu²⁺	53.6	½ SO_4^{2-}	80.0(5)
⅓ La³⁺	69.7	⅓ $Fe(CN)_6^{3-}$	100.9
Ag⁺	61.9(3)	¼ $Fe(CN)_6^{4-}$	110.6
NH_4^+	73.5(8)	F⁻	55.4

[a]Values in the SI units of S m²/mol, the *molar ionic conductivities*, are 1/10,000 this size; an appropriate value for aqueous hydrogen ion would be about 35 mS m²/mol. For details see Table A2.4 in Appendix 2.

$$D_i = RT/6\pi\nu N r_i$$

The Stokes–Einstein equation can be used for calculation of the diffusion coefficients of uncharged species if the species diffusing is roughly spherical and much larger than the solvent molecules.

When an ion is in motion, the force upon it is not thermal buffeting but the result of interaction of the electric field and the ionic charge. The mobility u'_i is the electrical mobility obtained when an ion moves under unit potential gradient. The conductivity due to a single ion κ_i is, in molar units,

$$\kappa_i = zFcu'_i \tag{4.14}$$

which, since $\lambda_i = \kappa_i/c$, is more usefully expressed as

$$\lambda_i = zFu'_i \tag{4.15}$$

The electrical mobility under unit potential gradient u'_i is the product of the absolute mobility u_i, the absolute value of the ionic charge, and the absolute value of the charge on the electron, which, expressed in molar units, is

$$u_i = u'_i N/zF \tag{4.16}$$

After substituting the preceding equation

$$u_i = \lambda_i N / z^2 F^2 \tag{4.17}$$

Combination of this equation with Eq. (4.11) yields, at infinite dilution in the medium when ions conduct independently, an equation derived by Nernst (S2) for the diffusion coefficient of the ion:

$$D^0{}_i = (RT/z^2 F^2)\lambda^0{}_i \tag{4.18}$$

The equation derived by Nernst is useful for calculating the diffusion coefficients of ions in dilute solutions. Some care is required in its use because molar conductivities of ions and specific ionic conductances are often given in the form λ_i/z as indicated by species such as $\tfrac{1}{2}Ca^{2+}$ listed in the table; the redundant z must then be dropped from the equation. Alternatively, the values given in that form may be converted to those of real species; λ^0 (Ca^{2+}) is simply $2\lambda^0$ ($\tfrac{1}{2}Ca^{2+}$).

Combination of Eqs. (4.11), (4.15), and (4.16) yields, under conditions of infinite dilution where λ equals λ^0,

$$\lambda^0{}_i = zFe/6\pi\nu(z_i/r_i) \tag{4.19}$$

where e is simply the value of the electronic charge, F/N. The molar conductivity Λ^0 of any salt can be calculated by application of the equation immediately above; for a univalent electrolyte it is given by

$$\Lambda^0 = (zFe/6\pi\nu)[(z(+)/r(+)) + (z(-)/r(-))] \tag{4.20}$$

Of the terms on the right-hand side of this equation, only the viscosity ν is strongly dependent upon the medium. Hence as an approximation for all media in which ions may move:

$$\Lambda^0\nu = \text{constant} \tag{4.21}$$

This relationship, first suggested empirically by Walden, is known as *Walden's rule*. It is more often honored in the breach than in the observance, because the actual or *solvated* radius of an ion is medium-dependent. Good agreement is found only for larger ions, whose solvated radii change less markedly with the medium in which they conduct.

4.2   THEORETICAL CONCEPTS IN IONIC SOLUTIONS

The chemical potential of a solute, μ_i, is the sum of a standard chemical potential $\mu^0{}_i$ and an activity term:

$$\mu_i = \mu^o{}_i + RT \ln (a_i) \tag{4.22}$$

The standard chemical potential will be taken up later, in Chapter 7. The chemical potential of a solute whose degree of dissociation does not change with concentration is the product of the solute concentration and an *activity coefficient*. In dilute solutions, Henry's law is obeyed, which means that the chemical activity of a solute is proportional to its concentration. The proportionality constant, or activity coefficient, is considered to be unity at infinite dilution. This may be written mathematically as

$$a_i = \gamma_i c_i, \ \lim \gamma_i(c_i \to 0) = 1 \tag{4.23}$$

Since for any equilibrium in which any species is dissociated into ions the activities of the species on both sides are equal, the uni-univalent example is

$$a(MX) = a(M^+)a(X^-) \tag{4.24}$$

$$a(MX) = \gamma(M^+)c(M^+)\gamma(X^-)c(X^-) \tag{4.25}$$

This simplifies to

$$a(MX) = c^2\gamma(M^+)\gamma(X^-) \tag{4.26}$$

because the concentrations of positive and negative charges must be equal in solution.

Unfortunately the single-ion activity coefficients $\gamma(M^+)$ and $\gamma(X^-)$ are not experimentally measurable quantities even for uni-univalent electrolytes, so a *mean ionic activity coefficient* $\gamma^\pm$ is used in their place. The mean ionic activity coefficient is the geometric mean of the individual ionic activity coefficients. For a uni-univalent electrolyte this is given by

$$\gamma^\pm = (\gamma(M^+)\gamma(X^-))^{1/2} \tag{4.27}$$

This mean ionic activity coefficient, which can be measured experimentally, does go to zero in the limit of infinite dilution. Similar but more complicated expressions for the mean ionic activity coefficient are obtained for salts which dissociate to give ions greater in number or charge than a simple uni-univalent electrolyte.

Real solutions may contain more than one dissociated solute and the solute may dissociate to give ions of different charge. It was for this reason that in 1921 G. N. Lewis (Lewis and Randall, S3) defined the empirical concept of *ionic strength, I*:

$$I = \tfrac{1}{2} \, \Sigma_i c_i z^2{}_i \tag{4.28}$$

where c_i is the ionic concentration and z_i is the ionic charge. Thus the ionic strength of a uni-univalent salt such as NaCl is equal to its concentration, while that of a di-divalent salt such as magnesium sulfate is equal to four times its concentration.

The concept of ionic strength was given a theoretical basis in the Debye–Huckel theory of ionic solutions two years later (S4). The activity coefficient of any ion depends primarily upon two factors: ionic charge and ionic concentration. This is true because in dilute solutions the primary interaction between ions is a simple coulombic attraction or repulsion. This interaction extends considerably further into the solution than do other intermolecular forces. The problem addressed by Debye and Huckel was to calculate the difference in chemical potential that exists as a result of these electrostatic interactions. Assuming a Boltzmann distribution of the ions, using the law of electroneutrality, and expanding exponentials while retaining only the first one or two terms of the expansion power series, such calculation is possible. The resulting *Debye–Huckel equation* is

$$\ln (\gamma_i) = -Az^2_i \, I^{\frac{1}{2}} \tag{4.29}$$

where the parameter A is given by

$$A = (2\pi N)^{\frac{1}{2}}(10\epsilon kT/e^2)^{3/2} \tag{4.30}$$

In this equation N is Avogadro's number, e is the electronic charge, ϵ is the dielectric constant of the solvent, k is the Boltzmann constant, and T is the absolute temperature. For water at 25°C, the Debye–Huckel equation becomes

$$\log \gamma_i = -0.509z_i^2 I^{\frac{1}{2}} \tag{4.31}$$

where the constant has the units of $kg^{\frac{1}{2}} \, mol^{-\frac{1}{2}}$.

Since the activity coefficient of a single ion is not experimentally accessible, the Debye–Huckel equation is more frequently seen in a form appropriate for the mean ionic activity coefficient:

$$\log \gamma\pm = -0.509z(+)z(-)I^{\frac{1}{2}} \tag{4.32}$$

The Debye–Huckel equation is a limiting law at low concentrations and, for uni-univalent electrolytes in aqueous solution, is in excellent agreement with experiment up to an ionic strength of 0.01. Since conductance is usually employed for measurements at or below this ionic strength, the use of Debye–Huckel estimations of mean ionic activity coefficients is common.

Inherent in the Debye–Huckel discussion is the concept of an *ionic atmosphere*, since the Debye–Huckel expression is concerned with the energy difference between charging an ion in the absence of other ions and charging the same ion in their presence. The atmosphere may be visualized as a sphere of oppositely charged ions whose apparent radius r is given by

$$1/r^2 = 8\pi Ne^2 I^2/1000\epsilon kT \tag{4.33}$$

In solutions of high ionic strength, the ionic atmosphere is closer to the ion it surrounds and interacts more strongly with it. As the ionic strength decreases, the radius of the sphere of oppositely charged ions increases. The interaction with the central ion decreases to negligible as the solution becomes infinitely dilute, since the radius becomes infinite.

Onsager used the Debye–Huckel idea of an ionic atmosphere to quantitatively explain two effects that become significant as the solution becomes less dilute. Both operate to reduce the conductance to values lower than would be expected. The first is an *asymmetry effect* in which the motion of an ion is retarded by the electrical attraction of ions of opposite charge in the solution. The asymmetry effect arises because, in the absence of an applied electric field and hence of net motion of an ion, the ionic atmosphere is symmetric about the ion. As the ion moves under the influence of the field, the ionic atmosphere drags behind it, retarding the motion of the ion.

The second, the *electrophoretic effect*, arises from the fact that the positive and negative ions, moving in opposite directions, each carry some solvent molecules with them. The electrophoretic effect operates to retard the ion because the solvent molecules pulled along by the ion are going in the opposite direction to those pulled along by the oppositely charged ionic atmosphere. In effect, each is swimming upstream through the solvent pulled along by the other. Onsager developed expressions in terms of the parameters A and B for the asymmetry effect,

$$A = 82.4/(\epsilon T)^{1/2}\nu \tag{4.34}$$

and the electrophoretic effect,

$$B = 8.20 \times 10^5 \, \Lambda^0/(\epsilon T)^{3/2} \tag{4.35}$$

These are usually combined into the *Onsager equation*,

$$\Lambda = \Lambda^0 - (A + B\Lambda^0)c^{1/2} \tag{4.36}$$

The Onsager equation gives good values for the limiting slopes of plots of $\Lambda - \Lambda^0$ against $c^{1/2}$ in aqueous solutions, and is in excellent agreement with experimental values of conductivity up to about $0.02M$.

Ionic conductivity in aqueous solution is observed to increase in electric fields greater than about 10^5 V/cm (the *Wien effect*). An increase in conductivity is also observed at AC frequencies greater than about 3 MHz (the *Debye–Falkenhagen effect*). Neither effect is of analytical significance.

4.3   MEASUREMENT OF CONDUCTANCE

Conductance as such is not measured, although its reciprocal, resistance, is; such bridges as are calibrated in conductance units are actually resistance or impedance bridges. The measurement of resistance and impedance is discussed in Section 2.5; only such techniques as are specific to solution conductance measurements are included here. Additional references are given in the useful discussion of Braunstein and Robbins (S5).

Conductance is usually measured using an impedance bridge and an AC signal of small amplitude at about 1 kHz, since the impedance due to the double layer capacitance is then a negligible part of the cell resistance. Signals of larger amplitude produce effects other than simple conductance (Feates, Ives, and Pryor, S6), some of which are of analytical significance and are discussed in Chapter 16. As the frequency of the AC signal is changed, the impedance of the cell changes somewhat; use is made of this in AC methods (Chapter 16). Conductance can also be measured using DC signals if two unpolarizable electrodes are used rather than platinized platinum; other than for experimental simplicity there is little to commend such practice, although it has been used in conductometric titrations.

4.3.1   Conductance Cells

Conductance is measured by connecting an impedance bridge to the two electrodes of a *conductance cell*, which consists of a container, usually glass, and two plane parallel electrodes. These electrodes are usually made of platinum. Some of the types of cells used are shown in Figure 4.1; many of these are commercially available. The classical Kohlrausch cell with variable spacing is no longer used, because the proximity of the electrode leads introduces excessive capacitance. Precise work employs cells of fixed spacing (4.1*A*). If sufficient rigidity and fixed position of electrodes can be provided by standard taper glass joints, the conductometric titration cell (4.1*B*) is convenient. The dip-type model (4.1*C*) can be placed directly in the solution being titrated in a beaker. It is adequate for measurements where accuracy not greater than 0.5% is acceptable.

The planar electrodes need be platinized only for precision measurements of conductance; bright platinum suffices for conductometric titrations and most other electroanalytical measurements. Platinization gives coatings that can adsorb impurities; light, rather than heavy, coatings are preferable.

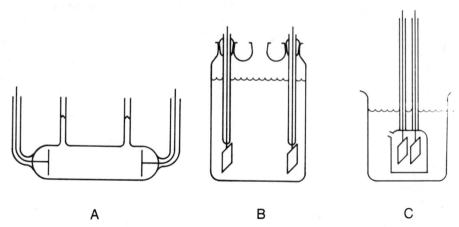

Figure 4.1   Conductance cells. (*A*) precision conductance cell; (*B*) conductometric titration cell; (*C*) dip-type cell.

4.3.2   Conductance Cell Constants

It is possible in principle to calculate the conductance to be observed for any solution in any cell if the geometry of the conductance cell, the specific molar ionic conductances, and the ionic concentrations are all known with sufficient precision. In practice, this cannot readily be achieved, and thus the *cell constant* is determined for each cell by measuring the conductance when the cell is filled with a solution whose specific conductance is very accurately known. Since the units of conductance are siemens and those of conductivity are siemens per meter (S/m), the SI units of cell constants are meters (cgs, cm).

Very precise values of the conductivity of aqueous potassium chloride solutions have been determined by Jones and Bradshaw (S7), whose values are generally accepted as the standard for conductance measurements (see Table 4.3). It is preferable to calibrate the cell using a solution whose conductance is of the same order of magnitude as the solution whose conductance is to be measured.

4.3.3   Control of the Temperature

All conductance values change significantly with temperature, and for precise work a constant-temperature bath is essential. A water thermostat will suffice for an accuracy of 0.1%, but for more precise work an oil bath is preferable. In conductometric titrations the constancy of the temperature rather than its absolute magnitude is important, and a large pan of water at room temperature will serve adequately as a thermostat. Titration without this precaution is not recommended because of possible interference from the heat of the titration reaction as well as from environmental variations.

TABLE 4.3   Conductance Cell Calibration Data. Specific Conductances of Aqueous Potassium Chloride Solutions[a]

Nominal mol kg	Actual g KCl/ kg solution	κ 0°C	κ 18°C	κ 25°C
1.0	71.1352	0.06517(6)	0.09783(8)	0.11134(2)
0.1	7.41913	0.007137(9)	0.011166(7)	0.012856(0)
0.01	0.745263	0.0007736(4)	0.0012205(2)	0.001408(7)

[a]Values of G. Jones and B. C. Bradshaw, *J. Amer. Chem. Soc.* **55**, 1780 (1933). Conductances are given in *international* ohms per centimeter; see Appendix 3.

4.3.4   Conductance of the Solvent

Measured values of conductance include the contributions to conductance from all ions present in solution as well as any conductance due to the solvent. For aqueous work, the conductivity of ordinary distilled water in equilibrium with the atmosphere is often negligible, being about 1×10^{-6} S/m (1×10^{-4} Ω^1 cm^1), most of which is due to the ionization of dissolved carbon dioxide. Specially prepared conductivity water (Kohlrausch and Heydweiller, S8) can reach 6×10^{-6} S/m (0.06×10^{-6} Ω^{-1} cm^{-1}) but can be obtained only with considerable effort and is rarely required. Solvent corrections are not as simple as it might appear, and thus use of a solvent whose conductivity is negligible is the usual, and preferred, practice. In nonaqueous work, purification of a solvent to conductivity standards is a major problem.

4.4   APPLICATIONS OF DIRECT CONDUCTANCE MEASUREMENTS

Most of the applications of direct conductance measurements relate to the measurement of dissociation equilibrium constants. Since conductance is a general property of ions, conductance measurements are not sufficiently selective to be significant in the analytical determination of species in natural media.

4.4.1   Ionization Constants

Ionic association is important for weak electrolytes. The Λ values of acetic acid deviate markedly from the Λ^0 values as the concentration of acetic acid is increased because acetic acid molecules are formed, and such molecules, being uncharged, contribute nothing to the ionic conductance. If the degree of dissociation of a univalent electrolyte is symbolized by α, then its dissociation constant K_d is given by the *Ostwald dilution law*

$$K_d = c\,\alpha^2/(1-\alpha) \qquad (4.37)$$

Since, as suggested by Ostwald, the experimental value of α at any concentration c is Λ/Λ^0, the value of a dissociation constant can be obtained from conductance measurements and the resulting formula:

$$K_d = c\,\Lambda^2/[\Lambda^0(\Lambda^0 - \Lambda)] \tag{4.38}$$

This is a useful method of obtaining the dissociation constants of weak electrolytes, including the acid ionization constants of weak acids such as acetic acid. The approach is useful for any solution equilibrium between ions and uncharged species, or in which charged species react to form other species of different charge.

When a weak acid or base is dissolved, a small fraction of the molecules dissociate into ions. The remaining molecules, being uncharged, cannot contribute to the observed conductance, which will be the sum of the individual ionic conductances of the constituent ions at their equilibrium concentrations in the solution of the weak acid or base. While the actual equivalent ionic conductances are not known, those at infinite dilution are known and can be used as reasonable approximations since the concentrations of the ions cannot be very high if the electrolyte is truly weak. For both strong and weak electrolytes the value of λ^0_i is greater than that of λ_i at any concentration because the attraction between ions decreases λ_i as concentration increases. For weak electrolytes, the increasing concentration also reduces the degree of molecular dissociation, and this factor is of more importance than the effect of ion–ion interactions.

Conductance measurements alone can give good values of ionization constants of weak acids and bases only if they are monoprotic and if no other ionic species are involved, since conductance measurements cannot distinguish among different ions. Measurement of the total ionic conductance is not sufficient to allow calculation of the concentrations of several different species.

4.4.2 Solubility Products

Direct measurement of conductance is well suited for measurement of the solubility product constant of a slightly soluble salt in a pure solvent. If the ionic concentrations are millimolar or less, the actual equivalent conductances of the ions can again, as a good approximation, be taken as equal to their values at infinite dilution, λ^0_i; since these values of λ^0_i are known, the concentration in the saturated solution is calculable.

This method will not be useful if specific molar ionic conductance values for the ions of the salt are not available, and, particularly in nonaqueous work, their separate determination may be necessary. It further assumes that the dissolved salt is completely dissociated. Thus phenomena such as ion pairing, found particularly in nonaqueous solvents, will give erroneous values of solubility product constants.

4.4.3   Ionic Composition of Salts

One function of conductometric studies, of use in the early Werner–Jorgenson controversies of inorganic complex bonding, is the determination of the number of ions released when one molecule of a salt of a complex ion dissolves in a solvent. This requires knowledge of the molecular weight and the conductance, which can then be interpreted as a sum of ionic conductances. In the series of cobalt-ammine complexes of structure $Co(NH_3)_{6-n}(NO_2)_nCl_{3-n}$, where n goes from 0 to 4, conductance measurements clearly indicate that cobalt is hexacoordinate throughout the series.

4.4.4   Ion Association Constants

Association of ions causes a reduction in the net ionic charge and may produce larger ionic species moving through the solution. The effect of ionic association is thus to reduce the conductance of a solution in which association occurs to a value below that calculated from limiting ionic conductances. Measurement of solution conductance as a function of concentration can permit the calculation of the association constant, or constants, of any pair of ions. This technique is useful in solvents of low dielectric constant, in which ion association is often significant. For example, the association of Cu(I) with tetrafluoroborate and hexafluorophosphate in acetonitrile was found by Yeager and Kratochvil (S9).

4.5   CONDUCTOMETRIC TITRATIONS

By far the most common analytical use of conductance measurements is their application as endpoint detectors in titrations. Although in principle conductometric end-point detection may be employed in any titration reaction, in practice it is limited to acid–base and precipitation reactions. Casual electrolytes, or ions that do not participate in the reaction, are a serious interference in redox titrations. These are better followed potentiometrically or amperometrically. Complexation reactions with neutral ligands produce only a small change in conductance with complex formation and cannot be effectively used, and there are usually better methods for complexometric titrations even when a charged ligand is involved.

4.5.1   Precipitation Reactions

A simple example of a precipitation titration is the aqueous titration of silver nitrate with sodium chloride:

$$(Ag^+ + NO_3^-) + (Na^+ + Cl^-) \rightarrow AgCl(s) + (Na^+ + NO_3^-)$$

$$AgCl(s) \rightarrow Ag^+ + Cl^-$$

For reactions of this general type when a slight dissociation of the precipitate

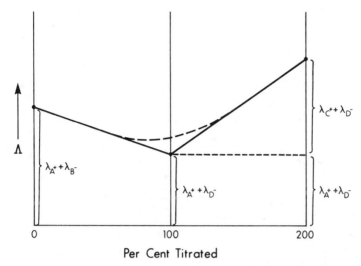

Figure 4.2 Conductometric titration curve for precipitation titrations. Solid straight lines are obtained if the solid compound has zero solubility; dashed curve is obtained if its solubility is appreciable. Vertical axis is conductance, increasing upward.

occurs,

$$(A^+ + B^-) + (C^+ + D^-) \rightarrow CB + A^+ + D^-$$

$$CB \rightarrow C^+ + B^-$$

The conductometric curve is as shown in Figure 4.2.

4.5.2 Acid–Base Reactions

Conductometric titrations in aqueous solution using acid–base reactions are probably the most common group of conductometric titrations. The primary reason for this is the anomalously high equivalent ionic conductances of hydrogen ion and hydroxide ion in aqueous solution, as shown in Table 4.2. These two ions travel through solution by a different and considerably faster mechanism than do ions that are not components of the solvent. For the titration of hydrochloric acid with sodium hydroxide, the reaction and limiting ionic conductances are

Reaction: $H^+ + Cl^- + Na^+ + OH^- \rightarrow Na^+ + Cl^- + H_2O$

λ^0_i: $350 + 76 + 50 + 198 \rightarrow 50 + 76 + 0$

and the titration curve obtained is shown in Figure 4.3.

For strong acids, the contribution of the acid is far greater in the early stages of the titration, but for weak acids its contribution is of course less, and for very

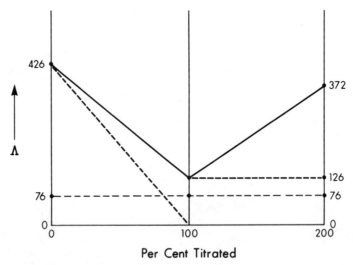

Figure 4.3 Conductometric titration curve for titration of HCl with NaOH. Solid lines represent total measured conductance, dashed lines the contribution to conductance from acid, base, and salt, respectively. Dilution not considered.

weak acids the salt contribution is dominant, as shown in Figure 4.4.

If the acid is strong or very weak, linear portions of the curve are generally sufficient to permit accurate extrapolation to the equivalence point. If, however, the acid is about 50% ionized, the line is significantly curved everywhere, and its extrapolation to the equivalence point is not possible. Under these circumstances, weakening the acid by the use of mixed solvents may help; strengthening of the acid is not generally a useful procedure.

Conductometric titrations are particularly useful in acid–base chemistry for very dilute strong acids and bases, for mixtures of strong and weak acids or strong and weak bases, for very weak acids or bases, or for ampholytes such as amino acids; in the latter case multiple breaks are obtained corresponding to the ionized forms and to the zwitterion.

The important measurements in conductometric titrations are those made well away from the equivalence point of the titration. These are then extrapolated graphically or by computation to intersect at the equivalence point. The points in the region of the equivalence point will not give linear plots if the reactions are incomplete; greater curvature is observed for less complete reactions.

Dilution corrections in conductometric titrations are necessary to avoid serious error. The volume must be corrected by multiplication by $(V_1 + V_2)/V_1$ where V_1 is the original or initial volume and V_2 is the volume of titrant added. This assumes that conductance is a linear function of dilution, which is not exactly true for real solutions, but it is approximately true if the dilution factor is small. For this reason the concentration of the titrant solution is often 10 or more times greater than the concentration of the solution being titrated.

Conductometric measurements and titrations have a decreasing importance in electroanalytical work, because conductance is a nonselective technique responding to all ions present in the solution. This disadvantage cannot be overcome.

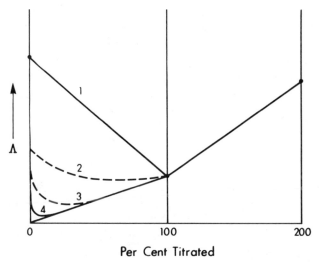

Per Cent Titrated

Figure 4.4 Effect of acid strength upon conductometric titration curves. (1) strong acid; (2) 50% ionized; (3) weak acid; (4) very weak acid. Dilution not considered.

On the other hand, conductometric titrations have some advantages in special areas of application. If a strong acid or base is titrated conductometrically, the *relative change* in conductance is virtually independent of the original concentration; very dilute solutions can be titrated with nearly the same precision as more concentrated ones. Thus for very dilute solutions conductometric titration is a useful method of standardization. Conductometric titration sometimes permits use of incomplete titration reactions because, at the extremes of the curve where measurements are made, the reaction is forced further toward completion by the presence of common ions. Conductometric rather than potentiometric titration of very weak acids and bases is therefore often useful.

4.6   CONDUCTOMETRIC DETECTORS

Since conductance is a general property of ionic substances in solution, a general detector can be made by placing a conductance cell in the effluent stream of a liquid chromatograph. To employ conductometric detection with a gas chromatograph, the gaseous effluent must pass through a contactor containing a liquid solvent in which the effluent can dissolve to give an ionic solution. This solution must then be passed through a conductance cell. It is preferable to split the solvent and pass the splits, one of which has contacted the gaseous effluent and one of which has not, through two cells operated differentially. The reference cell corrects for variation in temperature and flow rate. The application to gas chromatography, though more complex, originated in 1962 (Piringer and Pascalau   S10) for carbon-containing compounds. The output of the gas chromatograph was directed through a high-temperature reactor in which oxygen was injected to combust the compounds to carbon dioxide, with subsequent absorption in basic solution and conductometric detection.

Significant use of conductometric detectors in an element-selective form developed from the work of Coulson (Coulson, S11; Natusch and Thorpe, S12). They have included nitrogen, halogen, and sulfur (Jones and Nickless, S13; Hall, S14; Pope, Rodgers, and Flynn, S15) compounds, especially halogen-containing pesticides; other detectors are preferable for carbon compounds. Nitrogen is detected by reaction with hydrogen to form ammonia followed by absorption in basic aqueous solution. In the presence of a nickel catalyst, the reaction can be made to proceed for nitrogen alone, and the detector is nitrogen-specific. More often high-temperature reaction with hydrogen is carried out without a catalyst, converting nitrogen to ammonia and chlorine to HCl. Detection of nitrogen at the 100 pg level and chlorine at the 500 pg level is possible with these detectors. If oxygen is used for reaction, sulfur is also evolved as sulfur dioxide, which can be detected at the 1-ng level. Direct pyrolysis (no reactant gas) can give responses to nitrogen, chlorine, and sulfur (Cochrane, Wiilson, and Greenhalgh, S16), since the necessary hydrogen and oxygen are generally available in the organic molecules. Detectors of the Coulson type are commercially available.

Liquid chromatography has adapted the Coulson detectors for aqueous high-pressure liquid chromatography (Dolan and Sieber, S17; Poppe and Kuysten, S18) when aqueous eluent solutions are used. The response of the detector is sensitive to temperature and, in some designs, to the flow rate of the mobile phase. The response to different compounds is predictable from conductance data, and a linear region of response to concentration can usually be obtained. In these applications the eluent contains a low concentration of ions, since the response is to all ionic species in the solutiion.

In ion-exchange chromatography, the concentration of ions is high in the eluent aqueous solution. For ions that do not absorb light in the UV region, conductometric detection is the method of choice. A succession of two columns must be used (Small, Stevens, and Baumann, S19; Stevens, Turkelson, and Albe, S20), with the analysis column followed by a stripper column of opposite type that removes the ions of the eluent. Anion mixtures can be analyzed with an anion-exchange column followed by a cation-exchange stripper column when sodium hydroxide is used as eluent. Cation mixture analysis requires a cation-exchange column followed by an anion-exchange stripper column, with HCl used as eluent. Periodic regeneration of the stripper column is necessary in these methods.

4.7   HIGH-FREQUENCY METHODS

High-frequency conductance methods were suggested by Blake (S21, S22) in 1933 but were not used until the work of Jensen and Parrock (S23) in 1946; the methods enjoyed some popularity for 20 years thereafter, but are now comparatively rare. High-frequency methods are those in which a cell containing a solution is placed either between two metal sheets to form a capacitive circuit or within a coil of wire to form an inductive circuit. Although in principle one could employ high-frequency methods with the usual conductance cell containing two electrodes dipping into the solution, there are no advantages and some disadvantages in so doing. The cell is subjected to a high-frequency sinusoidal voltage, usually in the radiofrequency (rf) range of 1 to 400 MHz, and the rf current is measured with an rf bridge. The response of the cell to this voltage depends on the cell itself, the dielectric constant of the cell contents, and the conductance of the contents. The quantitative interpretation of these readings is difficult, as is their direct application to analysis. They have, however, received some use in following titrations, using methodology very similar to normal conductometric titrations, and can also be used to follow variations in the output of chemical process streams.

The sole advantage of high-frequency methods is their ability to operate without having electrodes within the cell itself. In some cases, this makes them the method of choice. In theory and in practice, they are more difficult to employ than are ordinary conductance measurements and are therefore comparatively uncommon. Reviews of the applications of high-frequency methods have been given by Reilley (S24) and by Ladd and Lee (S25, S26).

Measurements of the dielectric constant of a medium are very similar to high-frequency methods in experimental operations and in the scope of their analytical application (Thomas and Pertel, S27).

4.8 REFERENCES

General

G1   R. A. Robinson and R. H. Stokes, *Electrolyte Solutions*, 2nd ed., Butterworths, London, 1959.

G2   H. S. Harned and B. Y. Owen, *The Physical Chemistry of Electrolytic Solutions*, 3rd ed., Reinhold, New York, 1958.

G3   J. W. Loveland, "Conductometry and Oscillometry," in *Treatise on Analytical Chemistry*, Part 1, I. M. Kolthoff and P. J. Elving, Eds.; John Wiley, New York, Volume 4, 1963, pp. 2569–2627.

G4   T. Shedlovsky and L. Shedlovsky, "Conductometry," in *Physical Methods of Chemistry*, Part IIA, A. Weissberger and B. W. Rossiter, Eds., John Wiley, New York, 1971, pp. 163–204.

G5   M. Spiro, "Transference Numbers," in *Physical Methods of Chemistry*, Part IIA, A. Weissberger and B. W. Rossiter, Eds., John Wiley, New York, 1971, pp. 163–204.

G6   G. H. Nancollas, *Interactions in Electrolyte Solutions*, Elsevier, Amsterdam, 1966, 214 pp. General thermodynamic treatment of ionic association in complexes or ion pairs.

G7   C. B. Monk, *Electrolytic Dissociation*, Academic Press, New York, 1961, 320 pp. Rather readable discussion of theory and experimental approaches to ions in solution.

G8   R. M. Fuoss and F. Accascina, *Electrolytic Conductance*, Interscience, New York, 1959, 279 pp. More advanced theory of conductance.

Specific

S1   G. G. Stokes, *Trans. Camb. phil. Soc.* **8**, 287 (1845).

S2   W. Nernst, *Z. phys. Chem.* **2**, 613 (1888).

S3   G. N. Lewis and M. Randall, *J. Amer. Chem. Soc.* **43**, 1112 (1921).

S4   P. Debye and E. Huckel, *Phys. Z.* **24**, 185, 305 (1923).

S5   J. Braunstein and G. D. Robbins, *J. Chem. Educ.* **48**, 52 (1971).

S6   F. S. Feates, D. J. G. Ives, and J. H. Pryor, *J. Electrochem. Soc.* **103**, 580 (1956).

S7   G. Jones and B. C. Bradshaw, *J. Amer. Chem. Soc.* **55**, 1780 (1933).

S8   F. Kohlrausch and A. Heydweiller, *Wied. Ann.* **53**, 209 (1894); *Z. phys. Chem.* **14**, 317 (1894).

S9   H. L. Yeager and B. Kratochvil, *J. Phys. Chem.* **73**, 1963 (1969).

S10 O. Piringer and M. Pascalau, *J. Chromatogr.* **8**, 410(1962).

S11 D. M. Coulson, *J. Gas Chromatogr.* **3**, 134 (1965).

S12 D. F. S. Natusch and T. M. Thorpe, *Anal. Chem.* **45**, 1185A (1973).

S13 P. Jones and G. Nickless, *J. Chromatogr.* **73**, 19 (1972).

S14 R. C. Hall, *J. Chromatogr. Sci.* **12**, 152 (1974).

S15 B. E. Pope, D. H. Rodgers, and T. C. Flynn, *J. Chromatogr.* **134**, 1 (1977).

S16 W. P. Cochrane, B. P. Wilson, and R. Greenhalgh, *J. Chromatogr.* **75**, 207 (1973).

S17 J. W. Dolan and J. N. Seiber, *Anal. Chem.* **49**, 326 (1977).

S18 H. Poppe and J. Kuysten, *J. Chromatogr.* **132**, 369 (1977)

S19 H. Small, T. Stevens, and W. C. Baumann, *Anal Chem.* **47**, 1801 (1975).

S20 T. S. Stevens, V. T. Turkelson, and W. R. Albe, *Anal. Chem.* **49**, 1176 (1977).

S21 G. G. Blake, *J. Roy. Soc. Arts* **82**, 154 (1933).

S22 G. G. Blake, *Conductometric Measurements at Radio Frequencies*, Chapman and Hall, London, 1950.

S23 F. W. Jensen and A. L. Parrack, *Anal. Chem.* **18**, 595 (1946).

S24 C. N. Reilley, in P. Delahay, *New Instrumental Methods in Electrochemistry*, Interscience, New York, 1954, pp. 319ff.

S25 M. F. C. Ladd and W. H. Lee, *Talanta* **4**, 274 (1960).

S26 M. F. C. Ladd and W. H. Lee, *Talanta* **12**, 941 (1965).

S27 B. W. Thomas and R. Pertel, "Measurement of Capacity: Analytical Uses of the Dielectric Constant," in *Treatise on Analytical Chemistry,* Part 1, I. M. Kolthoff and P. J. Elving, Eds., John Wiley, New York, Volume 4, 1963, pp. 2631–2672.

4.9 STUDY PROBLEMS

4.1 The conductance of a solution of nitric acid was measured as 0.02273 S in a cell whose constant was measured in a separate experiment as 0.0053 m^{-1}. Calculate the concentration of nitric acid in the solution.

4.2 Given that the solubility of carbon dioxide in water at 25°C is 1.45 g/dm^3 under a pressure of 1 atm of CO_2, use the appropriate limiting ionic mobilities and the conductivity of water in equilibrium with the air to calculate the extent of dissociation of carbon dioxide in water at 25°C. The mole fraction of CO_2 in dry air is 0.000314.

4.3 A conductometric titration of 10 cm^3 of 0.01 mol/dm^3 BaCl$_2$ solution was carried out with 0.01 mol/dm^3 standard aqueous sulfuric acid. Compute the conductivity of the solution (a) at the beginning of the titration, before any sulfuric acid has been added, (b) at the equivalence point of the titration, and (c) after twice the stoichiometrically required amount of sulfuric acid has been added to the barium chloride solution. Neglect the common ion effect. Comment on the practicality of the titration.

4.4 The solubility product of lead(II) sulfate is about 2×10^{-8} mol^2/dm^6. Calculate the conductivity expected for a saturated solution of lead(II) sulfate, assuming the ions exist in solution as simple ions at neutral pH.

4.5 The ionization constant of ammonium ion is about 5.5×10^{-10} mol/dm^3. Calculate the conductivity expected for a solution nominally 0.01 mol/dm^3 in ammonium chloride. Comment upon the reasonability of the use of conductance in determining the ionization constant of ammonium ion. (Note: Think before calculating.)

4.6 A conductometric titration of dilute NaOH solution is to be carried out with standard HCl solution. If the original concentrations were millimolar for both solutions, calculate the conductivity expected at 50%, 100%, and 150% titrated. Repeat these calculations for micromolar and nanomolar solutions. (Ignore any effects of carbonate equilibria.) Comment upon the analytical significance of these calculations.

4.7 Calculate the diffusion coefficients of H$^+$, Ba^{2+}, K$^+$, and Cl$^-$ in aqueous solution at 25°C from their limiting ionic mobilities. Does this calculation support the oft-assumed value of 1×10^{-5} cm^2/s for aqueous ions? If not, why not?

4.8 A sulfuric acid bath used in metal treatment is found to function best when its conductivity is 0.0517 S/cm. The available technical-grade sulfuric acid has a conductivity of 0.18 S/cm. Calculate the number of kilograms of acid that must be added to 500 kg of water to produce the desired bath.

4.9 Measurement at 25°C of the resistance of a saturated solution of aqueous silver chloride gave the value of 47.25 ± 0.05 Ω; water alone gave a value of 100.8 ± 0.05 Ω. The cell constant measured in a separate experiment was 0.0020 ± 0.0002 m^{-1}. Calculate the solubility product of silver chloride and the relative error in the solubility product. What is the source of the largest component of this error?

CHAPTER
5
ELECTRODE–ELECTROLYTE INTERPHASES

In the previous chapter the topics treated could have been called *ionics*—those processes that occur generally in solutions of electrolytes, or ionic solutions. In contrast, the area of physical electrochemistry taken up in this chapter may be called *electrodics* (Bockris and Reddy, G3) because it centers on the electrode rather than the solution.

5.1 MODELS OF THE INTERPHASE REGION

Electrodics is the study of those processes that take place within the anomalous interphase region surrounding the electrode. The most useful way to enter the area of electrodics is to consider the electrode itself as a giant single ion, of variable charge, in the ionic electrolyte solution. Any charge placed on such an ion-electrode must then cause a readjustment of the solution in its immediate vicinity. The following sections give three attempts made to describe the outcome of such a readjustment.

5.1.1 Helmholtz Model

If any net charge exists on a metal electrode conceived as a giant ion, such a charge must draw out from the electrolyte solution a *counterlayer* of charge of opposite sign. This counterlayer will be made up of ions whose charges are opposite in sign to that on the electrode. Such an electrified interface of two sheets of charge of opposite sign is called the *electrical double layer*, or simply the *double layer*. The sheets of charge are equal in total charge, equal also in charge density, and opposite in sign. This is equivalent to the description of a simple parallel-plate capacitor, and thus this model, the Helmholtz (S1) model of the interphase, is that the interphase region is both in structure and in behavior like an electrostatic parallel-plate capacitor. For such a capacitor, the potential difference across it is

$$E = 4\pi Q x/\epsilon \qquad (5.1)$$

where Q is the charge on the capacitor, x is the distance separating the plates, and is the dielectric constant of the medium separating the plates.

Such a model does give a reasonable fit to some electrochemical data. For example, in studies of electrocapillary maxima, the slope of the curve of surface tension γ against cell potential E for a cell of constant electrolyte composition is given (Lippmann, S2) by the *Lippmann equation*:

$$d\gamma/dE = -Q \tag{5.2}$$

Differentiation of the Helmholtz model equation with respect to charge gives

$$dE = (4\pi x/\epsilon)\, dQ \tag{5.3}$$

which when substituted into the Lippmann equation is

$$d\gamma = -(4\pi x/\epsilon)\, Q\, dQ \tag{5.4}$$

Integration of this equation gives

$$\gamma + \text{const} = -(4\pi x/2\epsilon)Q^2 \tag{5.5}$$

The peak of the curve, corresponding to a maximum in surface tension, is called the *electrocapillary maximum*. The electrocapillary maximum is taken as the *point* or *potential of zero charge*, that is, where $Q = 0$ and no net charge exists in the interphase region. At this potential E_{pzc} the potential difference across the capacitor is necessarily zero. The constant of integration in the above equation is simply γ_{max}, the maximum value of γ, when $Q = 0$. Using again the Helmholtz model equation, this yields

$$\gamma = \gamma_{max} - \epsilon(E - E_{pzc})^2/8\pi x \tag{5.6}$$

This equation gives a parabola symmetric about the potential of the electrocapillary maximum. Actual experimental curves (Grahame, S3) of surface tension against potential are approximately, but not exactly, of this shape as shown in Figure 5.1. Deviations of the experimental curves from the theoretical parabola are usually greater on the positive side and are greater with some solutions than with others. These deviations depend significantly upon the ions in solution, and more strongly upon the anions in the positive region because anions form the counterlayer to an electrode positive to the electrocapillary maximum (point of zero charge) while cations form the counterlayer to an electrode negative to the point of zero charge.

Differentiation of the Helmholtz model equation with respect to potential yields

$$dQ/dE = \epsilon/4\pi x = C_{dl} \tag{5.7}$$

where C_{dl} is the capacitance of the double layer. As in any other parallel-plate

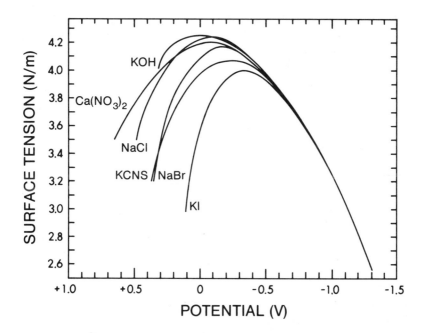

Figure 5.1 Experimental electrocapillary curves. Mercury electrode, aqueous solutions, 18°C. Potential scale arbitrarily set to zero at potential of electrocapillary maximum. Redrawn from Grahame (S3).

capacitor where the areas of the plates do not change, this capacitance is constant, that is, independent of potential, unless E or x is taken as variable, which in the simple Helmholtz model they are not. Experimental values generally range between 200 and 400 mF/m² (cgs: 20–40 μF/cm²), which is in rough agreement with the value calculated for a parallel-plate capacitor with plate spacing equal to the diameter of a water molecule and a dielectric constant for water of 10. The use of the normal bulk dielectric constant of water, 78.54, does not give values of capacitance close to those observed experimentally when used in the simple Helmholtz parallel-plate capacitor model; this is ascribed to inadequacy in either the model or in the use of bulk dielectric constants at the molecular level. Experimentally, it is found that the double-layer capacitance does vary with potential and with the concentration of the electrolyte, dependences that the Helmholtz model is unable to explain.

Double-layer capacitance, the surface tension of mercury, and the potential of the electrocapillary maximum all change with the potential of the electrode surface. The reason for the change in surface tension with potential is that the added charges on the electrode surface repel each other, reducing the surface tension. If electrolytes such as KNO_3, $KClO_4$, or NaF are added the curve does not change significantly since these ions are not specifically adsorbed. But if a negative ion such as iodide which is specifically adsorbed is added to the electrolyte, the electrocapillary curve will move toward a more negative charge; if a positive ion is adsorbed the curve will move toward a positive charge. Addition of an adsorbed nonelectrolyte will not shift the curve but will tend to flatten it out.

Specific adsorption, shown in Figure 5.2, will affect the double-layer capacitance as well. Concentration changes not involving specific adsorption will do so to a much lesser degree, and can usually be ignored. A value of 0.20 to 0.25 F/m^2 is a reasonable average for the double-layer capacitance at most potentials of interest in aqueous solution. Double-layer capacitance is simply a ratio of change in charge to change in potential, both per unit area. Since dQ/dE is the product of the area and the double-layer capacitance per unit area, integration yields

$$Q = C_{dl}A(E_{pzc} - E) \tag{5.8}$$

where Q is the charge on the electrode and E_{pzc} is the potential of the electrocapillary maximum.

Although the simple parallel-plate capacitor model of Helmholtz has inadequacies and more advanced models have been developed, much of the terminology of physical electrochemistry is still based on the Helmholtz model, and it is probably adequate for all *electroanalytical* purposes. Let us therefore look in some detail at this physical model and at the terminology associated with it.

Remember that the electrode was originally considered as a giant ion. Like all real ions in aqueous solution, it is solvated by a layer of oriented water molecules. This solvation layer, normally considered to be only one water molecule thick, is oriented by the charge on the electrode, with the positive ends of the water dipoles oriented toward a negative electrode and the negative ends of the water dipoles oriented toward a positive electrode. *Outside* this solvation layer, and separated from it by their individual solvation sheaths, is the layer of ions of opposite charge to that on the electrode itself, as postulated by Helmholtz. The plane parallel to the electrode that constitutes the locus of the centers of charge of this layer of ions is called the *outer Helmholtz plane*. Ions can, under certain circumstances, *adsorb* or attach themselves directly to the surface of the electrode, displacing the water molecules that would normally occupy that position. If such ions are present, the plane parallel to the electrode that constitutes the locus of centers of charge of this layer or partial layer of adsorbed ions is called the *inner Helmholtz plane*. Except in cases of specific adsorption, the bulk of the ions are believed to be in the outer, rather than the inner, Helmholtz plane. The entire double-layer structure thus described, from the outer Helmholtz plane inward to the electrode surface, is referred to as the *compact double layer*. Within the compact double layer, the potential difference between the electrode and the outer Helmholtz plane is considered to decrease linearly with distance as it does in an ordinary electric parallel-plate capacitor.

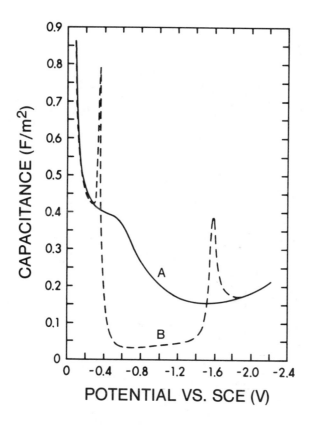

Figure 5.2   Effect of adsorbed heptanol upon differential double layer capacity. Mercury electrode, 0.5 mol/dm³ aqueous Na_2SO_4. Curve A is obtained in the absence of heptanol and curve B in its presence. Redrawn from Grahame (S3).

The schematic physical representation of the Helmholtz model is shown in Figure 5.3.

5.1.2   Gouy–Chapman Model

The inability of the simple Helmholtz model to account totally for the observations related to double-layer capacitance and to surface-tension measurements led to independent postulation of an alternative model by Gouy (S4) and Chapman (S5) about 1910. The fundamental idea was that the ions should be free of the rigid sheet parallel to the electrode surface. Once free, the ions are then affected by thermal buffeting from other particles of the solution and move under the combined thermal and electrical forces until a time-average or equilibrium ionic distribution is achieved in the vicinity of the electrode. This distribution is called the *diffuse double layer* as opposed to the compact double layer of the Helmholtz model.

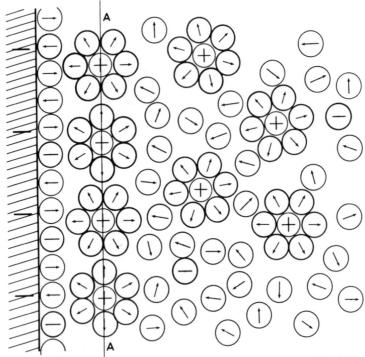

Figure 5.3   Schematic representation of Helmholtz model. Line A–A, outer Helmholtz plane. Water dipoles represented by circles containing arrows, ions indicated by circles containing + (cations) and – (anions) symbols. Additional ions beyond A–A are not part of the Helmholtz model.

Further quantitative discussion of the diffuse double layer can be found in the works cited in the General section of the references of this chapter. It is of some consequence to note, however, that in such a diffuse double layer the potential difference between the electrode and the extreme outer edge of the diffuse layer, which is the bulk solution, falls off exponentially with distance from the electrode. Experimentally, the Gouy–Chapman approach accounts for some of the properties of interphases that the Helmholtz model does not explain, but at the cost of loss of explanation of much of the behavior that is dealt with by the Helmholtz model. The Gouy–Chapman model is not of great importance in itself, but it is a necessary step to the synthesis of Stern.

5.1.3   Stern Synthesis of Models

The most satisfactory present theory of structure of the double layer is the synthesis of Stern (S6) of the two models just described. In the original Gouy–Chapman approach, the ions were taken as point charges. If the ions are taken as having real dimensions instead of point charges, there exists of necessity a distance of closest approach to the electrode; using hydrated radii, the model then becomes that of a Helmholtz plane with scattered ions beyond it, so that the actual double layer is made up of a compact (Helmholtz) double layer plus a diffuse (Gouy–Chapman) double layer beyond it. The charge Q in the interphase may be considered equal to $Q_H + Q_G$, where Q_H is the charge in the compact double layer, and Q_G is the charge in the diffuse double layer. The total capacity for charge of the interphase C is then made up of two capacitances, C_H and C_G. These capacitances are additive in the form $1/C = 1/C_H + 1/C_G$, which is the usual form for the total capacitance of two capacitors in a *series* circuit. This electrical model is equivalent to the Stern synthesis.

The relative contributions of C_G and C_H to the overall capacitance C depend upon the concentration and other properties of the electrolyte solution. With increasing concentration C approaches C_H, while with decreasing concentration C approaches C_G. Physically, this means that with increasing concentration the observed capacitance approaches that of the Helmholtz model; at most experimental concentrations the latter is the case. The charge on the double layer is then squeezed by the ionic concentration in the solution into the outer Helmholtz plane, and thus the Helmholtz model is the more significant component of the Stern synthesis. (See Figure 5.4.)

The magnitude of the capacitance of the double layer is often quite large. Values on smooth metal surfaces such as bright platinum or mercury are normally 0.1 to 0.4 F/m². Platinized platinum may have capacitance up to three orders of magnitude higher with respect to its *geometric* area.

5.1.4   Electrical Analogue Models

The foregoing models are all physical models, but it is sometimes useful to employ an electrical analogue model of an electrochemical cell. An electrical analogue model is one in which the cell is replaced by a circuit whose response to an external electrical signal is identical to the response of an actual electrochemical cell. It is actually impossible to obtain an exact electrical analogue model of the response of any cell to the full range of possible external signals, but quite simple models suffice for most purposes. Assuming that the cell has been properly designed such that the external circuit is responding essentially to the properties of only the desired electrode, the simple Helmholtz physical model of an electrode often provides the requisite quantitative insight. In its simplest form, this model is that of a parallel-plate capacitor. Thus, the simplest useful electrical analogue model of an electrochemical cell that can deal with any of the theory of this chapter requires the addition of the double-layer capacitance C_{dl} in series with the cell resistance discussed in the previous chapter. These constitute a series RC circuit whose properties will depend on the frequency or frequencies of any signal applied to the cell. For some electroanalytical techniques, particularly AC techniques, this model is inadequate, and more complex combinations of resistance and capacitance must be employed.

5.2   ADSORPTION

Adsorption is the attachment of entities such as ions or molecules to surfaces. In electrochemistry, *adsorption* generally means the attachment of ions and/or molecules to the surface of electrodes. Adsorption is divided into *specific adsorption* and *nonspecific adsorption*. Nonspecific adsorption is the general tendency of ions and molecules to attach themselves to the surfaces of electrodes. The water molecules in the immediate vicinity of an electrode are, at least weakly, attached to the electrode surface, and ions in solution may be attached as well. This nonspecific adsorption, though real, is of no significant interest to the *electroanalytical* chemist.

Specific adsorption, on the other hand, is a matter of concern. Specific adsorption is adsorption that cannot be accounted for by the diffuse double-layer part of the Stern synthesis. It is the tendency of specific types of ions or molecules to attach themselves to the electrode surface. Specific adsorption takes place in the compact double layer, often at the inner Helmholtz plane. It is usually much stronger than nonspecific adsorption and is often comparable to the strength of chemical bonds; it is also called *chemisorption*. It is usually significantly dependent on the electrode potential and the electrode material, as well as on the nature of the adsorbing species. Specific adsorption will often produce a monolayer of adsorbed material on the electrode surface, although availability of the adsorbed species may permit only partial coverage of the electrode surface by the monolayer.

Specific adsorption can influence electroanalytical measurements whether or not the species adsorbed is electrochemically active. Adsorption of an inactive substance can cause the electrode surface to become less accessible to electrochemically active species, and also alters the double-layer capacitance of the electrode since the substance is present in the compact double layer. This alteration of the double-layer capacitance can be significant, especially when the adsorption–desorption process depends upon the potential of the electrode.

Specific adsorption of species that are electrochemically active, by which is meant species which can be oxidized or reduced within the range of potential used, can give rise to other problems in interpretation. Most of the theoretical treatments of electroanalytical techniques assume an initial uniform solution phase; this will not exist if specific adsorption is present, since the adsorbed material is then preferentially collected at the electrode surface. Another problem is that the free energy of adsorption is often significant, so that different potentials are observed for the redox processes of the adsorbed and bulk solution species.

The effect of adsorption of electrochemically active species upon electroanalytical observations is complex. It depends on whether the reactants, products, or both are adsorbed; on the strength and potential dependence of the adsorption; on the nature of the adsorbed species; on the components of the electrolyte solution; and on the technique being employed. The effect cannot be

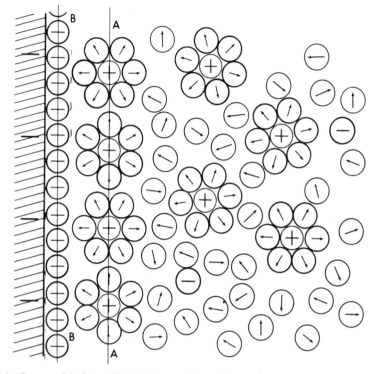

Figure 5.4 Stern model of electrified interface, with specific anion adsorption. Shaded area, electrode; A–A, outer Helmholtz plane; B–B, inner Helmholtz plane. Water dipoles indicated by arrows in circles.

usefully discussed generally here; some discussion devoted to particular techniques is included in Parts 2 and 3. Some comments on the effect of concentrations in the electrolyte solution can, however, be made at this point.

5.2.1 Adsorption Isotherms

An *adsorption isotherm* is a relationship between the quantity of material adsorbed on a surface from a solution or the gas phase (under isothermal conditions) and the properties of the particular phase solution and surface. The adsorption isotherms of electroanalytical interest are those in which the solution is an electrolyte and the surface is an electrode.

At very low surface coverages a simple *linear adsorption isotherm* may be adequate. The amount of a species adsorbed is then directly proportional to its bulk concentration. The proportionality constant B is the adsorption coefficient and may or may not specifically incorporate the maximum possible amount adsorbed, Γ_{max}.

$$\Gamma = \Gamma_{max} \, Bc \tag{5.9}$$

The adsorption isotherm most useful in electrochemistry is the *Langmuir adsorption isotherm*:

$$\Gamma = \Gamma_{max}\,(Bc/(1 + Bc)) \tag{5.10}$$

where Γ is the equilibrium number of moles, Γ_{max} is the maximum possible number of moles, c is the concentration of the adsorbed species in the bulk of the solution, and B is an adsorption coefficient.

The adsorption coefficient in the Langmuir (S7) adsorption isotherm is the equilibrium constant for the equilibrium between the solution species A, the sites upon which it could be adsorbed S, and the species adsorbed upon sites AS. Upon incorporating activity coefficients into the constant and expressing the surface concentrations in mole fractions:

$$A(soln) + S(surf) \rightleftharpoons AS(surf)$$

$$B = X(AS)/X(S)c(A)$$

Since the fractional surface coverage cannot exceed coverage of the total surface area, which is unity, the ratio $X(AS)/X(S)$ is also the ratio of (surface coverage)/(1 − surface coverage). Substitution and rearrangement gives

$$surface\ coverage = Bc/(1 + Bc)$$

which is identical to Eq. (5.10), since Γ/Γ_{max} is also the fractional surface coverage.

The Langmuir adsorption isotherm is useful because it corresponds to a physical model in which the maximum coverage is that of a complete layer, generally a monolayer, upon the electrode surface. Both Γ and Γ_{max} are directly proportional to the surface area, so the *surface excess* Γ or Γ_{max} is usually given in units of moles per square meter (cgs: mol/cm^2) of *geometric* electrode surface area. With increasing concentration, the surface excess always increases until the maximum surface excess is reached.

One of the additional factors that has been taken into account in adsorption is the potential dependence of B as used by Frumkin (S8, S9). Frumkin postulated that the potential dependence is given by

$$B = B_0 \exp(-A(E - E'')^2) \tag{5.11}$$

In this equation E'' is the potential at which the adsorption is a maximum, B_0 is the adsorption coefficient at that potential, and A is a quantity given by

$$A = [C_{dl}(free) - C_{dl}(ads)]/2RT\,\Gamma_{max} \tag{5.12}$$

In the above equation, $C_{dl}(free)$ is the capacitance of the double layer in the absence of any material that is specifically adsorbed, and $C_{dl}(ads)$ is the capacitance of the double layer when the electrode is fully covered with a monolayer of the specifically adsorbed substance.

Other factors are the orientation of molecules on the electrode surface, including the possibility of multiple potential-dependent orientations; and the rate of the adsorption process itself, since unless the rate of adsorption is much more rapid than the rate of a measurement technique no adsorption isotherm can be used to describe the adsorption seen by that technique.

5.3   CONCEPTS AND DEFINITIONS OF ELECTRODE KINETICS

Electrode kinetics is part of that branch of kinetics in which chemical reactions are stoichiometrically linked to electrons through Faraday's laws, and thus the rates of these processes are directly linked to the passage of current. When the electrode kinetics being referred to include only those of the actual electrode process in which the form of charge movement changes between ionic conduction and electronic conduction they are called *charge-transfer kinetics*.

5.3.1   Polarization

It is an experimental fact that when a net current flows through the interphase region that is an electrode, the potential of that electrode, or more accurately the potential difference between that electrode and some reference electrode not affected by the net current flowing, will be changed. If the potential of an electrode through which zero net current if flowing is written as $E(0)$ to denote zero current flow, the difference between the potential in general E, where E is the actual potential of the electrode whether net current is flowing or not, and $E(0)$ is called the *polarization* of the electrode:

$$\eta = E - E(0) \tag{5.13}$$

Polarization of the electrode is by definition zero when E equals $E(0)$, and normally not otherwise, although the polarization may be too small to be measured. Polarization is a function of the current density j and generally increases with increasing current density.

The term polarization and its various derived adjective forms such as *polarized, polarizable*, and *unpolarizable* are all applicable regardless of the cause of the deviation of E from $E(0)$.

5.3.2   Overvoltage

Polarization is a measurable quantity and is independent of the origin of $E(0)$, which need not be an equilibrium potential in the sense of being determined by a single electrode couple. If the origin of $E(0)$ is a single electrode reaction and there is no interference with the development of its equilibrium potential, the equilibrium potential E^0 will be identical to $E(0)$. The equilibrium potential due to a single couple E^0 used in Part 1 is not the same as the standard thermodynamic potential of a couple, also symbolized by E^0, taken up in Part 2; no account of standard states or reference electrodes has been introduced as yet. All thermodynamic standard potentials are equilibrium potentials due to a single couple, but the converse is not true.

Under these circumstances

$$\eta = E - E^0 \tag{5.14}$$

and η is now called the *overvoltage* rather than the polarization. Overvoltage is thus a subdivision of the more general term polarization.

Overvoltage and polarization are often used in confusing and overlapping ways in the electrochemical literature; the distinction drawn here follows Vetter (G8). The terms *anodic polarization (anodic overvoltage)* and *cathodic polarization (cathodic overvoltage)* refer to the polarization (overvoltage) due to a net anodic current and a net cathodic current, respectively. The signs of overvoltages have not achieved complete consensus in the literature. For this work, the recommendations of the International Union of Pure and Applied Chemistry (S11) will be followed. These are, fundamentally, that positive values are assigned to anodic currents (net oxidation) and negative values to cathodic currents (net reduction). Thus the existence of a net anodic current causes an anodic overvoltage, whose sign is *positive*, and the existence of a net cathodic current causes a cathodic overvoltage, whose sign is *negative*. The sign of the overvoltage is related to the direction of the net current flow at the electrode and not to its actual equilibrium potential in sign or otherwise. For example, if a battery is discharging, the effect of overvoltage is to decrease the potential difference between the electrodes, but if a battery is charging, the effect of overpotential is to increase the potential difference between the electrodes. In either case, overpotential is deleterious to its performance!

Overvoltage can be assigned to various steps in the overall reaction that occurs at an electrode, and hence terms subdividing total overvoltage arise. The most important are *charge-transfer overvoltage, reaction overvoltage, diffusion overvoltage*, and *crystallization overvoltage* (Vetter, G8). The total overvoltage is the sum of these. Each arises from a single *type* of step in the overall electrode reaction. *Reaction overvoltage* is that overvoltage that arises solely from the reaction steps in which charge transfer through the interphase region is not directly involved, as would be the case for a step $Pb(OH)_3^- \longrightarrow 3OH^- + Pb^{2+}$, which might precede the reduction of Pb^{2+}. *Diffusion overvoltage* is the result of slow mass transfer to the electrode by diffusion, and *crystallization overvoltage* is the result of slow incorporation of ions into a crystal lattice or of slow removal of them from it.

Certain other descriptions of the types of overvoltage have been used. Agar and Bowden (S12,S13) defined their term *concentration overvoltage* as exactly equivalent to diffusion overvoltage. Their term *resistance overvoltage* is not an overvoltage at all and corresponds to *resistance polarization* (see below). Their term *activation overvoltage* is the sum of the charge-transfer overvoltage, the reaction overvoltage, and the crystallization overvoltage, or, in other words, the total overvoltage less the diffusion overvoltage.

The term *overpotential*, occasionally found in the electrochemical literature, is synonymous with *overvoltage*. The term *resistance polarization*, introduced as resistance overvoltage by Agar and Bowden, is not an overvoltage, because it does not correspond to any electrode reaction or step in an electrode reaction. It is the sum of the ohmic drops that, according to Ohm's law $E = IR$, must arise when current is passed through any conductor. Resistance polarization includes as components the resistance of the bulk electrolyte; the resistance of the electrolyte within a diffusion layer, which is normally not the same as, and may be considerably greater than, the resistance of the bulk electrolyte; and the resistance of surface films such as oxides, which may far exceed that of both of the other types. Resistance polarization is an undesirable component of the overall polarization, and steps such as the addition of a supporting electrolyte are normally taken to reduce its experimental magnitude to the greatest extent possible.

5.4   FUNDAMENTALS OF CHARGE TRANSFER OVERVOLTAGE

Conventionally, the rate of any chemical reaction such as the generalized redox reaction $Ox + ze^- \rightleftharpoons Red$, where Ox represents the oxidized species and Red the reduced species, is given by an equation such as

$$\text{Rate} = -dn_{Ox}/dt = dn_{Red}/dt$$

where n_{Ox} represents moles of oxidized species and n_{Red} moles of reduced species. In electrochemical reactions direct access to amounts of substance reacting is often denied us, but through Faraday's laws the measured current I will always be proportional to the rate of change in the amounts of products and reactants, which is the reaction rate. Since the units of dn/dt are moles per second,

$$-I_c = zF(-dn_{Ox}/dt)$$

maintains consistent units, where z is the number of electrons per mole of oxidized species used and F, the Faraday constant, converts from moles to the unit of charge, coulombs. Cathodic current is taken as negative, as recommended by IUPAC.

5.4.1   Heterogeneous Rate Constants

Electron transfer, or charge transfer, is necessarily a heterogeneous process, and so the rate constant for it is of necessity a heterogeneous rate constant, that is, a rate constant valid only on the surface of the electrode. Moreover, the molecularity of the reaction is one, in the species Ox, and since molecularity of a step—here the charge transfer step—necessarily also implies order of that step, it follows that

$$-dn_{Ox}/dt = k_{c,h} A c^0_{Ox} \tag{5.16}$$

where $k_{c,h}$ is the *cathodic heterogeneous rate constant*, A is the electrode area, and c^0_{Ox} is the *surface* concentration of the species Ox. Only the surface concentration of Ox is relevant because only the quantity of species Ox at the electrode surface participates in the heterogeneous process that takes place there. While it is true that relations do exist between c^0_{Ox} and the bulk concentrations, these are for our present purposes irrelevant. It has been assumed in this equation that the surface concentration can be used in place of the surface activity, or is at any rate proportional to it so that the proportionality constant can be incorporated into the heterogeneous rate constant. Combination of the two preceding equations yields:

$$-I_c = zFAk_{c,h}c^0_{Ox} \tag{5.17}$$

where the units of $Ak_{c,h}c^0_{Ox}$ are moles per second. Since the concentration c^0_{Ox} is in moles per cubic meter, while the units of the heterogeneous rate constant $k_{c,h}$ are meters per second, the area of the electrode A, in meters squared, must be introduced. This gives the cathodic component of the current. The units of *heterogeneous* first-order rate constants are meters per second, while those of *homogeneous* first-order rate constants are per second, s^{-1}.

Treatment of the anodic component of the current in the same fashion yields

$$I_a = zFAk_{a,h}c^0_{Red} \tag{5.18}$$

for the anodic current, which is positive, and $k_{a,h}$ is then the *anodic heterogeneous rate constant*. The number of electrons z in the balanced forward reaction must equal those in the balanced reverse reaction, so the value of z in both equations is the same, while the values of the surface concentrations and heterogeneous rate constants need not be. The actual net current I is simply the sum of the anodic and cathodic components $I_a + I_c$ taken with regard to sign:

$$I = zFAc^0_{Red}k_{a,h} - zFAc^0_{Ox}k_{c,h} \tag{5.19}$$

This equation can be used as it stands for reaction overvoltage, that is, overvoltage due to the slow kinetics of a *chemical* reaction. By a chemical reaction is meant a reaction whose rate constant does not, as charge-transfer overvoltage does, depend on the potential of the electrode, and thus a *chemical* reaction is formally distinguished from an *electrochemical* one.

A chemical reaction, whether electrochemical or not, is found to have a temperature-dependent rate constant k_f. The temperature dependence is expressed by a relation proposed by Svante Arrhenius in 1889, which bears his name as the *Arrhenius equation*:

$$k_f = A \exp(-E^*/RT) \tag{5.20}$$

In the Arrhenius equation, A is the preexponential factor, or *frequency factor*, and E^* is the *activation energy*. Alternatively, the empirical Arrhenius equation can be derived from absolute rate theory as proposed by Eyring in 1935. The rate constant is then expressed as a function of the *free energy of activation* ΔG^*:

$$k_f = A \exp(-\Delta G^*/RT) \tag{5.21}$$

The free energy of activation in ordinary kinetics is viewed as the height of the energy barrier that the reacting species must cross in order for the reaction to take place. In electrochemical kinetics, it is viewed, in an analogous way, as the height of the energy barrier that the electron must cross as it moves between the reacting species and the electrode. The frequency factor A is, in the absolute rate theory, given as a simple function of the Boltzmann constant k and the Planck constant h:

$$A = kT/h \tag{5.22}$$

For an electrochemical reaction it is most convenient to express free energy of activation in terms of potential:

$$\Delta G^* = -zFE \tag{5.23}$$

$$k_h = A \exp(zFE/RT) \tag{5.24}$$

In an electrochemical situation the reactions at the electrode are necessarily heterogeneous. It is convenient to distinguish them from homogeneous solution kinetic reactions by the subscript h. The potential is simply the actual potential of the electrode. Since the potential of a single electrode is not measurable, it is necessary to define the rate constant at some fixed standard physical electrode potential E^0 as the *standard heterogeneous rate constant* k_h^0:

$$k_h^0 = A \exp(zFE^0/RT) \tag{5.25}$$

The standard potential may be taken with respect to a standard physical reference electrode, such as the standard hydrogen electrode (SHE). It is more commonly taken with respect to the standard potential of the electrochemical reaction involved, which is itself taken with respect to the potential of a standard physical reference electrode. In terms of the standard heterogeneous rate constant, the measured rate constant is then:

$$k_h = k^0_h \exp(zFE/RT) \tag{5.26}$$

Heterogeneous rate constants need not be the same for an anodic reaction as for the same reaction occurring in a cathodic direction, and the direction must therefore be distinguished by the subscripts a and c. The anodic heterogeneous rate constant $k_{a,h}$ is then

$$k_{a,h} = k^0_{a,h} \exp(zFE/RT) \tag{5.27}$$

The value of the heterogeneous rate constant $k_{a,h}$ depends upon the electrode potential E and approaches the value of $k^0_{a,h}$ as the potential difference $(E - E^0)$ goes to zero. This equation then becomes

$$k_{a,h} = k^0_{a,h} \exp(zF\eta/RT) \tag{5.28}$$

since $\eta = E - E^0$, where E^0 is the true equilibrium value of the potential at which $k_{a,h} = k^0_{a,h}$.

5.4.2   Exchange Current

At any equilibrium potential no *net* current can flow, and no overvoltage or polarization can exist. The condition is, however, dynamic rather than static and is thus similar to other types of dynamic chemical equilibrium such as precipitation. When a slightly soluble material is in equilibrium with a solution of its ions, as in the classic example $AgCl(s) \rightleftharpoons Ag^+ + Cl^-$, there is continuous precipitation and continuous redissolution, but, since the rates are equal, no net change is observed. At an electrode at equilibrium there is significant and continuous movement of charge carriers in both directions through the interphase region so that identical anodic and cathodic currents, and thus a *net* zero current, flow through the interphase region and thus through the external circuit. Thus, no *net* current flows externally, and no *net* chemical reaction occurs.

The magnitude of these real and mutually compensating currents flowing at any zero-current potential $E(0)$, taken as an absolute value and therefore positive in sign, is called the *exchange current* I^0. The exchange current is a measurable quantity for any electrode, but is characteristic of a particular electrode reaction only when $E(0)$ is determined by a single electrode reaction and is then equal to E^0. By the use of E^0 throughout this chapter we mean the potential of an electrode reaction under conditions of zero current, that is, an unpolarized electrode potential established by a single oxidation or reduction reaction.

When exchange current is measured as a current per unit of electrode area it is called the *exchange current density*, j^0. Since the exchange current I^0 is the current when η is zero,

$$I^0 = zFAc^0_{Red}k^0_{a,h} = zFAc^0_{Ox}k^0_{c,h} \tag{5.29}$$

At zero net current the surface concentrations of both will be the equilibrium, and normally the equal, concentrations.

5.4.3 Transfer Coefficient

The effect of overvoltage on the two reaction directions is a complicating factor that has not yet been considered. If the overvoltage is positive (anodic), it would accelerate the movement of charge in the direction so as to increase anodic current. It would also, however, necessarily *decelerate* the movement of charge in the direction of cathodic current. Thus the effects are twofold: an increase in anodic current and a decrease in cathodic current. These two effects need not be equal. Their difference will depend on the symmetry of the electrochemical activation energy barrier that exists in the electrode interphase region, in the compact double layer. The possibility of asymmetry is taken into consideration by introduction of the *transfer coefficient* α. The transfer coefficient is that fraction of the overvoltage that goes to increase the anodic current. Since $0 < \alpha < 1$, the fraction $1 - \alpha$ must therefore be the fraction of the overvoltage that goes to decrease the cathodic current. The transfer coefficient α normally does not change significantly with potential over small potential ranges, and is frequently taken equal to 0.5. The assumption is then that the activation energy surface is symmetric. Measured values of transfer coefficients in aqueous solution generally range from 0.2 to 0.8.

Introducing transfer coefficients with due regard to sign, for the anodic reaction,

$$k_{a,h} = k^0_{a,h} \exp(\alpha z F \eta / RT) \tag{5.30}$$

and for the cathodic reaction

$$k_{c,h} = k^0_{c,h} \exp(-(1-\alpha)zF\eta/RT) \tag{5.31}$$

Substitution of both of these rate constants into the equation for the net current I derived earlier in this section, and use of the equation for exchange current given above, yields the *Butler–Volmer equation*:

$$I = I^0[\exp(\alpha z F \eta / RT) - \exp(-(1-\alpha)zF\eta/RT)] \tag{5.32}$$

Further details of the origin (Erdey-Gruz and Volmer, S14; Butler, S15) and derivation of the Butler–Volmer equation can be found in the General references at the end of this chapter.

5.5   CONSEQUENCES OF CHARGE-TRANSFER OVERVOLTAGE

In many electroanalytical measurements, observed currents are limited by charge transfer and processes of interest may include charge transfer. Some consequences of the Butler–Volmer equation and its limiting cases are given below. In all of these cases it is assumed that only charge transfer will limit the current. If this is not the case, the equations will become more complex because then the processes of diffusion, reaction, and crystallization must also be considered.

5.5.1   Condition at Zero Current or Overvoltage

There is no overpotential when the net current I is zero, and no net current when the overpotential is zero.

5.5.2   Condition at Large Overvoltages

When the overpotential becomes large in either the anodic or the cathodic direction, then one of the two exponential terms becomes very small relative to the other. For example, at high positive (anodic) overvoltages,

$$I = I^0 \exp(\alpha z F \eta / RT) \tag{5.33}$$

or

$$\ln (I/I^0) = \alpha z F \eta / RT \tag{5.34}$$

which is more commonly seen in the form

$$\eta = -RT/\alpha z F \ln I^0 + RT/\alpha z F \ln I_a \tag{5.35}$$

This equation has the same form as the empirical *Tafel equation* (Butler, S16):

$$\eta = A + B \log I \tag{5.36}$$

obtained for electrode reactions under conditions where the observed current is limited only by the rate of charge transfer, and permits interpretation of the Tafel slopes and intercepts in terms of transfer coefficient and exchange current. A plot of η against $\log I$ yields a straight line whose slope gives α and whose intercept then gives I^0. Use of the Tafel equation is quantitatively valid whenever the overpotential exceeds 120 mV, for a one-electron reaction and $\alpha = 0.5$, since then the current due to the process being retarded is less than 1% of the current due to the process being accelerated; in general, this is true when the absolute value of the overpotential is greater than twice $2.303RT/zF$. For a one-electron reduction with $\alpha = 0.5$, an overpotential of 120 mV should be sufficiently large that Tafel behavior is observed at 25°C; for a two-electron reaction only 60 mV should be needed.

The Tafel equation and Tafel plots are commonly used in physical electrochemistry; an example can be seen in the work of Conway and Dzieciuch (S17) on Kolbe electrooxidation.

5.5.3   Condition at Small Overvoltages

Taking the derivative of the Butler–Volmer equation with respect to overvoltage, and applying the rules that

$$d(\exp(bx))/dx = b\exp(bx) \text{ and } d(u + v + w)/dx = du/dx + dv/dx + dw/dx,$$

gives equation 5.37:

$$dI/d\eta = I^0[\alpha zF/RT \exp(\alpha zF\eta/RT) + (1 - \alpha)zF/RT \exp(-(1 - \alpha)zF\eta/RT)]$$

In the limit of values of η, as $\eta \rightarrow 0$, $\alpha zF\eta/RT$ also approaches zero, and for small values the exponential term $\exp(\alpha zF\eta/RT)$ can be approximated by $1 + \alpha zF\eta/RT$. Hence the equations simplify to

$$dI/d\eta = I^0[\alpha zF/RT + (1 - \alpha)zF/RT] \tag{5.38}$$

$$dI/d\eta = I^0[zF/RT] = zFI^0/RT \tag{5.39}$$

By analogy with Ohm's law, from which the ratios I/E and dI/dE have the units of reciprocal resistance, it is convenient to define the *polarization resistance* R_p as the derivative of overpotential with respect to current. At low overpotentials,

$$R_p = RT/zFI^0 = RT/zFAj^0 \tag{5.40}$$

The polarization resistance is a measure of the voltage change or overvoltage η arising from a current I, and is a useful alternative to expressing the rate of an electrochemical reaction and its sensitivity to passage of current. The polarization resistance should not be confused with the *resistance polarization* discussed earlier. High values of polarization resistance indicate that comparatively little current flows when the overvoltage increases, or that a comparatively small change in current produces a large change in overvoltage.

If a measuring instrument draws a momentary current that will produce a small overpotential, the effect can be seen from the integral of the Butler–Volmer equation:

$$\eta = RT/zF (I/I^0) \tag{5.41}$$

Clearly, a larger overpotential will be produced by the same current drain I if the exchange current of the reaction is smaller, hence a reaction for which I^0 is small is said to be more easily polarized. It is convenient to think of this as a statement of Ohm's law; the equation $-\eta = (RT/zFI^0)I$ is analogous to the statement of Ohm's law $E = RI$. The ratio of overpotential to current, the *polarization resistance* R_p, is high if the electrode is easily polarized and low if the electrode is not easily polarized.

Using the simplest equivalent circuit for an electrochemical cell to which an overpotential may be applied, which is the series combination of the double-layer capacitance C_{dl} and the polarization resistance R_p, the overpotential will decay exponentially as a function of time according to the relation

$$\eta(t) = \eta(0) \exp(-t/\tau) \tag{5.42}$$

where τ is the *time constant* of the RC equivalent circuit. The quantity 2.3 times the time constant is the time required for the overpotential to decay by one order of magnitude. This time constant is R_pC_{dl}; on platinum, a value of C_{dl} would be about 1.0 F/m².

Example
The reaction $O_2 + 4H^+ + 4e^- \rightarrow 2H_2O$ has a value of j^0 of about 1×10^{-6} A/m² on bright platinum; $R_p = RT/4Fj^0 = 6 \times 10^7 \ \Omega$ cm². For oxygen on bright platinum, the time constant τ is $(6 \times 10^3 \ \Omega \ \text{m}^2) \times 1.0$ F/m² $= 6 \times 10^3$ s, and thus $2.3\, \tau$ is 3.8 h.

Example
The much faster reaction $Zn^{2+}(c = 0.020) + 2e^- \rightleftharpoons Zn(Hg)$ has an exchange current density of 54 A/m² on mercury; for this reaction R_p is $RT/2Fj^0 = 0.00024 \ \Omega$ m². For Zn^{2+} on zinc amalgam, the time constant τ is $0.00024 \ \Omega$ m² $\times$ 1.0 F/m² $= 2.4 \times 10^{-4}$ s, and thus 2.3τ is 0.55 ms.

It is clear that the zinc electrode is far more difficult to polarize and that its polarization disappears much faster than does that of an oxygen electrode, even if C_{dl} of bright platinum is used for the amalgamated surface.

5.6 TERMS ARISING FROM ELECTROCHEMICAL KINETIC CONCEPTS

Terminology used with different electroanalytical techniques has its roots in the concepts of electrochemical kinetics, which is closely related to ordinary chemical kinetics. However, the additional variable of electrode potential at which the reaction of interest is occurring has very significant influence on the results obtained.

5.6.1 Fast and Slow Reactions

As applied in electrochemical kinetics, *fast* and *slow* electrode reactions are relative and not absolute characterizations; they are relative to the rate of the electrochemical technique used to study the reactions. The rate of a measuring technique can have two distinct meanings. One of these is the response speed of the measuring technique, which is a time phenomenon. An electrode reaction is then said to be *fast* when its rate is much greater than the response speed, or resolving time, of the technique by which it is being studied. A reaction that is fast when studied polarographically may well be slow when studied by a potential-step technique because the resolving time of a polarograph is the drop time, usually 3 s or more, while the resolving time of a potential-step technique is of the order of milliseconds.

The second meaning of the rate of a measuring technique is in terms of current. Here the term "fast" is taken to mean a situation at the electrode in which the actual current density demanded by the measuring technique j is much less than the exchange current density j^0. If j approaches or exceeds j^0 the reaction is said to be slow. To avoid confusion the terms "fast" and "slow" should not be used in this way; the concept of polarization resistance should be used instead.

5.6.2 Reversible and Irreversible Reactions

The terms *reversible* and *irreversible* are commonly used in kinetics in two ways, one of which can lead to serious confusion. The acceptable meaning of an irreversible reaction is that the rate of the forward reaction is so much greater than that of the backward reaction that the backward reaction can be neglected; such reactions are termed irreversible in electrochemical kinetics just as in ordinary chemical kinetics. If these rates have rate constants k_f and k_b, then a reaction may be termed *totally irreversible* when $k_f \geq 100k_b$; *partially irreversible* or *quasi-reversible* when $100\,k_f > k_b > k_f$; and *reversible* when $k_f \leq k_b$. The direction of k_f and k_b depends upon the direction of reaction giving rise to the net current flow. This meaning is not capable of serious misinterpretation.

Unfortunately, however, the rates of charge-transfer reactions are potential-dependent. Thus the meaning of reversible and irreversible, if interpreted in terms of current, depends upon the current demanded and the overvoltage thereby produced. A reaction for which $k_f \leq k_b$ at one potential may well be totally irreversible at another, since both k_f and k_b are potential-dependent and change in opposite directions with overvoltage. Thus the use of reversible and irreversible in the literature is confusing; in some cases *reversible* is taken to mean an absence of detectable charge-transfer overvoltage, or an observed current limited by some process other than charge transfer, while *irreversible* is used in the sense of charge-transfer-limited current being observed. Care in the use of these terms is recommended.

5.7 MECHANISMS OF ELECTROCHEMICAL REACTIONS

Mechanisms of reactions that take place at electrodes, or *electrochemical mechanisms*, can be quite complex. In general, a mechanism may be written as a sequence of steps in which one substance is converted into another:

A → B, B → C, C → D, ...

In an electrochemical mechanism the steps can be of two types, called *chemical steps* (C) and *electrochemical steps* (E). A chemical step is a step in which no electron transfer to or from the electrode takes place. Such a step does not of itself produce a charge flow into or out of the electrode and thus is not directly observable by an external measuring circuit. It may, however, influence charge flow occurring because of other steps in the mechanism and can then be detected indirectly. Since chemical steps involve no electron transfer, they are governed by ordinary chemical kinetics. The rates of chemical steps are not strongly dependent upon the potential of the electrode at which they occur, although some influence of the potential is felt because the double-layer structure of the interphase region is determined in part by the potential of the electrode. The chemical steps in an electrochemical mechanism may or may not take place in the interphase region.

An electrochemical step, on the other hand, involves electron transfer to or from the electrode. Such a step does produce a charge flow measurable by an external circuit. Electrochemical steps are governed by electrode kinetics as described earlier in this chapter, and their rates are strongly dependent on the potential of the electrode at which they occur. An electrochemical step probably always involves the transfer of only one electron. Many systems are known, however, in which two, and less often even more, electrons appear to be transferred in a single electrochemical step because the experimental technique used is not capable of resolving the separate steps.

It is therefore necessary to distinguish between the stoichiometric number of electrons involved in a reaction, z, and the number of electrons involved in one kinetic step of that reaction. In an overall electrode reaction, a certain number of electrons appear; two, in the reaction $Fe^{2+} + 2e^- \rightarrow Fe(0)$. The presence of these electrons follows from the stoichiometric necessity of balanced charge, and their number z is fixed for a given overall reaction. This number of electrons arises in various fundamental electrochemical equations that refer to equilibrium. One example is $\Delta G = -zFE$ for the free-energy change in an electrochemical cell whose equilibrium potential difference is E; another is its derivative, the Nernst equation.

When the net overall electrode reaction is considered to occur in several steps, however, the rate of the overall electrode reaction may well be determined by only one of those steps. As in normal chemical kinetics, when a mechanism involves several successive steps the rate of the reaction proceeding by the overall mechanism is determined by the rate of the slowest step. While the number of electrons involved in the overall reaction stoichiometry is z, the number of electrons involved in the rate-determining step may well not be equal to z and, in practice, turns out to be unity most of the time. The number of electrons involved in any step of the overall net reaction is indicated by v, the *stoichiometric factor* of the step. It has the same meaning for the step as does z for the overall net electrode reaction. When the step is the only reaction being considered and no ambiguity can arise, z is the preferred symbol.

Rates of chemical reactions are functions of the systems being studied and the concentrations of the reacting species in the interphase region, if double-layer effects are neglected. Rates of electrochemical steps are functions of these and of the electrode potential as well. Concentrations and electrode potentials may vary slowly or rapidly with time, as may electrical signals applied to an electrode at which reactions are occurring. A general solution of the appropriate differential equations for each mechanism is not possible. However, solutions under the specific conditions appropriate to different electroanalytical techniques have been extensively developed. For the sake of clarity, chemical steps are given as first order, and electrochemical steps are given as reductions in the general mechanism descriptions below.

5.7.1   Single Steps of Electrochemical Mechanisms

Two different *single steps of electrochemical mechanisms* are possible.

The C step or mechanism

$$A \rightleftarrows B$$

consists of a single chemical step occurring at the electrode surface in which $A \rightarrow B$ has a forward rate constant $k_{f,h}$. If the value of $k_{f,h}$ is large, the reaction is said to be fast, while if the value of $k_{f,h}$ is small, the reaction is said to be slow. Both of these terms are relative, and may be relative to each other, relative to another step, or relative to the technique that is being used to study the mechanism. The reverse or back rate constant k_b is that of the reverse reaction, $B \rightarrow A$. A mechanism consisting of a single chemical step cannot be studied directly by any electroanalytical technique, but electroanalytical techniques are routinely used to monitor time-dependent concentrations of either A or B if they are electrochemically active. The use of polarography to follow the rates of hydrolysis of nitrosoureas is but one of many such applications.

The E step or mechanism

$$A + e^- \rightleftarrows B$$

consists of a single electrochemical step governed by charge-transfer kinetics, as described by the Butler–Volmer equation.

5.7.2   CC Mechanism

Four different *two-step mechanisms* are possible. Two successive chemical steps constitute the *CC mechanism*:

$$A \rightleftarrows B, B \rightleftarrows C$$

When two successive kinetic steps take place, the overall or observed kinetics will depend on the relative magnitudes of their rate constants and on the species monitored. If the concentration of the final product C is monitored and the rates are widely disparate, the concentration of C will be governed virtually completely by the rate of the slowest, or *rate-limiting*, step of the sequence whether first or second. If the two rates are comparable, both must be considered in order to obtain the overall rate.

Both of the chemical steps involved may be heterogeneous or homogeneous, depending on whether they take place in the bulk solution or at the electrode surface. Again, such a mechanism can only be followed by electrochemical methods.

5.7.3   EE Mechanism

Two successive electrochemical steps constitute the *EE mechanism*:

$$A + e^- \rightleftharpoons B, \quad B + e^- \rightleftharpoons C$$

When two successive electrochemical steps take place, the observed kinetics will depend on the relative magnitudes of their forward rate constants, which in turn depend on the potential of the electrode. Such a mechanism may appear as a single multiple-electron transfer process at all experimentally accessible electrode potentials. Usually only the forward reactions need be considered, although the four possible limiting cases of R−R (reversible-reversible), I−R (irreversible-reversible), R−I (reversible-irreversible), and I−I (irreversible-irreversible) electrochemical steps give distinguishable effects with some electroanalytical techniques.

5.7.4   CE Mechanism

The *CE mechanism* includes both types of steps:

$$A \rightleftharpoons B, \quad B + e^- \rightleftharpoons C$$

and amounts to a preceding chemical equilibrium; the effect of the equilibrium depends on the rates of both forward and reverse reactions, whose quotient is the equilibrium constant.

The preceding equilibrium can be homogeneous or heterogeneous. The simplest case is a fast, first-order heterogeneous equilibrium, for which the equilibrium constant is

$$K = c^0{}_B / c^0{}_A$$

The current observed for such a reaction is given by the Butler–Volmer equation in which the term $c^0{}_{Ox} k^0{}_{c,h}$ is replaced by $c^0{}_A K k^0{}_{c,h}$. The current is thus a function of the surface concentration of A rather than B, since the surface concentration of

A governs that of B as long as the preceding equilibrium is fast enough to keep up with the current flowing.

When the preceding chemical equilibrium takes place in the bulk solution rather than at the electrode surface, the above replacement is possible only when c_{Ox} equals c^0_{Ox} (uniform concentration) or when a functional relationship linking c_{Ox} to c^0_{Ox} is known.

Analogous equations are obtained for anodic reactions. More complex equations can be derived for reactions of higher order, when the equilibrium is not much faster than the rate of charge transfer, or when processes such as diffusion govern the dependence of c^0_{Ox}.

The rate of the electrochemical step depends strongly on the electrode potential, while the rate of the chemical step does not. By appropriate variation of the electrode potential it is usually possible to shift from an overall rate controlled virtually completely by the chemical kinetics of the chemical step to one controlled virtually completely by the electrode kinetics of the electrochemical step. Techniques in which electrode potential is the independent variable can often distinguish this mechanism.

5.7.5   EC Mechanism

The EC mechanism

$$A + e^- \rightleftharpoons B, \quad B \rightleftharpoons C$$

is similar to the CE mechanism just discussed.

The effect of the following equilibrium appears through the Butler–Volmer equation. It affects now the term $c^0_{Red} k^0_{a,h}$, which is replaced by $c^0_C K k^0_{a,h}$, in the simple first-order, heterogeneous, fast-equilibrium case. Again, an analogous equation holds for the anodic reaction. More complex equations can be derived for other cases.

5.7.6   EC(R) Mechanism

The EC(R) mechanism

$$A + e^- \rightleftharpoons B, \quad B + X \rightleftharpoons A$$

is a special case of the EC mechanism in which the reactant A is catalytically regenerated from the product of the electrode reaction B. The regeneration of A may occur by reaction with a third component X present in essentially constant concentration in the solution and thus involves pseudo-first-order rather than first-order kinetics.

An example of the EC(R) mechanism is the reduction of Ti(IV) in aqueous acid solution on mercury in the presence of hydroxylamine. The Ti(III) produced by reversible reduction is reoxidized by hydroxylamine to Ti(IV), while hydroxylamine is not reducible on mercury at this potential. The mechanism has been studied by several different electroanalytics techniques (Blazek and Koryta, S18; Delahay, Mattax, and Berzins, S19; Herman and Bard, S20; Fischer, Bracka and Fischerova, S21; Saveant and Vianello, S22; Christie and Lauer, S23; Lingane and Christie, S24). The pseudo-first order catalytic rate constant is about 40 dm^3/mol s at 25°C.

The EC(R) mechanism with disproportionation

$$A + e^- \rightleftarrows B, \quad B + B \rightleftarrows A + C$$

is a variant of the EC(R) mechanism in which the third component X is another molecule of B; that is, the reduction product B reacts with itself, disproportionating to give a product C as well as the original reactant A.

Examples of the EC(R) mechanism with disproportionation include the sulfonephthalein acid–base indicator phenol red, which upon one-electron reduction in aqueous solution gives a free radical product that disproportionates to form the original sulfonephthalein and the reduced (two-electron) sulfonephthalin (Daum and McHalsky, S25). The rate constant is 340 ± 30 dm^3/mol s (Senne and Marple, S26). The two-electron reduction is observed, but only at more negative potentials. An inorganic example of disproportionation is the one-electron polarographic reduction of U(VI) to U(V); U(V) disproportionates to U(VI) and U(IV) in acidic aqueous solution (Kudirka and Nicholson, S27).

5.7.7   Three-Step Mechanisms

Most of the possible *three-step mechanisms* have not received serious study. The CCC and EEE mechanisms are simply extensions of their CC and EE two-step counterparts. The possible EEC, CEE, CCE, and ECC mechanisms can usually be analyzed only in two-step form, since the two steps of the same type are difficult to distinguish by varying the experimental conditions.

The ECE mechanism

$$A + e^- \rightleftarrows B, \quad B \rightleftarrows C, \quad C + e^- \rightleftarrows D$$

in which a chemical step separates two electrochemical steps, has been the subject of considerable study. The ECE mechanism may be considered a combination of the EC and CE mechanisms. Again, its observation and analysis depend on the fact that the rates of the electrochemical steps are strongly dependent upon the electrode potential while the rate of the chemical step is not. If the chemical step is very fast relative to the electrochemical ones under all conditions, this mechanism is indistinguishable from the EE mechanism. The reversibility of the electrochemical steps sometimes produces observable effects, depending on the experimental techniques and conditions.

The EC(R)E mechanism

$$A + e^- \rightleftharpoons B, \quad B + X \rightleftharpoons A, \quad B + e^- \rightleftharpoons C$$

is a special case of the ECE mechanism in which the second electrochemical step competes for B with the chemical step, and is an extension of the EC(R) mechanism. The reversibility of the electrochemical steps may need to be considered.

5.8 REFERENCES

General

G1 D. A. MacInnes, *The Principles of Electrochemistry*, Dover, New York, 1961; orig. ed. 1939; 478 pp. A classic general treatment of physical electrochemistry. Well worth having.

G2 H. R. Thirsk and J. A. Harrison, *A Guide to the Study of Electrode Kinetics*, Academic Press, New York, 1972, 174 pp. A brief but effective treatment of the interpretation of experimental electrochemical data from the standpoint of physical electrochemistry.

G3 J. O'M. Bockris and A. K. N. Reddy, *Modern Electrochemistry* (2 vols.), Plenum Press, New York, 1970, 1432 pp. Best modern text.

G4 J. O'M. Bockris and R. A. Fredlein, *A Workbook of Electrochemistry*, Plenum Press, New York, 1973. Set of problems to accompany Bockris and Reddy.

G5 B. E. Conway, *Theory and Principles of Electrode Processes*, Ronald Press, New York, 1965, 303 pp. Advanced monograph.

G6 P. Delahay. *Double Layer and Electrode Kinetics*, Interscience, New York, 1965, 320 p. Advanced monograph.

G7 T. Erdey-Gruz, *Kinetics of Electrode Processes*, Adam Hilger, London, 1972.

G8 K. J. Vetter, *Electrochemical Kinetics*, Academic Press, New York, 1967, 789 pp. Classic advanced monograph and text.

G9 B. B. Damaskin, *The Principles of Current Methods for the Study of Electrochemical Reactions*, G. Mamantov, trans., McGraw-Hill, New York, 1967.

G10 E. Yeager and A. J. Salkind, Eds. , *Techniques of Electrochemistry, Volume 1* (1972); *Volume 2* (1973); *Volume 3* (1977). Wiley-Interscience, New York. A series of review articles on specific techniques from the standpoint of physical electrochemistry and electrochemical engineering.

G11 P. Delahay and C. W. Tobias, Eds. , *Advances in Electrochemistry and Electrochemical Engineering* (from Volume 10, 1977, H. Gerischer and C. W. Tobias, Eds.), Wiley-Interscience, New York. A series of review articles on particular topics in both fundamental and applied electrochemistry.

G12 J. O'M. Bockris and B. E. Conway, *Modern Aspects of Electrochemistry*, Plenum Press, New York. A series of review articles on particular topics in physical electrochemistry. Less applied coverage than the Delahay and Tobias series.

G13 *Specialist Periodical Reports: Electrochemistry*, Chemical Society, London. Each volume contains several review articles.

Specific

S1   H. von Helmholtz, *Wied. Ann.* **7**, 337 (1879).

S2   G. Lippmann, *Ann. Chim. Phys. (Paris)* **5**, 494 (1875).

S3   D. C. Grahame, *Chem. Rev.* **41**, 441 (1947).

S4   G. Gouy, *J. Chim. Phys. (Paris)* **29**, 145 (1903); A. Gouy, *J. Phys.* **9**, 457 (1910); *Compt. Rend.* **149**, 654 (1909); *Ann. Phys.* **7**, 129 (1917).

S5   D. L. Chapman, *Phil. Mag.* **25**, 475 (1913).

S6   O. Stern, *Z. Elektrochem.* **30**, 509 (1924).

S7   I. Langmuir, *Phys. Rev.* **8**, 149 (1916).

S8   A. N. Frumkin, *Z. Phys.* **35**, 792 (1926).

S9   A. N. Frumkin, *Electrochim. Acta* **9**, 465 (1964).

S10 R. Parsons, *J. Electroanal. Chem.* **7**, 136 (1964); also following comment by A. N. Frumkin.

S11 IUPAC Information Bulletin Appendixes on Provisional Nomenclature, Symbols, Units, and Standards, Number 42, 1975.

S12 F. P. Bowden and J. N. Agar, *Ann. Rept. Progr. Chem.* **35**, 90 (1938).

S13 J. N. Agar and F. P. Bowden, *Proc. Roy. Soc.* **169A**, 206 (1939).

S14 T. Erdey-Gruz and M. Volmer, *Z. phys. Chem.* **150A**, 203 (1930).

S15 J. A. V. Butler, *Trans. Faraday Soc.* **19**, 729 (1924); **19**, 734 (1924).

S16 J. Tafel, *Z. phys. Chem.* **50**, 641 (1905).

S17 B. E. Conway and M. Dzieciuch, *Can. J. Chem.* **41**, 21 (1963).

S18 A. Blazek and J. Koryta, *Coll. Czech. Chem. Comm.* **18**, 326 (1953).

S19 P. Delahay, C. C. Mattax, and T. Berzins, *J. Amer. Chem. Soc.* **76**, 5319 (1959).

S20 H. B. Herman and A. J. Bard, *Anal. Chem.* **36**, 510 (1964).

S21 O. Fischer, O. Bracka, and E. Fischerova, *Coll. Czech. Chem. Comm.* **26**, 1505 (1961).

S22 J. M. Saveant and E. Vianello, *Electrochim. Acta* **10**, 905 (1965).

S23 J. H. Christie and G. Lauer, *Anal. Chem.* **36**, 2037 (1964).

S24 P. J. Lingane and J. H. Christie, *J. Electroanal. Chem.* **13**, 227 (1967).

S25 P. H. Daum and M. L. McHalsky, *Anal. Chem.* **52**, 340 (1980).

S26 J. K. Senne and L. W. Marple, *Anal. Chem.* **42**, 1147 (1970).

S27 P. J. Kudirka and R. S. Nicholson, *Anal. Chem.* **44**, 1786 (1972).

S28 I. M. Kolthoff and W. E. Harris, *J. Amer. Chem. Soc.* **68**, 1175 (1946).

S29 R. Tamamushi, *Kinetic Parameters of Electrode Reactions of Metallic Compounds*, Butterworths, London, 1975.

5.9   STUDY PROBLEMS

5.1   On solid platinum or carbon electrodes the Ce(IV)/Ce(III) couple has a transfer coefficient of 0.26 ± 0.05 and a (heterogeneous) rate constant at the equilibrium potential of $3.8 \pm 0.1 \times 10^{-6}$ m/s . Compute the exchange current density of this couple in a solution uniform at 0.1 mol/dm^3 in both Ce(IV) and Ce(III).

5.2   On mercury the reduction of Cd(II) to its amalgam depends strongly on the ions in solution; for aqueous sulfate solutions, transfer coefficients are around 0.25, and rate constants at the equilibrium potential are around 4×10^{-4} m/s. Compute the exchange current at a 0.02-cm^2 hanging-mercury-drop electrode in 1.0 mmol/dm^3 Cd(II) solution and the corresponding polarization resistance due to this couple.

5.3   In this chapter the author derives an equation for large anodic overpotentials as a limiting case of the Butler–Volmer equation. Derive the equation for the limiting case of large cathodic overpotentials from the Butler–Volmer equation. Remember that by convention cathodic currents are taken as negative.

5.4   The author states that use of the Tafel equation is quantitatively valid, that is, correct within 1%, whenever the absolute value of the overpotential is greater than twice $2.303RT/zF$. Prove him right or wrong, assuming the Butler–Volmer equation is correct.

5.5   For a hanging-mercury-drop electrode in a certain solution, 95% coverage was obtained for a millimolar bulk-solution concentration. Compute a numeric value, with units, for the adsorption coefficient, assuming the Langmuir adsorption isotherm. Do the same with the linear adsorption isotherm. When is it reasonable to use the linear adsorption isotherm?

CHAPTER
6

ELECTROCHEMICAL STOICHIOMETRY

Analytical applications of electrochemistry date from the beginning of the last century. In 1801, Cruikshank (S1) plated out silver and copper electrolytically in order to identify them, while in 1807, Sir Humphrey Davy employed electrolysis to generate and characterize sodium and, later, other alkali metals. These studies were qualitative. The quantitative relationships of electrochemistry were developed 30 years later by Michael Faraday. Those currents that correspond to passage of charge that effects a net chemical reaction are therefore called *Faradaic* currents. The quantitative relation between charge transferred and quantity of reaction carried out is the basis of *coulometry*, whose name is derived from that of the unit of charge.

The fundamental principle of coulometry was elucidated by Faraday (G1) around 1833–1834: the amount of chemical reaction (moles of chemical reaction) caused by the passage of electrical charge is directly proportional to the amount of charge (coulombs) passed through the cell. These laws were initially formulated as two statements:

1. The weight of a chemical substance liberated at an electrode is directly proportional to the amount of current [total amount of electrical charge] passed through the cell.

2. The weights of different substances produced by a given amount of current are proportional to the equivalent weights of the substances.

When an electrochemical cell is set up such that only one chemical reaction can occur when there is charge passage through that cell, then the stoichiometry of that one reaction can be directly and quantitatively linked to total charge passage. If, however, more than one pathway is available for the chemical use of the charge passed, the link between number of electrons (charge) and moles of a specific chemical reaction is no longer quantitative. Such situations are usually described as those in which the *current efficiency* for the reaction in question is less than 100%, and are not usually of analytical utility. The stoichiometric requirement is for 100% *overall* current efficiency, that is, for a relationship between *total* charge passed and *total* quantity of the final electrolysis product. The existence of multiple pathways of reaction schemes is irrelevant if, but only if, all yield the same final product and require the same number of electrons to obtain it.

In coulometry, the quantity directly linked to total moles of chemical reaction is the total charge passed Q. Although it is possible to measure total charge passed directly by putting a coulometer in series with the electrolytic cell, this is uncommon in modern analytical practice. Current, rather than charge, is the usual measured parameter. But since it is charge that must be determined, the definition of current as charge passed per unit of time, $I = Q/t$, is used to obtain Q. The instantaneous current is actually $I = dQ/dt$, and therefore the charge is the integral of the current over time.

The Faraday constant is itself an analytical standard, since its value is more precisely known than are the atomic weights of many elements (S2). Coulometry thus does not require chemical standards for quantitative determinations, and measurement of the electrical parameter of charge permits, when the reaction is stoichometric, quantitative calculation of the amount of a substance present.

6.1   STOICHIOMETRY OF ELECTRODE REACTIONS

Electrons can enter or leave electrolyte solutions only at electrodes, the points at which charge conduction can change between ionic and electronic. Since electrolyte solutions must possess overall electrical neutrality, introduction of electrons at one electrode in a reduction process must always match loss of electrons at another electrode in an oxidation process. Physically, the electrodes in a cell can be quite separate. For this reason one often concentrates on one electrode, even though two or more are present in the cell. It is useful to use *half-reactions*, which correspond to actual or potential *electrode reactions*. Electrode reactions, or half-reactions, explicitly show electrons as products or as reactants. These electrons are supplied or removed by electronic conduction through the metallic wires leading to the electrode, and can be measured as a flow of electrical current by electrical instruments.

Chemical stoichiometry is normally expressed in *moles*, the SI base unit for quantity of substance, while electrical charge is normally expressed in *coulombs*, which are ampere seconds. These units are simply related by the *Faraday constant F*, which is the product of the electronic charge (1.6021×10^{-19} C/electron) and the Avogadro number (6.0225×10^{23} electrons/mol electrons) to give 96,487 coulombs per mole electrons. The Faraday constant is used whenever conversion between the electrical units of coulombs and the chemical units of moles is necessary.

6.1.1   Writing Electrode Reactions

An electrode reaction or electrochemical half-reaction is a reaction written so that all *atoms* appearing on one side are balanced, in type and number, by atoms appearing on the other. The total *charge* on one side of an electrode reaction is equal to the total charge on the other, but at least one *electron* appearing on one side is not balanced by an electron appearing on the other. In other words, an electrode reaction must explicitly involve a net loss of free electrons or a net gain of them. A reaction that does not explicitly involve one or more net free electrons is not an electrode reaction; it may be an electrochemical cell reaction, or it may not be electrochemical at all.

Electrode reactions are of two types, *oxidation* and *reduction*. There are several alternative and equivalent definitions of these terms. Oxidation is that form of electrode reaction:

1.   in which the oxidation state(s) of the element(s) involved *increases*, that is, becomes more positive.

2.   that occurs at, and thereby defines, the *anode*.

3.   in which electrons are *lost*; that is, in which "free" electrons appear as net *products* in the electrode reaction equation. The electrons are lost *from* the species undergoing oxidation and thereby appear as net products in the reaction. They are, of course, *gained* by the electrode (or by another species, which is thereby reduced). It is traditional to relate to the species rather than to the electrode and thus oxidation is considered a *loss of electrons*, despite the apparent production of "free" electrons in oxidation reactions. The gain of electrons upon reduction is viewed in the same way.

The opposite process of *reduction* is that form of electrode reaction:

1.   in which the oxidation state(s) of the element(s) involved *decreases*, that is, becomes less positive.

2.   that occurs at, and thereby defines, the *cathode*.

3.   in which electrons are *gained*; that is, in which "free" electrons appear as net reactants in the electrode reaction equation.

The terms *oxidizing agent* or *oxidant* and *reducing agent* or *reductant* follow from the definitions of oxidation and reduction given above. An oxidizing agent oxidizes something else, and is itself reduced in so doing, while a reducing agent reduces something else, and is itself oxidized in so doing.

Electrode reactions take place either at the surface of an electrode, in which case the "free" electrons are lost to or gained from the metallic conductors of the external circuit, or in the electrolyte itself, in which case the "free" electrons are gained or lost by species undergoing the opposite form of electrode reaction. Full

```
R                       O              C A N A D A
E                       X              A           N
D                       I              T           O
U                       D              H           D
C A T H O D E           A N O D E      O           E
T                       T              D
I                       I              E
O                       O
N                       N
```

Figure 6.1   Useful mnemonic devices.

reactions that take place either at a combination of two electrodes in a cell, or directly without electrodes in the electrolyte, are therefore called *redox reactions* since both reduction and oxidation are necessarily involved. (See Figure 6.1.)

6.1.2   Balancing Electrode Reactions

Since electrode (or redox) reactions take place in the electrolyte or at an electrode–electrolyte interface, the constituents of the electrolyte can be used freely to balance either type of reaction. In aqueous solutions, the simplest procedure is shown by the following example:

STEP 1. Write the product(s) and reactant(s) known.
$$H_2C_2O_4 \rightarrow CO_2$$
$$MnO_4^- \rightarrow Mn^{2+}$$

STEP 2. Balance all of the elements *except* those of water (H and O) by appropriate multiplication.
$$H_2C_2O_4 \rightarrow 2CO_2$$
$$MnO_4^- \rightarrow Mn^{2+}$$

STEP 3. Balance oxygen by adding H_2O as necessary.
$$H_2C_2O_4 \rightarrow 2CO_2$$
$$MnO_4^- \rightarrow Mn^{2+} + 4H_2O$$

STEP 4. Balance hydrogen by adding H^+ as neccessary.
$$H_2C_2O_4 \rightarrow 2CO_2 + 2H^+$$
$$8H^+ + MnO_4^- \rightarrow Mn^{2+} + 4H_2O$$

STEP 5. Balance charge by adding electrons as necessary.
$$H_2C_2O_4 \rightarrow 2CO_2 + 2H^+ + 2e^-$$
$$8H^+ + MnO_4^- + 5e^- \rightarrow Mn^{2+} + 4H_2O$$

Should the reaction, which is now a balanced electrode reaction, actually take place in basic solution where protons are not available, the steps above should be followed and the balanced acid solution result obtained. Then by (a) adding the same number of OH^- to *both* sides as H^+ ions appear on *one* side, (b) combining the OH^- and H^+ together to form H_2O, and (c) removing the H_2O present on both sides (if any) equally, one will obtain the reaction that *formally* exists in basic solution:

$$H_2C_2O_4 + 2OH^- \rightarrow 2CO_2 + 2H_2O + 2e^-$$

$$MnO_4^- + 4H_2O + 5e^- \rightarrow Mn^{2+} + 4OH^-$$

However, neither $H_2C_2O_4$ nor Mn^{2+} *actually* exists in basic solution, thus the reactions do not correspond to the true processes taking place. No purely formal procedure can give this information; knowledge of the species actually or probably present in basic solution is essential.

In solutions that are not aqueous, analogous procedures can be used to balance electrode reactions.

6.2   STOICHIOMETRY OF REDOX REACTIONS

The stoichiometry of redox reactions is simply the combined stoichiometry of the two electrode or half-reactions that make up a full redox reaction. To obtain a balanced redox reaction, the two component half-reactions are first balanced separately and then are combined. The combination is effected by:

1.   multiplying each equation by the number of electrons appearing in the other,

2.   adding the resulting equations, and

3.   simplifying the result to the extent possible.

The equations of the previous example would be combined as follows:

$$H_2C_2O_4 \rightarrow 2CO_2 + 2H^+ + 2e^-$$

$$8H^+ + MnO_4^- + 5e^- \rightarrow Mn^{2+} + 4H_2O$$

1. $5H_2C_2O_4 \rightarrow 10CO_2 + 10H^+ + 10e^-$
   $16H^+ + 2MnO_4^- + 10e^- \rightarrow 2Mn^{2+} + 8H_2O$

2. $16H^+ + 2MnO_4^- + 5H_2C_2O_4 \rightarrow 2Mn^{2+} + 8H_2O + 10CO_2 + 10H^+$

3. $6H^+ + 2MnO_4^- + 5H_2C_2O_4 \rightarrow 2Mn^{2+} + 8H_2O + 10CO_2$

A redox reaction, unlike an electrode reaction, must not include any "free" electrons. It must therefore be a combination of one oxidation half-reaction and one reduction half-reaction. A redox reaction may occur in a cell, which contains two electrodes, as well as in a simple beaker in which the reactants are mixed. Redox reactions are therefore also called *cell reactions*.

6.3 REFERENCES

General

G1 M. Faraday, *Experimental Researches in Electricity* (A collection of his papers from Phil. Trans. Roy. Soc.); selections from his work can be found in *A Source Book in Physics*, Magie, Ed., pp. 473ff. and *A Source Book in Chemistry*, Leicester and Klickstein, Eds., pp. 280ff.

G2 J. J. Lingane, *Electroanalytical Chemistry,* 2nd Ed., Interscience, New York, 1958, pp. 296–616.

G3 G. W. C. Milner and G. Phillips, *Coulometry in Analytical Chemistry*, Pergamon Press, New York, 1967.

G4 G. A. Rechnitz, *Controlled Potential Analysis*, MacMillan, New York, 1963.

G4 K. Abresch and I. Claassen, *Coulometric Analysis*, L. L. Leveson, trans., Chapman and Hall, London, 1965.

Specific

S1 W. Cruikshank, *Ann. Physik* **7,** 105 (1801)

S2 *Pure and Appl. Chem.* **45,** 125 (1976).

S3 E. P. Przybylowicz and L. B. Rogers, *Anal. Chim. Acta* **18,** 596 (1958).

S4 D. S. Tutundzic and S. Mladenovic, *Anal. Chim. Acta* **12,** 282 (1955).

6.4 STUDY PROBLEMS

6.1 Complete and balance the following reactions in aqueous acid solution:
   (a) $Cd^{2+} + Pb(s) \rightarrow$?
   (b) $HNO_2 + H_2SO_3 \rightarrow NH_4^+ +$?
   (c) $MnO_4^- + IO_3^- \rightarrow$?
   (d) $Cr_2O_7^{2-} + Fe^{2+} \rightarrow$?
   (e) $Cr^{2+} + HClO \rightarrow$?
   (f) $VO_2^+ + Fe^{2+} \rightarrow$?

6.2 In electrogravimetric analysis of a brass sample, a sample weighing 3.7411 g was dissolved and the copper content plated onto a platinum electrode whose weight before deposition was 32.1962 g and after deposition was 35.0841 g. Compute the percentage of copper in the brass.

6.3 A constant current source used in a coulometric titration of cysteine with Hg(II) generated from a mercury-pool electrode provided a current of 10.00 mA. If a solution of 10.00 ml volume required 215 s for completion of the titration, what is the concentration of cysteine in the solution? Cysteine and Hg(II) react in a 2:1 ratio (Przybylowicz and Rogers, S3).

6.4 The classical permanganate titration of Fe(II) can be carried out coulometrically using permanganate generated coulometrically from $MnSO_4$ in 2 mol/dm³ aqueous sulfuric acid, which also contains the sample. Given a 10.00-mA constant current source, how long would it take to titrate the Fe(II) in a 10-cm³ aliquot sample of 0.062 mol/dm³ Fe(II)? (Tutundzic and Mladenovic, S4)

6.5 A controlled potential electrolysis of a certain quinone in aqueous solution gave 642 C for a 25-cm³ aliquot of a solution of the quinone. If the reduction followed the usual two-electron, two-proton quinone reduction reaction, what was the concentration of the quinone in the solution?

6.6 A solution of unknown hydrogen-ion content was titrated using coulometric generation of hydroxide ion at a constant current of 300 mA; a total of 467 s was required to reach the end point. How many moles of hydrogen ion were present in the unknown solution?

6.7 A cadmium amalgam was prepared by passing a current of 500 mA through an aqueous solution of cadmium salt using a 10.0-g mercury pool as cathode. Compute the percentage by weight of cadmium in the amalgam produced as a function of time. How long would it take to produce an amalgam of exactly 10.0 wt % cadmium?

6.8 An electrode is to be plated with 0.75 mm of copper from a solution of a copper(II) cyanide complex. How long should a current of 5.0 A be left on if the electrode area is 25.0 cm²?

PART
2
EQUILIBRIUM ELECTROANALYTICAL TECHNIQUES

CHAPTER
7
EQUILIBRIUM POTENTIALS

An equilibrium potential is a potential difference observed in a system that is in equilibrium with respect to electrons, that is, one in which there is no or essentially no movement of electrons in the conductors whose potential is being measured. This is precisely the same thing as saying that no current is drawn from, or introduced into, the electrochemical system by the measurement technique. It does not mean that no current flows in the system, but simply that there is no net flow of electrical current into or out of it. The techniques covered in this part thus differ from those covered in Part 3, in which a net current must flow into or out of the electrochemical system.

Since all of the potentials taken up in Part 2 are measured at zero current, or at currents so low that they do not change the potential from its zero current value, the notation $E(0)$ is superfluous. As used in this part, the superscript zero carries its usual meaning of the thermodynamic standard state.

7.1 ORIGIN OF ELECTROCHEMICAL POTENTIALS

An electrical potential is the tendency of electrons to leave a homogeneous phase, normally a metallic conductor, and enter some other homogeneous phase. Such an "absolute" electrical potential is without significance in analytical electrochemistry, however great its significance may be elsewhere. By *electrical potential* we mean a measurable difference in the tendency of electrons to leave different electronic conductors (wires) made of the same metal—almost invariably copper. An electrical potential, then, is that potential difference measured by a suitable measuring device (meter) connected to two copper wires.

When two *metallic wires* are connected through a meter, electrons will spontaneously flow through the meter from the most negative metal, which has the highest "electron pressure," to the least negative metal, which has the lowest "electron pressure." Such a flow of electrons constitutes an electric current. The flow would be brief in such a circuit, however, for the excess electrons moved into the least negative metal would give it a negative charge, repelling further electrons and terminating the current flow. For this reason a complete electrical circuit must provide a return path for electrons so that a continuous current may flow, as shown in Figures 3.1 and 7.1.

It is instructive to think of the flow of electricity as analogous to the flow of water: the wire is analogous to a pipe, the electrons are analogous to the molecules of water, the sources/sinks of electrons are analogous to lakes or basins containing water, and the electrical potential is analogous to the height or pressure head of the water, a spontaneous direction of flow being from more negative potential to less negative, just as the spontaneous direction of flow of water is from a higher level to a lower level. When water flows spontaneously, it can be made to do useful work; when electrons flow spontaneously, they can be made to do useful work. By application of energy from an external source, water can be made to flow in a nonspontaneous direction (uphill); so also can electrons be made to flow in a nonspontaneous direction by application of sufficient external energy.

The analogy can be carried somewhat further. For example, the unit of current is really number of electrons per unit time (1 A = 1 C/s; coulombs are related by the value of the electronic charge to the number of electrons) just as the unit of water flow is really number of molecules per unit time (1 mol/s can easily be related to cubic meters per second or kilograms per second using the molar volume or molecular weight). The flow of water through any pipe or water system is determined by the applied pressure difference driving it and the resistance of the system. So also the flow of electrons through any electrical or electrochemical system is determined by the potential difference driving it and also the resistance of the system.

A *chemical potential*, or more properly a difference in chemical potential, is a measurable difference in the tendency of some component to leave a homogeneous chemical system. This difference is a difference in free energy of one or more components of the systems. When two chemical systems are connected such that at least some types of matter and energy can flow from one to the other, spontaneous flows of matter and energy from one system to the other take place until the free energies of the two systems are the same, the condition of being in chemical equilibrium with each other.

It is possible to connect together two chemical systems by means of a metallic wire; in such a case the wire, which is an electronic conductor, can carry a flow of electronic charge and thus provide a pathway through which the two systems may achieve equilibrium. The resulting flow of electrons increases the negative charge of the system that receives them, however, and this increasing negative charge increasingly repels additional electrons back along the wire toward the originating system. In practice, very few electrons need transfer before the electrostatic charge difference prevents further current flow, so that such a connective electron pathway alone is without practical significance.

The potential differences that can arise in a system of two different metallic conductors are usually referred to as *contact potentials*. When a system composed of two different metallic conductors is connected to an external device, the potential difference is measured if and only if the input leads of the external device are of dissimilar metals. If they are of the same metal, normally copper

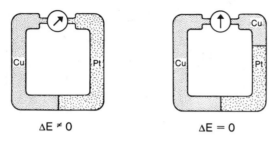

$\Delta E \neq 0$ $\Delta E = 0$

Figure 7.1   Potential differences in system of two metallic phases.

wires, no potential difference is measurable, because equal and opposite contact potentials exist at the two metal–metal contacts as shown in Figure 7.1. The total electric potential difference sensed by an external device is therefore the *net* electric potential difference that exists across all of the phases and interphase regions in the circuit, regardless of the nature of the phase or the interphase region.

If in addition to the electronic pathway a second and ionic pathway is provided for charge transfer, the electrostatic charge difference can be removed by movement of ions, and thus a continuous current can flow under a continuous difference in electrical potential. The continuous electric potential provided by a difference in chemical potential between two systems is called an *electrochemical potential* or an *electrochemical potential difference*, and the configuration of two linked chemical systems in which it arises is called an *electrochemical cell*. If the two systems are in chemical equilibrium with each other, the difference in chemical potential between them must be zero, and the consequent electrical potential is also zero. Such cells are normally of little interest. It is more frequently the case that the two systems are each in internal chemical equilibrium, though not in equilibrium with each other. There is then a difference in chemical potential between them that can result in a difference in electrical potential.

The simplest possible electrochemical cell consists of a single *ionic conductor* that interfaces with two *electronic conductors* to give two *electrodes*, as shown in Figure 7.2. The electrodes A and C may be made of any metal as long as it is the same for both, since contact potentials will then either not arise at the input leads or be equal and opposite if they do. The net electric potential E then arises in the electrodes A and C, in the electrolyte B, or at the interphase boundaries AB and BC between them. A more general electrochemical cell is shown in Figure 7.3. Again, the net measured electric potential is the sum of the potentials in the electrodes, electrolytes, separating ionic conductor, and the interphase regions. Physically, electric potentials due to the establishment of arrangements of charge, electronic or ionic, can occur at any or all of them.

The chemical potential of a species i in a homogeneous phase, μ_i, is simply the work required to bring a particle i from a location at which its chemical potential

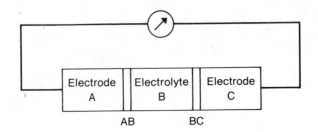

Figure 7.2 Simplest possible electrochemical cell. The electrodes (A, C) and electrolyte (B) are usually assumed to be internally uniform. The interphasial regions AB and BC may or may not be identical even if the electrodes A and C are identical.

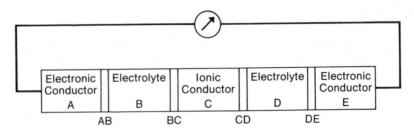

Figure 7.3 Electrochemical cell showing phases and interphases. Ionic conductor C may be identical to electrolytes B and D, but often consists of a physically distinct salt bridge.

is zero, that is to say at infinity, to any point in the homogeneous interior of the phase. For a charged species, electrical work is possible as the species being brought in is moved through the charged regions which include ions, dipoles, and double layer between infinity and the interior of the phase, while for an uncharged species no such work is possible. The chemical potential μ_i of an uncharged species i will therefore be identical to its electrochemical potential μ'_i while for a charged species they will differ. The difference is usually written as

$$\mu_i = \mu'_i - z_i F \Phi \tag{7.1}$$

in molar units, where the quantity Φ is called the *inner electric potential* of the phase.

Conceptually, this equation divides each phase into two divisions: a homogeneous interior volume deprived of its charge and double layer, and a thin outer shell into which the charge and double layer are compressed. The inner electric potential is then a difference in potential seen by particle i as it enters or leaves the bulk homogeneous phase. The division is conceptual rather than real, and so although the electrochemical potential μ'_i is in principle a measurable quantity, the value of neither Φ nor the chemical potential μ_i for a charged species in the bulk uncharged phase is independently measurable. In principle, the inner electric potential can be calculated from a theoretical model of the structure of a phase, but it cannot be measured.

The *inner electric potential* Φ can be considered as the sum of two terms called the *outer electric potential* Ψ and the *surface electric potential* χ:

$$\Phi = \Psi + \chi \tag{7.2}$$

The outer electric potential is, in principle, both measurable and calculable. It represents the work required to bring a particle i from infinity to a point *outside the phase but situated in the immediate vicinity of its surface*. The surface potential then represents the work required to move the particle i across the surface shell into the homogeneous interior of the phase, and is neither measurable nor calculable. The surface potential is often regarded as being due to a distribution of dipoles at the surface while the outer electric potential is regarded as due to distribution of ions, though there is no need to make such a distinction, and it is not universally applicable. The outer electron potential is sometimes used to give a *real potential* α_i:

$$\alpha_i = \mu'_i - zF\Psi \tag{7.3}$$

The real potential can be regarded as the value of the electrochemical potential when the phase is uncharged, since the surface electric potential is then zero.

The potentials of particles in phases are more of theoretical interest than of real interest, but differences between them can produce real and measurable phenomena. The units of μ_i, μ'_i and α_i are those of energy, joules per mole, while the units of Ψ, Φ, and χ are those of potential difference, volts.

The difference between the outer electric potentials of two phases is called the *Volta potential difference*:

$$\Delta\Psi = \Psi(1) - \Psi(2) \tag{7.4}$$

When a species i is in equilibrium between two phases, the difference in its electrochemical potential is zero and then

$$\Delta\Psi_i = (\alpha_i(1) - \alpha_i(2)/z_iF \tag{7.5}$$

Likewise, the difference between the inner electric potentials of two phases is called the *Galvani potential difference*:

$$\Delta\Phi = \Phi(1) - \Phi(2) \tag{7.6}$$

The Galvani potential difference, in terms of electrochemical potential, is given for a species i by

$$\Delta\mu'_i = \Delta\mu_i + z_iF\Delta\Phi \tag{7.7}$$

The values of Volta potential differences are in principle always measurable quantities. The values of Galvani potential differences are not, unless the bulk phases between which they are measured are the same, in which case the Galvani potential difference is simply the measurable electric potential difference. Thus the net electric potential, which is the sum of all of the Galvani potentials, is measurable by an external device with identical metallic leads, but the component Galvani potentials are not experimentally accessible. For the overall cell,

$$E = \Delta\Phi = \Sigma\Sigma_i(\Delta\mu'_i - \Delta\mu_i)/z_iF \qquad (7.8)$$

where the first summation is the summation over all phases and interphase regions and the second is the summation over all the species i. Only the difference between μ'_i and μ_i within a single phase or at its boundary can give rise to an electrical potential.

In the absence of a net current flow, as in all of the methods considered in Part 2, the electronic conductors can be considered uniform except in the interphase regions. If these regions are not identical, a net electrical potential can arise between them that will be sensed by the external circuit. If they are identical, a net electrical potential can be sensed only if it appears also as a Galvani potential difference at BC, in C, or at CD, which will occur only if the chemical and electrochemical potential differences at one or all of these points are unequal. In other words, a net difference in Galvani potentials must appear somewhere in the cell in order for it to be sensed by an external circuit, which means that at some point the electrochemical and chemical potentials must differ.

In many cells it is desirable to concentrate solely on the interphase regions associated with the electrodes, AB and DE. This can be achieved by minimizing the *net* difference in Galvani potentials at BC, in C, and at CD in Figure 7.3. When the interphase region BC is very similar to that at CD, but constructed in an opposite sense, and when all ionic charges are transported identically through C so that charge is not relevant there and μ_i and μ'_i are the same, the net difference in Galvani potentials is reduced to insignificance. This can be accomplished by means of a *salt bridge*.

In other cells it is desirable to concentrate solely on the ionic conductor C and its associated interphase regions BC and CD. This can be achieved by minimizing the *net* difference in Galvani potentials at AB and at DE, which is to say the net difference in electrode potentials. This is achieved by making the electrodes identical, and preferably unpolarizable, since they are necessarily constructed in opposite senses. Identical, opposed reference electrodes are usually employed for this purpose. The electrodes then act as sensors for the difference in Galvani potential that appears across the ionic conductor C and its associated interphase regions BC and CD.

7.2   BASIC THERMODYNAMICS

The basic equations of thermodynamics constitute much of an elementary course in physical chemistry and are discussed at length in many excellent texts (G1–G6). Only a brief summary is given here.

The *law of conservation of energy*, also known as the *first law of thermodynamics*, can be written as

$$dU = dq + dw \tag{7.9}$$

or

$$dq + dw - dU = 0 \tag{7.10}$$

where U is *internal energy*, q is *heat*, and w is *work*. Usually work is taken to mean only pressure–volume or expansion work $p\,dV$; all other types of work are neglected as will be done here. Then, taking work absorbed as negative;

$$dU = dq - p\,dV \tag{7.11}$$

expresses the fact that energy is conserved. Any difference between heat and work must appear as a change in internal energy.

The *law of entropy*, also known as the *second law of thermodynamics*, can be written as

$$dq = T\,dS \tag{7.12}$$

where T is the absolute temperature and S is a *state function* called *entropy*. Other state functions are also useful to us. The *enthalpy H* is defined as

$$H = U + pV \tag{7.13}$$

The *free energy* or *Gibbs free energy G* is defined as

$$G = H - TS \tag{7.14}$$

The two laws of thermodynamics, when combined, constitute the *fundamental equation of thermodynamics*:

$$dU = T\,dS - p\,dV \tag{7.15}$$

This equation, together with the definitions of enthalpy and free energy, gives for the derivative of the free energy

$$dG = -S\,dT + V\,dp \tag{7.16}$$

The derivative of the free energy contains two terms. The first of these can be considered to be the variation of free energy with temperature and the second the variation of free energy with pressure. At constant pressure the derivative of the free energy is then given simply by the *Gibbs–Helmholtz equation*:

$$dG/dT = -S \qquad (7.17)$$

The general equations apply to changes in free energies, as in chemical reactions, just as they do to the state functions themselves. At constant temperature

$$\Delta G = \Delta H - T\,\Delta S \qquad (7.18)$$

and

$$d\Delta G/dT = -\Delta S \qquad (7.19)$$

hence

$$\Delta H = \Delta G + d\Delta G/dT \qquad (7.20)$$

This relationship permits us to obtain values of free energies from measured cell potentials and to calculate potentials of cells from tabulated free energies. The temperature coefficient of the cell potential gives the entropy change of the cell reaction:

$$\Delta S = zF(dE/dT) \qquad (7.21)$$

and thus its enthalpy change as well:

$$\Delta H = -zFE + zFT(dE/dT) \qquad (7.22)$$

7.3   ELECTROCHEMICAL THERMODYNAMICS

The most basic relationship of electrochemical thermodynamics is

$$-dG = zFE \qquad (7.23)$$

where dG is the free energy change of an electrochemical reaction, z is the number of electrons involved, F is the Faraday constant, and E is the potential difference between the electrodes when the reaction takes place reversibly in an electrochemical cell. This equation is simply an expression of the usual unit of free energy, the joule, as its equivalent the volt coulomb since the Faraday constant and z relate moles of reaction to coulombs passed.

For any general chemical reaction

$$qA + rB + ... \rightarrow sC + tD + ...$$

the free energy of the reaction is given by the difference between the chemical potentials μ_i of the reactants and those of the products. This is the signed or algebraic sum of the chemical potentials, those of the products being taken as positive and those of the reactants as negative. Since the chemical potential is, as stated earlier, given by

$$\mu_i = \mu°_i + RT \ln a_i \tag{7.24}$$

the free energy change in the reaction is

$$\Delta G = \Sigma_i\mu°_i + RT \ln Q \tag{7.25}$$

The quotient Q is simply the quotient of the *actual* activities of the products to those of the reactants:

$$Q = a(C)^s a(D)^t ... /a(A)^q a(B)^r ... \tag{7.26}$$

The general relationship holds also at chemical equilibrium, where Q is simply the ratio of the *equilibrium* activities, which constitutes the *equilibrium constant*:

$$K = a(C)^s a(D)^t ... /a(A)^q a(B)^r ... \tag{7.27}$$

At equilibrium

$$\Delta G = \Sigma_i\mu°_i + RT \ln K \tag{7.28}$$

but since at equilibrium the free energy change is zero,

$$-\Sigma_i\mu°_i = RT \ln K \tag{7.29}$$

The summation of the free energies of reactants and products under standard conditions, with due regard to the signs associated with reactants and products, is simply the free energy change under standard conditions, $\Delta G°$:

$$-\Delta G° = RT \ln K \tag{7.30}$$

Substitution of $\Delta G°$ back into the quotient expression gives:

$$\Delta G = -\Delta G° + RT \ln Q \tag{7.31}$$

Expressing the values of ΔG in terms of $-zFE$,

$$E = E° - (RT/zF) \ln Q \tag{7.32}$$

where E is the actual measured potential difference across a cell in which the quotient of the actual activities of products and reactants is Q. The quantity E^0 is the potential difference measured across the cell under standard thermodynamic conditions, that are normally *not* those which prevail at equilibrium. This equation is known as the *Nernst equation* (S1). The Nernst equation will correctly give the equilibrium cell potential if the appropriate value of the standard cell potential E^0 and the actual activities of the species involved in the cell reaction are used.

7.3.1   Nernst Equation: Cell Potentials

A *cell potential* is the potential difference between two electrodes in a full cell. If the cell is operating under standard conditions, a *standard cell potential* E^0 will be observed as the cell potential. For a cell, the Nernst equation in the above form gives the observed potential difference. Since equations of reactions are normally written from left to right, the sign of a spontaneous reaction, which is positive potential difference and negative free energy change, should correspond to the physical sign of the charge on the right-hand electrode of the cell as written.

Example
The Daniell cell:

$$Zn/Zn^{2+}, (0.1\ mol/dm^3)\ //Cu^{2+}, 0.01\ mol/dm^3)/Cu$$

has a potential E calculable from the Nernst equation (when activities are taken equal to concentrations) of

$$E = [+0.34 - (-0.76)] - 0.03\ \log(0.1/0.01)$$

$$E = +1.04\ V$$

so that the cell is written above correctly for the reaction

$$Zn + Cu^{2+} \rightarrow Cu + Zn^{2+}$$

with products on the right and the most positive electrode, copper, on the right. The configuration

$$Cu/Cu^{2+}(0.01\ mol/dm^3)//Zn^{2+}(0.1\ mol/dm^3)/Zn$$

would be incorrect, because the products would be on the left as would the most positive electrode. When reduction standard electrode potentials are used to obtain standard cell potentials, they are always taken as right-hand electrode minus left-hand electrode as in the example above. The potential calculated for the incorrect configuration would be −1.04 V, *correct in magnitude but incorrect in sign.*

The writer prefers to use electrode potentials rather than cell potentials. Any physical voltage measuring device is connected between two physically separate electrodes. Calculation of the separate electrode potentials immediately shows which electrode is physically the more positive; the separate Nernst equations are then of the same form, in which the activities of the reduced species are in the numerator and those of the oxidized species are in the denominator. Moreover, when the potential of an electrode is externally (potentiostatically) controlled the activities of all species *at the electrode surface* are forced to the ratios given by the Nernst equation using the actual potential at which the electrode is controlled.

7.3.2   Nernst Equation: Electrode Potentials

In electroanalytical work, it is more common to separate the Nernst equation for a full cell into two equations, each of which corresponds to a half-reaction or electrode reaction. The separate equations have the form

$$E = E^0 - RT/zF \ln [a_{Red}/a_{Ox}]$$

where E is the observed potential of the electrode, E^0 is the potential the electrode would have under standard conditions, R is the universal gas constant, T is the thermodynamic temperature, z is the actual number of electrons involved in the electrode reaction responsible for the equilibrium potential, F is the Faraday constant, and a_{Red} and a_{Ox} are the activities of the oxidized and reduced forms participating in the electrode reaction responsible for the equilibrium potential. At 25°C (298.15 K) this equation becomes

$$E = E^0 - 0.05915/z \log [a_{Red}/a_{Ox}] \tag{7.34}$$

In dilute solutions this equation is often used with concentration replacing activity since in dilute solution activity can be taken as approximately equal to concentration. The equation will now give correctly the equilibrium *electrode potential* if the appropriate value of the *standard electrode potential E^0* and the actual activities of the species involved in the electrode reaction are used.

When Nernst equations are used for full cells, the question of reference electrode is irrelevant because the cell potential is the difference in potential between the two electrodes of the cell and no reference potential is involved. When Nernst equations involve half-cells, the reference electrode potential does not cancel. Electrode potentials are always related to some reference electrode, usually the aqueous SHE, whether standard electrode potentials or not. This dependence comes in with the value of the standard electrode potential and rarely causes difficulty if standard electrode potentials are all taken with respect to the same reference.

Calculation of the separate electrode potentials of a cell will give values whose difference (taken with regard to sign) is the measurable cell potential. The most positive electrode potential calculated will be the potential of the physically most positive electrode of the cell when reduction electrode potentials are used. (The reverse will be true when the obsolete oxidation electrode potentials are used, and the form of the simplified Nernst equations used in this text will be incorrect. The student is strongly advised to reverse the signs of oxidation electrode potentials, thereby making them reduction electrode potentials, wherever they are encountered.)

The Nernst equation for a half-cell (electrode) reaction uses the products of the *electrode* reaction in the numerator of Q regardless of what reaction takes place in the overall cell. For this reason electrode reactions have a direction that is related to the type of reaction; reduction, for reduction electrode potentials. The reduced forms are the products, and the oxidized forms are the reactants.

It is important to realize that potential differences are in *volts* while free energies are not; free energies have the normal energy units of *joules*, which are *volt coulombs* as well as *newton meters*. Thus when potentials of half-reactions are used to obtain the potentials of other half-reactions one must add the volt coulombs rather than volts, as shown in the following example:

$$Fe^{2+} + 2e^- \rightarrow Fe(0), \; E^0 = -0.44 \text{ V, but } zFE^0 = -0.88F \text{ V C}$$

$$Fe^{3+} + e^- \rightarrow Fe^{2+}, \; E^0 = +0.77 \text{ V, but } zFE^0 = +0.77F \text{ V C}$$

$$Fe^{3+} + 3e^- \rightarrow Fe(0), \; E^0 = -0.04 \text{ V, but } zFE^0 = -0.11F \text{ V C}$$

so that when equations are added algebraically the volt coulombs are additive, but the potentials of the couples are not. When half-reactions are added to give a full cell reaction, however, one oxidation and one reduction (which must each involve the same number of electrons in a balanced full cell reaction) are added to obtain the overall cell potential E. The addition is again algebraic but is now of *volts* and not *volt coulombs*; using the above example:

$$2Fe^{3+} + Fe(0) \rightarrow 3Fe^{2+}, \; E^0 = +0.77 \text{ V} - (-0.44 \text{ V}) = +1.21 \text{ V}.$$

The most positive electrode of a cell in which this reaction would occur is the electrode involving the Fe^{3+}/Fe^{2+} couple, which would physically be written as

$$//Fe^{3+}(aq), Fe^{2+}(aq)/Pt$$

whose standard potential is $+0.77$ V. The most negative electrode of the cell would be the electrode involving the $Fe^{2+}/Fe(0)$ couple, which would physically be written as

$$//Fe^{2+}(aq)/Fe$$

whose standard potential is -0.44 V. The cell could then be written as a linear combination of these two half-cells, either as

$Pt/Fe^{2+}(aq),Fe^{3+}(aq)//Fe^{2+}(aq)/Fe$

with the positive terminal on the *left*, or as

$Fe/Fe^{2+}(aq)//Fe^{3+}(aq),Fe^{2+}(aq)/Pt$

with the positive terminal on the *right*. By convention, a cell is normally written with the overall cell potential positive when the reaction proceeds from the left to right spontaneously. Since $-\Delta G = zFE$, the free-energy change is then negative as should be the case for a spontaneous reaction. Thus the sign of the right-hand electrode is the same as the sign of the overall cell potential when the reaction proceeds spontaneously from left to right as written, which is the case here. If the spontaneous reaction were the reverse, so also would be the sign of the overall cell potential, as must be the case from free energy considerations.

For most purposes, the potential of a full cell or a full reaction is of little interest to an electroanalytical chemist, since it is always clear from the two half-cells considered separately. Since each half-cell contains one electrode, potentials of half-cells are often referred to as *electrode potentials*. The signs of electrode potentials and their physical meaning have been the subject of controversy for many years (Licht and DeBethune, S2). The matter was resolved at the 17th Conference of the International Union of Pure and Applied Chemistry (IUPAC) at Stockholm in 1953. This agreement consists in using the word *potential* exclusively for that quantity associated with an electrode whose sign corresponds to the actual electrostatic charge on the metal. When reactions at electrodes are written as reductions, this is actually the case. If the reactions are written as oxidations (with concomitant reversal of signs), the sign is reversed from that of the electrostatic charge on the metal, so the term oxidation potentials is unacceptable. *Reduction potentials* are acceptable, and a partial table of them is given in Table 7.1. A more extensive table is given in Appendix 2.

7.4   STANDARD STATE DEFINITIONS

The problem of the definition of thermodynamic standard states (states in which the activity of a substance is unity) is one of the problems often overlooked when electrochemical thermodynamic data are being used. It becomes particularly difficult when dealing with nonaqueous media.

A thermodynamic quantity, such as free-energy change or potential, is a quantity referred to standard conditions when a superscript zero is added; for example, ΔG^0 or E^0. A subscript or following parenthesis is used to denote the temperature to which this quantity refers, for example, ΔG^0_{298} or E^0_{25}, if desired; an omitted subscript with a zero superscript refers to the usual temperature, 25°C or 298.15 K. The temperature is usually in kelvins; the only common exception is the use of 25 degrees Celsius rather than 298.15 Kelvin.

TABLE 7.1   Standard (Reduction) Electrode Potentials.[a] Aqueous Acid Solution, 25°C

Electrode Couple	E^0 (V)	dE/dT (mV/K)
$Na^+ + e^- \rightarrow Na$	−2.714	−0.772
$Mg^{2+} + 2e^- \rightarrow Mg$	−2.363	+0.103
$Al^{3+} + 3e^- \rightarrow Al$	−1.662	+0.504
$Zn^{2+} + 2e^- \rightarrow Zn$	−0.7628	+0.091
$Fe^{2+} + 2e^- \rightarrow Fe$	−0.4402	+0.052
$Cd^{2+} + 2e^- \rightarrow Cd$	−0.4029	−0.093
$Sn^{2+} + 2e^- \rightarrow Sn$	−0.136	−0.282
$Pb^{2+} + 2e^- \rightarrow Pb$	−0.126	−0.451
$2H^+ + 2e^- \rightarrow H_2(SHE)$	±0.0000	±0.000
$Cu^{2+} + e^- \rightarrow Cu^+$	+0.153	+0.073
$AgCl + e^- \rightarrow Ag + Cl^-$	+0.2223	−0.653
$Hg_2Cl_2 + 2e^- \rightarrow 2Cl^- + 2Hg(SCE)$	+0.2676	−0.317
$Cu^{2+} + 2e^- \rightarrow Cu$	+0.337	+0.008
$Cu^+ + e^- \rightarrow Cu$	+0.521	−0.058
$I_2 + 2e^- \rightarrow 2I^-$	+0.5355	−0.148
$Hg_2SO_4 + 2e^- \rightarrow 2Hg + SO_4^{2-}$	+0.6151	−0.826
$Fe^{3+} + e^- \rightarrow Fe^{2+}$	+0.771	+1.188
$Ag^+ + e^- \rightarrow Ag$	+0.7991	−1.000
$Br_2(aq) + 2e^- \rightarrow 2Br^-$	+1.087	−0.478
$O_2 + 4H^+ + 4e^- \rightarrow 2H_2O$	+1.229	−0.846
$MnO_2 + 4H^+ + 2e^- \rightarrow Mn^{2+} + 2H_2O$	+1.23	−0.661
$Cl_2 + 2e^- \rightarrow 2Cl^-$	+1.3595	−1.260

[a]Taken from DeBethune, Licht, and Swendeman (S7); for details and additional potentials see Appendix 2.5.

The standard state of an element is defined as the pure element, in its most stable form, under one atmosphere pressure at the specified temperature. Similarly, the standard state of a compound is defined as the pure compound under one atmosphere pressure at the specified temperature. These definitions are the usual thermodynamic definitions.

7.4.1   Solvents

An *ideal solution* is a solution in which the properties of each component are directly proportional to its mole fraction. In other words, an ideal solution is a solution that obeys *Raoult's law*:

$$a_i = a^0_i X_i \tag{7.35}$$

where a^0_i is the activity of the pure substance i and X_i is its mole fraction in the solution.

Raoult's law is obeyed, at least reasonably well, by solvents because the activity of any component of a solution must approach the activity of the pure component as the mole fraction of the component approaches unity. Solvent molecules constitute the vast majority of the molecules making up the dilute solutions with which electroanalytical chemists are usually concerned. For this reason the standard state of a solvent is generally taken to be the same as that of the pure element(s) or compound(s) of which it is composed.

7.4.2   Solutes

Minor components of solutions, including *solutes*, do not usually obey Raoult's law. Their behavior is better described by *Henry's law*:

$$a_i = k_{X,i}X_i \tag{7.36}$$

where $k_{X,i}$ is the Henry's law constant. The value of this constant depends both upon the solute and the solvent.

Henry's law is obeyed by all solutes when the solution is dilute enough. The limit of such solutions is called the *infinitely dilute solution*. Real solutions are found to obey Henry's law only in the limit of infinite dilution. As concentration increases, deviations from Henry's law become experimentally significant, and the value of $k_{X,i}$ does not remain constant.

Concentration scales other than mole fraction are not usually used with Raoult's law but are frequently used with Henry's law. For concentration generally the law would be written as

$$a_i = k_{c,i}c_i \tag{7.37}$$

and for molality it would be written as

$$a_i = k_{m,i}m_i \tag{7.38}$$

The values of $k_{X,i}$, $k_{c,i}$, and $k_{m,i}$ are not the same for the same solute and solvent.

It is convenient to consider the constant in Henry's law as the product of a strictly empirical constant k'_i and an activity coefficient γ_i. The activity coefficient then accounts for the changes in the Henry's law constant with increasing concentration while the empirical constant remains unique to the solute and solvent:

$$a_i = k'_{c,i}\gamma_i c_i \tag{7.39}$$

The activity coefficient is unity in the infinitely dilute solution where $k'_{c,i} = k_{c,i}$ and deviates from unity with increasing concentration. Similar expressions are written for the other concentration scales.

The activity of any component of a solution is directly related to its partial free energy or *chemical potential* μ,

$$\mu_i = RT \ln a_i \tag{7.40}$$

under all conditions including under standard conditions whatever they may be:

$$\mu^0{}_i = RT \ln a^0{}_i \tag{7.41}$$

The difference between chemical potential and chemical potential under standard conditions is

$$\mu_i - \mu^0{}_i = RT \ln k'_{c,i}\gamma_i c_i - RT \ln k'_{c,i}\gamma_i c^0{}_i \tag{7.42}$$

Provided the solution follows the behavior of the infinitely dilute solution, $k'_{c,i}$ is a constant and so is γ_i. The equation then simplifies to

$$\mu_i - \mu^0{}_i = RT \ln c_i - RT \ln c^0{}_i \tag{7.43}$$

Selection of $c^0{}_i$ as unity for the standard state eliminates the second term:

$$\mu_i - \mu^0{}_i = RT \ln c_i \tag{7.44}$$

When the potential of an electrode is determined by a single species, then

$$\mu_i - \mu^0{}_i = \Delta G - \Delta G^0 \tag{7.45}$$

$$\mu_i - \mu^0{}_i = (E - E^0)/-zF \tag{7.46}$$

$$E - E^0 = -(RT/zF) \ln c_i \tag{7.48}$$

$$E = E^0 - (RT/zF) \ln c_i \tag{7.48}$$

If the other species taking part in the electrode reaction are included, this is simply the Nernst equation. What is important for the present purpose is that a linear relationship between the measured values of E and $\ln c$ is generally observed, within experimental error, whenever c is less than approximately 0.1 m, which indicates that γ remains unity over this concentration range. The calculation of E^0 from this relationship requires that $E = E^0$ when c is unity *on the concentration scale chosen.* Thus the standard state of solute species is one that is physically unreal, for it is a solution of unit concentration (e. g., mole fraction) that has the thermodynamic properties of an infinitely dilute solution. The value of E^0 is usually calculated from several sets of $E - \ln c$ data. These E^0

values are experimentally significant only in solutions of concentrations below about 0.1 m, since the activity coefficients cannot remain unity at high concentrations.

The U.S. National Bureau of Standards (S4) uses the molality scale of concentration for the standard state of aqueous solutions and the mole fraction scale of concentration for the standard state of all nonaqueous solutions.

The standard state of a charged entity or ion such as I^- or Fe^{2+} in solution is the same as that of an uncharged species insofar as these are specified above; however, it is physically impossible to make a solution that contains only one type of ion. Solutions must be electrically neutral, and so they must contain at least two different types of ion—one positive and one negative. It is therefore impossible to have a solution of, say, Fe^{2+} alone. One must also have a negative ion present, say Cl^-, in equal concentration. In an infinitely dilute solution the Cl^- will have no effect on the Fe^{2+}, and vice versa, so that the activity coefficient γ of each ion will be constant and can be defined as unity.

7.4.3 Materials Linked to the Solvent Through a Chemical Equilibrium

In dealing with a solution, when the solvent has been defined to be in its standard state, the activity of the solvent is unity by definition. Since species in equilibrium with each other are of the same activity, it follows that in, for example, molten LiCl:

$$LiCl(l) \rightleftharpoons Li^+(solvated) + Cl^-(solvated)$$

Thus, if $a(LiCl) = 1$, the product $a(Li^+)a(Cl^-) = 1$. We can then define $a(Li^+) = 1$, in which case $a(Cl^-) = 1$ also. Thus the choice of a standard state for any species that is included in the solvent by a chemical equilibrium (e. g., chloride ions in molten alkali chlorides) cannot be independent of the choice of the solvent itself; the standard state for such a species is chosen as the actual state of the species in the specific solvent under 1 atm pressure at the specified temperature. Note that this means that, for example, in the LiCl—KCl eutectic the standard potentials reported for the Li^+/Li and Fe^{2+}/Fe couples are not strictly comparable, for in the former case the standard state of Li^+ is that state in which it exists in the pure LiCl—KCl solvent, while in the latter case the standard state of Fe^{2+} is the unit concentration—dilute solution hypothetical state. The chlorine–chloride and bromine–bromide couples are not strictly comparable for the same reason.

Consider two examples of standard—state problems in nonaqueous electrochemistry. The first arose in a study in fused LiCl-KCl eutectic involving sulfur and sulfide (Bodewig and Plambeck, S5). The problem was that the potential of the cell:

$$C/Li_2S(sat),S(sat),LiCl—KCl//LiCl—KCl,Ag^+/Ag$$

was stable while there was considerable drifting of the potential of the cell:

$$C/Li_2S(unsat),S(sat),LiCl—KCl//LiCl—KCl,Ag^+/Ag$$

Similar results were observed with K_2S. The problem appeared to be the volatilization of Li_2S (or Li_2S_2, etc.) from the melt. Nevertheless, it appears possible to calculate the standard potential of the S^{2-}/S couple by considering the following equilibrium:

$$Li_2S \rightleftarrows Li^+(solv) + S^{2-}(solv)$$

$$a(Li_2S) = a^2(Li^+)^2a(S^{2-})$$

In a saturated solution the activity of Li_2S is equal to the activity of pure solid Li_2S since these are in equilibrium, thus $a(Li_2S) = 1$. In fused LiCl—KCl, $a(Li^+) = 1$, and in the saturated solution $a(S^{2-}) = 1$ also. We are measuring the potential of the cell,

$$C/S^{2-}(a = 1),S(a = 1)//reference\ electrode$$

under standard conditions, thus obtaining its standard potential. It is a necessary consequence of this statement that the following experimental observations should prevail:

1. The cell potential will be independent of the amount of sulfur present.

2. The cell potential will be independent of the amount of Li_2S present as long as, and only as long as, the cell is maintained in a saturated condition.

3. The cell potential will be independent of the form in which the S^{2-} is added as long as that form is ionic.

This last point is less obvious. Consider that S^{2-} is added as CaS, which is ionic. Then $CaS \rightleftarrows Ca^{2+} + S^{2-}$, while $S^{2-} + Li^+ \rightleftarrows Li_2S$; and since Li^+ is present in far greater concentration than Ca^{2+}, the linked equilibria will be heavily shifted toward Li_2S, in accordance with Le Chatelier's principle, forming a saturated solution of Li_2S. The small amount of Ca^{2+} replacing Li^+ in the melt will presumably have a negligible effect on the observed potential.

Another case involved aluminum in chloroaluminate melts (Hames and Plambeck, S6). Here advantage was taken of the standard state of the solvent-linked aluminum ion to establish a standard reference electrode in $AlCl_3$—MCl melts. In such melts we have linked equilibria,

$$AlCl_3\text{—}MCl \rightleftharpoons M^+AlCl_4^- \rightleftharpoons M^+ + AlCl_4^-$$

and

$$AlCl_4^- \rightleftharpoons Al^{3+} + 4Cl^-$$

These equilibria operate such that the activity of aluminum ion is fixed (and equal to unity by definition) in any melt containing a fixed proportion of $AlCl_3$. Thus a bar of pure aluminum, whose activity is, by definition, unity in these solvents, will assume a fixed potential due to the Al^{3+}/Al couple, and this will also be the standard potential of the couple. This type of electrode should be less polarizable than other types of reference electrodes since the decrease (cathodic polarization) or increase (anodic polarization) in concentration of Al^{3+} due to electric current will be very small in comparison with the concentration of $AlCl_3$ and the chloroaluminate forms in equilibrium with it.

7.5 REFERENCES

General

G1   W. M. Latimer, *Oxidation Potentials,* 2nd ed., Prentice-Hall, Englewood Cliffs, N. J., 1952, especially Chapters 1 to 2.

G2   H. A. Laitinen, *Chemical Analysis*, McGraw-Hill, New York, 1960, Chapter 15.

G3   G. N. Lewis and M. Randall, *Thermodynamics,* 2nd ed., rev. Pitzer and Brewer, McGraw-Hill, New York, 1961, Chapter 24.

G4   G. W. Castellan, *Physical Chemistry,* 2nd ed., Addison-Wesley, Reading, Mass. 1971, 866 pp.

G5   W. J. Moore, *Physical Chemistry,* 4th ed., Prentice-Hall, Englewood Cliffs, N.Y., 1972, 977 pp.

G6   P. W. Atkins, *Physical Chemistry*, University Press, Oxford, 1978, 1018 pp.

Specific

S1   W. Nernst, *Z. Phys. Chem.* **4**, 129 (1889).

S2   T. S. Licht and A. J. DeBethune, *J. Chem. Educ.* **34**, 433 (1957); A. J. DeBethune, *J. Electrochem. Soc.* **102**, 288C (1955).

S3   A. J. DeBethune, T. S. Licht, and N. Swendeman, *J. Electrochem. Soc.* **106**, 616 (1959).

S4   U. S. Nat. Bur. Stds. Tech. Note 270-1, 1965, p. 3.

S5   F. G. Bodewig and J. A. Plambeck, *J. Electrochem. Soc.* **116**, 607 (1969).

S6   H. A. Laitinen and J. A. Plambeck, *J. Amer. Chem. Soc.* **87**, 1202 (1965).

S7   D. A. Hames and J. A. Plambeck, *Can. J. Chem.* **46**, 1727 (1968).

7.6 STUDY PROBLEMS

7.1 *Canadian Chemical Processing* reported in December 1976 that production of 1000 kg (1 metric ton or 1 Mg) of sodium chlorate required 565 kg of sodium chloride and 5600 kWh of electricity. The process is electrolytic, the oxidation reaction being carried out at a titanium anode. Calculate the number of coulombs required to produce 1 Mg of sodium chlorate. Then calculate the voltage of the electrolytic cell used. Does this correspond to the equilibrium potential difference for the couples involved? If not, why not?

7.2 *Canadian Chemical Processing* also reported, in the same article, that Canadian production capacity for chlorate, all of which is made by this process, is 205.4 Gg/yr. Ontario Hydro's Pickering nuclear electric generating plant is nominally capable of producing 100 MW(e) of electrical power, continuously in each reactor. Is one such reactor capable of supplying enough power for Canadian production capacity? If not, how many reactors would be required?

7.3 The standard potential for the half-cell reaction:

$$2HCl + 2e^- \rightarrow H_2 + 2Cl^-$$

was measured as -0.6935 ± 0.0047 V against the standard molar platinum electrode in fused LiCl—KCl eutectic at 723 K. In a separate study the standard potential of the half-cell reaction,

$$Cl_2 + 2e^- \rightarrow 2Cl^-$$

in the same solvent, against the same reference, and at the same temperature was $+0.3223 \pm 0.0016$ V. The USNBS thermodynamic tables give -47.246 kcal/mol, and the JANAF tables give -47.098 kcal/mol, for the free energy of the reaction:

$$H_2 + Cl_2 \rightarrow 2HCl$$

at 723 K, no error limits being given. With which, if either, thermodynamic value is the electrochemical measurement in agreement (Laitinen and Plambeck, S6)?

7.4 The standard molar aqueous potentials for the half-cell reactions $Hg_2^{2+} + 2e^- \rightarrow Hg$ and $2Hg^{2a} + 2e^- \rightarrow Hg_2^{2+}$ are, respectively, $+0.789$ V and $+0.920$ V against the SHE. The standard molar potentials for the same half-cell reactions were measured as $+1.028$ V and $+1.415$ V in fused $AlCl_3$ —NaCl—KCl against an aluminum reference electrode. In which solvent is Hg_2^{2+} more stable, and by how much (Hames and Plambeck, S7)?

7.5 From the values of the standard potentials of the appropriate couples, compute the numeric value of the equilibrium constant for the disproportionation of Fe(II) to Fe(III) and metallic iron in aqueous solution. Despite this value, it is difficult to maintain an aqueous standard solution of Fe(II) for any length of time. Why?

CHAPTER
8

EQUILIBRIUM ELECTROCHEMICAL COUPLES

There are three major classes of equilibrium potentials: *electrode potentials*, *redox potentials*, and *membrane potentials*. These major classes are distinguished in traditional ways.

8.1 ELECTRODE POTENTIALS

Electrode potentials are those equilibrium potentials that directly involve the electrode material. That is, they are equilibrium potentials that respond directly to the choice of the material in which the electronic conduction takes place. These potentials are thereby different from redox potentials, in which the choice of electrode material is irrelevant so long as it is chemically inert to all of the species present, and from membrane potentials, in which the actual potential difference across the membrane is measured using pairs of electrodes, inert or otherwise. Electrode potentials are divided into three traditional classes.

8.1.1 Electrodes of the First Kind

Electrodes of the first kind are those in which the electrode potential responds to a change in the activity of at least one of the species, oxidized or reduced, *directly*. The activity of the other species is often fixed, normally at unity, though it need not be. The most common type of electrode of the first kind is a metal in contact with a solution of its ions, such as $Ag(I)/Ag(0)$. For such electrodes, the Nernst equation becomes

$$E = E^0 - RT/zF \ln(1/a(\text{Ox})) \tag{8.1}$$

$$E = E^0(\text{Ag}^+/\text{Ag}) + 0.059 \log a(\text{Ag}^+) \tag{8.2}$$

The form in which the activity of the oxidized form is fixed while the activity of the reduced form varies is uncommon. The fixed activity is often unity, as in the example above involving silver metal; if it is not, the difference is usually incorporated into an $E^{0\prime}$ value. Such a value is known as a *formal potential*. A formal potential differs from a standard potential in that the concentrations rather than the activities of species appear in the Nernst equation. Thus the activity coefficients, and any other activity terms that remain constant such as free energies of complex formation, are incorporated into the numeric value of the formal potential.

Redox potentials are sometimes also described as belonging to "electrodes of the first kind" when the redox potential responds directly to changes in the activity of at least one of the species, as in the cases of $H^+(aq)/H_2(g),Pt$ and $I^-(aq), I_3^-$ (aq)/Pt.

8.1.2   Electrodes of the Second Kind

Electrodes of the second kind are those electrodes in which the relation between the electrode potential and the species activity is indirect, the species whose activity is of interest being linked to a species that is directly involved in the electrode reaction through *one* and only one chemical equilibrium. Examples of electrodes of the second kind are the silver chloride electrode:

$$//Cl^-(aq),AgCl(s)/Ag(0)$$

and the calomel electrode:

$$//Cl^-(aq),Hg_2Cl_2(s)/Hg(0)$$

which are of great practical use both as reference electrodes and as sensors responsive to the activity of chloride ions. The equilibrium involved is usually, although it need not be, a solubility product such as that of $AgCl$ or Hg_2Cl_2 in the examples above.

Nernst equations written for electrodes of the second kind may involve the equilibrium constant either explicitly or implicitly. For the first example, if the potential-determining reaction is considered to be $AgCl(s) + e^- \rightarrow Cl^-$ (ag) + Ag(s),

$$E = E^0 - RT/zF \ln[a(Ag)a(Cl^-)/a(AgCl)] \tag{8.3}$$

$$E = E^0(AgCl/Ag) - RT/F \ln a(Cl^-) \tag{8.4}$$

The simplified form is obtained since Ag and AgCl are present as pure solids and thus have unit activity.

Alternatively, if the potential-determining reaction is considered to be $Ag^+ + e^- \rightarrow Ag(s)$,

$$E = E^0(Ag^+/Ag) + RT/F \ln a(Ag^+) \tag{8.5}$$

It is necessary to use explicitly the solubility product of silver chloride:

$$K_s = a(Cl^-)a(Ag^+)/a(AgCl) \tag{8.6}$$

Substitution of this into the Nernst equation above gives

$$E = E^0(Ag^+/Ag) + RT/F \ln (K_s/a(Cl^-))$$

$$E = E^0(Ag^+/Ag) + RT/F \ln K_s - RT/F \ln a(Cl^-)$$

Comparison of this equation with the form of the Nernst equation above that implicitly incorporates the solubility product shows that the two equilibrium potentials are related directly through the solubility product constant:

$$E^0(AgCl/Ag) = E^0(Ag^+/Ag) + RT/F \ln K_s \tag{8.7}$$

Electrodes of the second kind are often used to measure equilibrium constants, as in the above example, as well as to measure ionic concentrations. As ion sensors, they are usually used for anions. Cation sensors of this type are uncommon because few of the well-behaved corresponding couples of electrodes of the first kind have oxidized forms of fixed activity.

8.1.3 Electrodes of the Third Kind

Electrodes of the third kind, like electrodes of the second kind, are electrodes for which the relation between the electrode potential and the species activity is indirect. In electrodes of the third kind, the species whose activity is of interest is linked to a species that is directly involved in the electrode reaction, through *two or more* chemical equilibria.

An example of electrodes of third kind is the oxalate electrode,

$$//Ca^{2+}(aq),CaC_2O_4(s),Ag_2C_2O_4(s)/Ag(0)$$

suggested by Reilley (S1) in which the two equilibria involved are the solubility equilibria governed by solubility products of silver oxalate and calcium oxalate. This electrode was used to follow the titration of calcium ion with oxalate ion. The electrode actually responds to the activity of the silver ion, which is dependent upon the activity of the oxalate anion through the solubility product of silver oxalate. The activity of oxalate is in turn governed by the activity of calcium ion through the solubility product of calcium oxalate:

$$E = E^0(Ag^+/Ag) - RT/2F \ln(1/a(Ag^+)^2)$$

$$K_{s,1}(Ag_2C_2O_4) = a(Ag^+)^2 a(C_2O_4^{2-})$$

$$K_{s,2}(CaC_2O_4) = a(Ca^{2+})a(C_2O_4^{2-})$$

$$E = E^0 + RT/2F \ln (K_{s,2}/K_{s,1}) - RT/2F \ln a(Ca^{2+})$$

Incorporation of the solubility product constants into a formal potential value $E^{0'}$ gives the useful potential of the electrode:

$$E = E^{0'} - RT/2F \ln a(Ca^{2+}) \tag{8.8}$$

Another example is the ethylenediaminetetraacetic acid electrode (EDTA is here indicated as H_4Y) responsive to metal ions:

$$//M^{2+}(aq),MY^{2-}(aq),HgY^{2-}(aq)/Hg(0)$$

in which both of the equilibria involved are complexation:

$$M^{2+}(aq) + Y^{4-}(aq) \rightleftarrows MY^{2-}(aq)$$

and

$$Y^{4-}(aq) + Hg^{2+}(aq) \rightleftarrows HgY^{2-}(aq).$$

The potential-determining couple is actually $Hg^{2+}(aq)/Hg(0)$. The ion $M^{2+}(aq)$ can be Zn^{2+}, Ca^{2+}, Pb^{2+}, Ca^{2+}, or mixtures of these; $Y^{4-}(aq)$ is the anion of EDTA.

8.2 REDOX POTENTIALS

The potential of a noble metal electrode, such as platinum, iridium, gold, palladium, and others, will respond to any redox couple present in solution, such as Fe(III)/Fe(II). In principle, such an electrode should give one and only one potential, which is the equilibrium redox potential of all of the couples present in the solution under the conditions in which they exist there, and such a redox potential should in principle be independent of the particular noble metal used. In practice, particularly in solutions that contain strongly oxidizing ions such as Co(III), MnO_4^-, $Cr_2O_7^{2-}$, and Ce(IV), the observed potential may not be the same, because the surface of the electrode is oxidized and the degree of oxidation depends on the noble metal used. The potential differences may range up to 200 mV in some cases, and can constitute a serious problem in potentiometric titrations using redox reactions.

It is often assumed that simply placing a noble metal electrode in a solution containing the two species that form an electrochemical redox couple will produce a voltage, measured against some reference electrode, that is characteristic of the species present in accordance with the Nernst equation. This assumption is frequently wrong. A noble metal electrode will respond correctly to the concentrations of species only if, first, the concentrations of species in solution are the true equilibrium concentrations, which means that the equilibria in solution must be rapid. Most, but by no means all, redox reactions are rapid; a slow reaction and thus a slow approach to equilibrium will produce slowly drifting redox potentials. Similar problems can arise from slow reactions with the oxygen of the air. A second problem that can arise with redox electrodes is the *mixed potential*, which occurs when multiple redox couples are possible and when equilibrium between them has not been achieved. The electrode then responds to the actual activities of the species of the redox couples with the highest exchange currents, which is itself a function of the rates of reaction with other species and cannot be used quantitatively. Thus, the response of the electrode is to couples with high exchange currents, and if these couples are not the desired couples interpretation of the measurements can be seriously awry.

8.3   MEMBRANE POTENTIALS

A *membrane potential* arises whenever a membrane separates two solutions of different composition. The basic structure of a membrane potential system is

reference 1/solution 1/*membrane*/solution 2/reference 2

in which two reference electrodes are required to measure the potential difference across the membrane. The potential difference will depend on the nature of the membrane and upon the concentrations of the ions making up solutions 1 and 2. Glass electrodes and ion-selective electrodes are both classed as membrane electrodes; the membranes can be solid or liquid.

Membranes are, from the point of view of an electrochemical cell, separators that exist in ionic conductors. They may separate solutions that are identical or different. They may be viewed as three-component systems: an interphase region with solution 1, a uniform bulk membrane region, and an interphase region with solution 2. It is usual to assume the interphase regions to be sharp phase boundaries in theoretical studies.

Membranes may be permeable to the solvent and to some, all, or none of the molecules and ions dissolved in it. For studies in dilute solution, permeability to the solvent is of little interest. Permeability to molecules, whether solvent or solute, does not directly produce electrochemically sensible phenomena and is therefore not usually considered. The permeability of the electrode to ions, however, is electrochemically significant.

When two solutions containing ions at different activities are separated, there exists between them a difference in chemical potential μ associated with each ion:

$$\mu_i = \mu^0_i + RT \ln a_i \tag{8.9}$$

Passage of one mole of electrons in the external circuit will cause a passage of t_i/z_i moles of ion i through the separator, where t_i is the transference number of the ion and z_i its charge. If the passage is carried out reversibly, then

$$dG = \Sigma_i(t_i/z_i) \, d\mu_i \tag{8.10}$$

$$dG = RT \, \Sigma_i(t_i/z_i) \, d \ln a_i \tag{8.11}$$

$$dE = -(RT/F) \, \Sigma_i(t_i/z_i) \, d \ln a_i \tag{8.12}$$

When the activities and transference numbers do not vary in the separation region, this equation is easily integrated to give

$$\Delta E = -(RT/F) \, \Sigma_i(t_i/z_i)\Delta\ln a_i \tag{8.13}$$

where ΔE is the observed potential difference across the separator and $\Delta \ln a_i = \ln (a_{i,2}/a_{i,1})$ reflects the difference in activity of ion i in the two bulk solutions that are separated. The difference ΔE is referred to as a *membrane potential* E(membrane) when the separator is a membrane and as a *liquid junction potential* E_j when the liquid electrolytes are in direct contact.

For a uni-univalent electrolyte such as the salt M^+X^-, two solutions of which can be designated as 1 and 2, the potential difference across the membrane is

$$E(\text{membrane}) = (RT/F)(t(-) - t(+))\Delta \ln [a(M^+)a(X^-)] \tag{8.14}$$

or, since $t(+) + t(-) = 1$,

$$E(\text{membrane}) = (RT/F)(2t(-) - 1)\Delta \ln [a(M^+)a(X^-)] \tag{8.15}$$

This equation gives the total potential difference present across a phase boundary, as would be measured by two identical electrodes in the two bulk electrolytes. Such boundary potentials are often called *Donnan potentials*, after F. G. Donnan, and arise whenever transport of all ions between the phases is not equally facilitated. The values of the transference numbers are those associated with transfer of the ions through the phase boundary, not those that prevail in either or both of the bulk electrolytes. The Donnan potential can be used either when the membrane is of zero thickness (and nonexistent) or when the membrane is of finite thickness but neither the activities nor transference numbers vary through it (uniform concentration throughout the membrane).

When the entire potential difference cannot be dealt with as a Donnan potential, it is possible to consider it as a sum of Donnan potentials, one at each membrane–solution phase boundary, and one *diffusion potential* arising from the differences in transport of ions in the bulk membrane. The separation is of theoretical rather than experimental significance, and it is usually sufficient to use either Donnan or diffusion potentials rather than both.

The potential difference between side 2 and side 1, ΔE, is:

$$\Delta E = (RT/F) (2t(-)-1)\ln [a(M^+,2)a(X^-,2)/a(M^+,1)a(X^-,1)] \tag{8.16}$$

The transference numbers t_i are the transference numbers of the ions *in the membrane* (usually assumed to be in the bulk membrane), which may be quite different from the transference numbers of the ions in either of the two bulk solutions. The magnitude of the potential difference ΔE depends upon the activity differences across the membrane *and* upon the transference numbers in the membrane. There are three special cases that are of interest.

If the membrane is equally permeable to both cation and anion, then $t(+) = t(-) = \frac{1}{2}$; the value of ΔE is necessarily zero. This is why a salt bridge, which is simply a bulk membrane between two interphases, should contain a cation and an anion whose transference numbers are nearly equal. The measured potential difference E across a cell is then

$$E = E_1 + E_2 + \Sigma E_j \tag{8.17}$$

where E_1 and E_2 are the potentials of the two electrodes between which the potential difference is measured, and ΣE_j is the sum of the liquid junction potentials in the cell. When the two electrodes are identical, only the junction potentials are measured because E_1 and E_2 are identical in magnitude but necessarily opposite in sign.

If the membrane is totally impermeable to the cation, then only the anion moves in it and $t(-)$ is unity. The value of ΔE is then RT/F times the logarithm of the activity terms. This is the greatest possible value of ΔE. The measured potential difference E across a cell is then

$$E = E_1 + E_2 + \Sigma E_j + (RT/F) \ln a_{\text{out}}/a_{\text{in}} \tag{8.18}$$

where the notation of activities a_{out} and a_{in} is a summary notation for the activity of all species across the separating membrane.

If the membrane is permeable to the cation and totally impermeable to the anion, only the cation moves in it, and $t(-)$ is zero. The value of ΔE is $-RT/F$ times the logarithm of the activity terms, equal to the preceding case but opposite in sign.

The discussion of membranes above applies also to salt bridges or other ionic conductors that have junctions with ionically conducting solutions. The potential difference calculated for salt bridge or membrane is that due to the bulk salt bridge or membrane and does *not* include the potentials of the interphase regions. These liquid junction potentials are assumed to cancel each other, which they will actually do only if the two interphase regions are identical.

Membrane potentials respond to the nature of the membrane as well as the reference electrodes on the two sides and to concentration of species in the test solution; in normal analytical practice the reference electrodes, the standard solution, and the membrane itself are all held constant so that only the variation in activity in the test solution is observed. These electrodes are discussed further in Chapters 10 and 11.

8.3.1   Liquid Junction Potentials

Consideration of physical separators such as membranes simply as discontinuous activity changes is reasonable, but when two electrolyte solutions are in direct contact to establish a *liquid junction* the discontinuous model is inadequate. Integration of the general Eq. (8.12) between solutions 1 and 2 gives

$$E_j = -(RT/F) \Sigma_i(t_i/z_i) \, d \ln a_i \tag{8.19}$$

Analytic solution of this general integral requires knowledge of the variation of t_i and a_i throughout the junction region between solutions 1 and 2. This knowledge is not available. Alternatively, since $a_i = \gamma_i c_i$, knowledge of the variation of the activity coefficient γ_i and the transference number t_i with concentration c_i, together with knowledge of the variation of c_i throughout the junction region, would suffice.

In deriving the equation that bears his name, Henderson (S2, S3) used this latter approach together with two simplifying assumptions. These were, first, that the activity coefficient γ_i was and remained unity throughout the junction region, and second, that the concentration c_i varied linearly from $c_{i,1}$ to $c_{i,2}$ across the junction region between the two bulk electrolyte solutions. This latter assumption is equivalent to taking the solution at a point a fractional distance x across the junction region as a mixture of fraction x of solution 2 together with fraction $1 - x$ of solution 1. The concentration c_i at that point x distance across the junction is then

$$c_i = xc_{i,2} + (1 - x) \, c_{i,1} \tag{8.20}$$

$$c_i = c_{i,1} + x(c_{i,2} - c_{i,1}) \tag{8.21}$$

Taking the derivative:

$$dc_i/dx = c_{i,2} - c_{i,1} \tag{8.22}$$

$$dc_i = (c_{i,2} - c_{i,1}) \, dx \tag{8.23}$$

and dividing both sides by c_i, an explicit statement of the variation of $d \ln c_i$ throughout the junction region is obtained because $d \ln c_i = d(c_i)/c_i$:

$$d \ln c_i = ((c_{i,2} - c_{i,1})/c_i) \, dx \tag{8.24}$$

The transport number t_i is the fraction of the conductivity due to the ith ion (Chapter 4):

$$t_i = \kappa_i / \Sigma_i \kappa_i$$

which in terms of the electrical mobilities of the ions μ'_i is

$$t_i = z_i c_i \mu'_i / \Sigma_i z_i c_i \mu'_i$$

The transport number t_i in the junction region is then given by

$$t_i = z_i \mu'_i c_i / (x \Sigma_i z_i \mu'_i c_{i,2} (1 - x) \Sigma_i z_i \mu'_i c_{i,1}) \tag{8.25}$$

The mobilities are taken as positive for cations and negative for anions, as is the charge. Multiplication of the two equations for concentration dependence and transport number gives

$$t_i \, d(\ln c_i)/z_i = \mu'_i(c_{i,2} - c_{i,1}) \, dx/(\Sigma_i c_{i,1} \mu'_i z_i + x \Sigma_i \mu'_i z_i(c_{i,2} - c_{i,1})) \tag{8.26}$$

Substitution of these relations back into the general relation followed by integration yields the *Henderson equation*:

$$E = (-RT/F)(\Sigma_i \mu'_i \, (c_{i,2} - c_{i,1})/\Sigma_i \mu'_i z_i(c_{i,2} - c_{i,1})) \ln (\Sigma_i \mu'_i z_i c_{i,2}/\Sigma_i \mu'_i z_i c_{i,1}) \tag{8.27}$$

For an equi-equivalent electrolyte the form of the Henderson equation simplifies to

$$E = (-RT/zF)((\mu'_+ - \mu'_-)/(\mu'_+ + \mu'_-)) \ln (c_2/c_1) \tag{8.28}$$

with charge and ionic mobilities of all ions taken as positive. According to the Henderson equation, the potential across the junction approaches zero as the mobilities of the positive and negative ions approach the same value. It is for this reason that aqueous KCl is the usual choice for the ionic conductor in salt bridges, whose purpose is to reduce the liquid junction potential to its lowest possible value.

8.4 ANALYTICAL APPLICATIONS OF ELECTRODE POTENTIALS

Analytical applications of electrode potentials will be covered here, with the exception of their use in titrations which is covered in Chapter 12. Electrodes of the first kind are excellent sensors for the ions that are directly involved in any electrochemical couple provided that the metal ion–metal couple has a reasonably high exchange current and that no other couples that are capable of chemical reaction with the electrode material itself exist in solution. The former limitation is occasional, but the latter is often severe; in the analyses of real solutions of analytical interest use of electrodes of the first kind is restricted for that reason. Electrodes of the second kind are more generally used as reference electrodes than as sensors, although they will serve well in the capacity of sensor. Electrodes of the third kind generally respond rather sluggishly and are rarely employed as sensors.

8.5 REFERENCES

S1   C.N. Reilley and R. W. Schmid, *Anal. Chem.* **30**, 947 (1958); C.N. Reilley, R.W. Schmid, and D.W. Lamson, *Anal. Chem.* **30**, 953 (1958).

S2   P. Henderson, *Z. phys. Chem.* **59**, 118 (1907).

S3   P. Henderson, *Z. phys. Chem.* **63**, 325 (1908).

8.6 STUDY PROBLEMS

8.1 Calculate the solubility product of silver bromide if the standard potential of the $AgBr/Ag$ electrode is $+0.095$ V against SHE.

8.2 Fluoride ion forms complex ions of the stoichiometry SnF_6^{2-} with Sn(IV). Calculate the formation constant of this complex ion, given that the SnF_6^{2-}/Sn electrode has a standard potential of -0.25 V against SHE.

8.3 Calculate the dependence of the formal potential of the $Cr_2O_7^{2-}/Cr^{3+}$ couple on pH, and on the basis of that calculation discuss the usefulness of dichromate as an oxidizing agent.

CHAPTER
9
pH AND THE HYDROGEN ELECTRODE

One of the most useful concepts in aqueous chemistry, often carried over into nonaqueous work, is the concept of solution pH. The quantity pH was originally defined by Sorensen (S1) as the negative logarithm of the *concentration* of the hydrogen ion in aqueous solution; this quantity varies between about 0 and 14 pH units. Strongly acidic solutions are found at the lower end of the scale, strongly basic solutions are found at the higher end, and neutral solutions have a pH of about 7.

The definition of pH was altered (S2) after the development of the thermodynamic concept of activity to

$$pH = -\log a(H^+) = -\log \gamma(H^+)c(H^+) \tag{9.1}$$

It is ironic that this useful concept was then defined in terms of a quantity that is in principle unmeasurable. The activity of hydrogen ion or any other single ionic species cannot be measured because a single ionic species cannot be added to a solution without at the same time adding another ion of the opposite charge. It is, however, possible to measure a quantity that, if not equal to the activity of the hydrogen ion, is very nearly so, and it is that quantity with which we are concerned.

Modern definitions of pH, as established by all national standardizing organizations, are operational rather than thermodynamic. They are based upon the measurement of potential differences between hydrogen electrodes in different solutions, which is relatable by use of standards to a quantity that is, very nearly, the activity of the hydrogen ion. Sensors for pH other than the hydrogen electrode can be, and for practical purposes usually are, employed in its place. By far the most important of these is the glass electrode. Other electrodes, such as the quinhydrone electrode, have also been used as have nonelectrochemical methods such as spectrophotometry with acid–base indicators. All of them, however, depend for their validity upon measurements made with the hydrogen electrode. It is convenient to consider the pH sensor, or measuring device, and the measuring procedure separately since essentially the same procedure and standards are used with all electrochemical sensors.

9.1   pH SENSORS: THE HYDROGEN ELECTRODE

The ultimate sensor of pH is the hydrogen electrode in either aqueous or nonaqueous solution. In protonic solvents, it is also the standard of the potential scale, the potential of a standard hydrogen electrode (SHE) being taken as the defined zero of potential. The schematic representation of a standard hydrogen electrode is

$$// \ H^+(\text{activity} = 1)/H_2(g, \text{fugacity} = 1), Pt$$

In practice, a normal hydrogen electrode (NHE) is virtually always used. Its schematic representation in aqueous solution is

$$// \ H^+(1.0M, aq)/H_2(g, p = 1 \ atm), Pt.$$

The potentials of the SHE and the NHE are virtually the same for all but the most precise work.

The construction of a normal hydrogen electrode (S3) is fairly simple, care being required only in platinizing the platinum surface. This is most effectively done by making the electrode the cathode of an electrolytic cell containing aqueous 1 to 3% chloroplatinic acid to which a very small amount of lead acetate is added and passing a current of 200 to 400 mA/cm^2 for 1 to 3 min, followed by extensive washing. In those rare cases where platinization produces intolerable catalysis, chloropalladous acid may be substituted, but the electrode response is not as rapid.

9.1.1   Measuring pH with the Hydrogen Electrode

There is no method of measuring an absolute pH without evaluating either a liquid junction potential or the activity coefficient (or activity) of a single ionic species. Separation of cationic and anionic contributions lies in the province of physical rather than analytical electrochemistry. The monograph of Bates (G1) or the general references of this chapter and Chapter 4 should be consulted for further details. Briefly, a cell without liquid junction is established:

$$Pt, H_2(g, p)/Cl^-(aq), H^+(aq)/AgCl(s), Ag$$

whose potential is given by the Nernst equation as

$$E = E(AgCl/Ag) - E(H^+/H_2)$$

$$E = E^0 - RT/F \ \ln[a(H^+)a(Cl\text{-})/p^{1/2}] \tag{9.2}$$

which rearranges to equation 9.3:

$$E = E^0 - 2.303RT/F \ \log[c(Cl\text{-})\gamma(Cl\text{-})/p^{1/2}] - 2.303RT/F \log a(H^+) \tag{9.3}$$

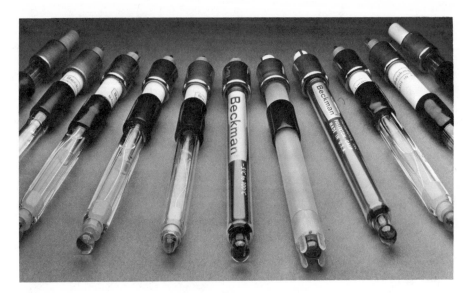

Figure 9.1   Modern commercial glass electrodes. Courtesy of Beckman Instruments, Inc.

in which E is the measurable cell potential, E^0 is the standard potential of the silver chloride reference (since the standard potential of the hydrogen electrode is by definition zero) and the pressure p of hydrogen and the concentration c of chloride ion are both measurable quantities. Only the activity coefficient $\gamma(Cl^-)$ of chloride ion and the activity of hydrogen ion remain, and of these log $\gamma(Cl^-)$ can be calculated using an extended form of the Debye–Huckel equation. Thus the remaining unknown, the activity of hydrogen ion, can be evaluated.

For analytical purposes, pH values of solutions are measured not by an absolute but by a relative method in which the pH of an unknown or test solution is compared with that of one or more reference pH standards, as discussed later in this chapter.

9.1.2   Usage of the Hydrogen Electrode

The hydrogen electrode is virtually obsolete for practical measurement purposes, although it remains the ultimate standard of hydrogen-ion activity measurement and, unlike any other electrode responsive to H^+, it is usable over the entire aqueous pH range and in nonaqueous solvents as well. For pH above 12, one might consider it in aqueous solutions, but never at lower pH.

The reason for the lack of use of hydrogen electrodes in practice is the prevalence of interferences. Both oxidizing ions and reducing ions in the solution interfere by causing the platinized platinum surface to act as a redox electrode. Since the exchange current of the $H^+/H_2(g)$ couple, even on platinized platinum, is low compared with the exchange currents of many redox couples at this type of an electrode, the interference is often severe. Moreover, many species in solution "poison" the catalytic platinized platinum surface so that the exchange current is further reduced, and the electrode becomes even less well-poised – a few examples being proteins, colloids, H_2S and other sulfides, arsine, and even oxygen.

9.2   pH SENSORS: THE GLASS ELECTRODE

The glass electrode is by far the most common electrochemical method of measuring the pH of aqueous, partially aqueous, and some nonaqueous systems. It is a type of membrane electrode, its general form of construction being:

$$Ag/AgCl(s), Cl\text{-}(aq), H^+(aq)/glass/test\ solution$$

The electrode response, first noted by Cremer in 1906 (S4) and systematically studied by 1919 (S5) can be given as

$$E = E' - RT/F \ln[a(H^+,1)/a(H^+,2)] \tag{9.4}$$

where E' is a potential at least nominally independent of hydrogen ion activity. Its components are given by

$$E' = E_{r1} - E_{r2} + E_{j1} - E_{j2} \tag{9.5}$$

where E_{r1} is the potential of the reference-electrode internal to the glass electrode, E_{r2} is the potential of a reference electrode external to the glass electrode, and the two values of E_j are the liquid junction potentials at the two interphases of the glass membrane. In aqueous solution at 25°, this simplifies to

$$E = E' + 0.05915 \log a(H^+,2) \tag{9.6}$$

where the E' term incorporates the difference in reference-electrode potentials, all liquid junction potentials in the cell, and the presumably constant term related to the activity of hydrogen ion in the internal solution.

It is more realistic to write this equation as

$$E = E' - E_{j2} + 0.05915 \log a(H^+,2) \tag{9.7}$$

thereby incorporating only the two reference-electrode potentials and the junction potential at the inner interphase of the glass membrane into the E' value, because the liquid junction potential at the external interphase will be affected even if only to a minor degree by the composition of the external solution. It should not be considered to be truly constant.

The mechanism of operation of the glass electrode has long been a subject of dispute. Experimentally, it has been shown by hydrogen ion analysis (S6) and tritium labeling (S7) that hydrogen ion is *not* transported through the membrane, and thus the discussion of the previous chapter cannot adequately describe the glass pH sensor. In the case of unequal permeability (of which the limiting version of total impermeability to one ion gives the *Donnan potential* observed across a membrane), the membrane is itself irrelevant. The distribution of ions imposed by the existence of the membrane, which will persist briefly in its absence, establishes a potential capable of being sensed by reference electrodes in the solutions. Under these circumstances the transport numbers in question relate to the solutions rather than the (not necessarily existent) membrane.

In water, the pH range over which the glass electrode can be used depends upon the specific glass of which the electrode is made. Hard glasses, such as Pyrex, are inferior in response to soft glass. As a result of the studies of MacInnes and Dole (S9) the popular soft soda-lime glass known as Corning 015 was developed; its composition is 72% SiO_2, 22% Na_2O, and 6% CaO.

For glasses of this type, the pH range accessible with the glass electrode is limited. More correctly, the deviations from Nernstian behavior become significant both in basic and in strongly acidic solutions. The error in basic solutions is the more serious of the two; it is generally insignificant up to pH 9, or perhaps slightly below this, but above pH 9 the potentials observed become significantly lower than the expected Nernstian values. The error increases rapidly with further increase in pH. At any given pH, the error increases with increasing concentrations of other cations in the solution, particularly alkali metal ions and especially sodium ion. At pH 12, the error would be about –0.7 pH in the presence of 1.0 mol/dm^3 sodium ion, but only about –0.3 pH in the presence of 0.1 mol/dm^3 sodium ion; errors due to other ions are considerably less than these.

On the acid side, the error does not become noticeable until the pH is less than 2 and becomes appreciable only at pH values near zero. Similar errors may be found in solutions of high ionic strength. These acid or ion errors arise from the fact that the activity of water is no longer the same on both sides of the membrane, and thus a concentration cell in water is created that also contributes to the observed potential.

The acid or ion error is generally not significant, which is fortunate in that it cannot be easily avoided. The basic error can be reduced significantly by changing the type of glass. Studies have been made by Perley (S10) and others on various types of glasses, and some lithium glasses have been developed with basic errors an order of magnitude or more below the errors of Corning 015 glasses. These are commercially available from several different manufacturers; some can be used even in 0.1-mol/dm^3 KOH or NaOH solutions.

9.2.1   Usage of the Glass Electrode

The glass electrode has a very high resistance, of the order of 1 to 500 MΩ, and its potential against any reference cannot be measured with any measuring system of low input impedance such as a potentiometer or a galvanometer-type voltmeter. An electronic voltmeter of unusually high input impedance is required, and such devices frequently are called *pH meters*. The precision and accuracy of pH measurements are virtually never limited by the pH meter in modern laboratories, but only by the glass electrode itself and by the standards used in its calibration.

Glass electrodes can be manufactured in the research laboratory, but rarely are since commercial electrodes are available at reasonable cost, and the quality control on commercial electrodes is better than on electrodes of local manufacture. Commercial electrodes are not small enough for many biochemical studies, and investigators have designed glass microelectrodes for these studies. But except for these microelectrodes commercial design prevails. Proper usage precautions are few, and commercial glass electrodes are remarkably robust. The presence of water in the glass membrane is essential for proper operation, and although the exact reason for this is still a subject of debate, the empirical fact is undisputed. Electrodes of modern design may be allowed to dry out, but stable Nernstian response will thereafter be delayed by a few minutes until the outer layers of glass become hydrated when the electrode is next used. Hydrolysis is particularly pronounced in basic solution, and storage of glass electrodes in solutions of pH greater than 9 is unwise.

Electrode response is rapid, and stable potentials are generally achieved within 1 to 3 min in shifts from one aqueous solution to another. Electrode response to changes of solvent may take longer, and storage in such solvents is unwise since rehydration to an equilibrium layer of silica gel on the outside of the glass bulb takes a great deal longer, perhaps several days. With proper care, glass-electrode membranes will last from one to two years; life is decreased significantly by higher temperatures (S8) and increased by storage in water (S11), since the hydrous silica layer protects the glass from further attack. Electrodes whose response has deteriorated and become sluggish owing to an excessive layer of hydrous silica can sometimes be restored by etching (2 min in 5% aqueous HF).

Glass electrodes of commercial design can be used normally in most mixed solvents, although storage in these solvents is unwise. The pH scale is shifted, and the values do not compare easily with those obtained in aqueous solutions, but they are themselves consistent. Glass electrodes can also be used in totally nonaqueous solvents, such as alcohols, heavy water, hydrogen peroxide, acetonitrile, dimethylformamide, propylene carbonate, and liquid ammonia; the response is Nernstian in these solvents over a reasonable pH range.

Minor errors in measurement of pH with glass electrodes may arise in strongly acidic or basic solutions because of the liquid junction potential between the standard solution and the reference electrode or, more often, between the test solution and the reference elctrode. This occurs because the H^+ or OH^- are then present in sufficient concentration that, with the help of their greater mobility, they carry a significant fraction of the ionic current. The error can be 0.03 to 0.04 pH units, because one pH unit corresponds to 59 mV, and the liquid junction potential may amount to perhaps 1 mV. Liquid junction potentials do in fact limit the precision of measurement of pH; 0.001 pH corresponds to 0.059 mV, and junction potentials of this magnitude cannot be avoided. Massive errors due to liquid junction potentials, of the order of 250 mV, are known to arise in colloidal suspensions or sludges since the mobility of Na^+ is then very large relative to the mobility of the negatively charged colloidal particles. In a two-phase solution (sludge plus supernatant) no potential difference is measured between two identical glass electrodes, one in the sludge and one in the supernatant, but a potential of almost 250 mV may be measurable between two identical reference electrodes, one in the sludge and one in the supernatant.

9.3   pH SENSORS: OTHER ELECTRODES

Electrochemical pH sensors other than the glass electrode are no longer of great use in electroanalytical work; a brief treatment of them suffices.

9.3.1   Quinhydrone Electrodes

Quinhydrone (Q_2H_2) is a slightly soluble compound that is formed by the combination of one mole of benzoquinone (Q) with one mole of benzohydroquinone (H_2Q). The presence of solid quinhydrone establishes the equilibrium

$$Q_2H_2(s) \rightleftharpoons Q + H_2Q$$

whose solubility product constant is

$$K_s = a(Q)a(H_2Q)/a(H_2Q_2) \tag{9.8}$$

The quinone and hydroquinone are also linked by the redox equilibrium $H_2Q \rightleftharpoons Q + 2H^+ + 2e^-$, which involves two protons and two electrons. The Nernst equation for a saturated aqueous solution of quinhydrone is, at 25°C:

$$E = +0.6994 - 0.05915/2 \log [a(H_2Q)/a(Q)a(H^+)^2] \tag{9.9}$$

Because the solution was made up by dissolution of H_2Q_2, the *concentration* of quinone is equal to the *concentration* of hydroquinone whether the solution is saturated or not, so if activities are approximated by concentrations,

$$E = +0.6994 + 0.05915/2 \log a(H^+)^2 \tag{9.10}$$

$$E = +0.6994 + 0.05915 \log a(H^+). \tag{9.11}$$

The activity effects, which are usually negligible, can be corrected for if necessary, but the quinhydrone electrode is rarely used for high-precision work (S12). The potential of the quinhydrone electrode is easily measured at a piece of bright platinum or gold wire, and the exchange current is high. The resistance of the electrode is low, and voltmeters of high input impedance are not necessary as they are with the glass electrode. This is the sole advantage of the quinhydrone electrode, and with the development of modern voltmeters it has become uncommon.

Quinhydrone electrodes have several disadvantages in addition to the fact that using one necessarily adds a substance to the test solution. These are, first, that they are suitable only below pH 8 or 9, because atmospheric oxidation of quinone and the dissociation of hydroquinone (it is an acid whose pK_a is about 10, and above pH 8.5 the dissociation is not noticeable) render its response nonNernstian to hydrogen ion. Second, unsaturated solutions of quinhydrone, or solutions saturated with quinhydrone only, are subject to a "salt error" of ± 0.06 pH units and sometimes more, depending upon the nature and concentration of other salts such as sodium chloride or magnesium sulfate in the solution. Such a "salt error" arises from the change in the activities in the ratio $a(H_2Q)/a(Q)$, which was assumed to be unity. By saturation of the solution with *both quinhydrone and hydroquinone* the "salt error" is eliminated since the ratio of activities $a(H_2Q)/a(Q)$ is then fixed, but the value of E^0 is more positive than that given above. Third, the quinhydrone electrode is not useful in solutions that contain strongly oxidizing or reducing ions, because these will react with either quinone or hydroquinone and upset the ratio. The standard potential of the quinhydrone electrode is $+0.699$ V, thus weaker oxidizing agents such as Cu^{2+} do not interfere.

9.3.2   Metal–Metal Oxide Electrodes

Metal–metal oxide couples (S13) have a long history of use in pH measurement, but are now obsolete except for occasional uses in extremely basic solution or in following a titration. The most common examples are $//H^+,Sb_2O_3(s)/Sb$ and $//H^+, HgO(s)/Hg$. The potential of the antimony(III) oxide electrode is adversely affected by complexing agents such as citrates and tartrates, by oxygen, and by strong oxidizing or reducing agents. The mercury(II) oxide electrode is less sensitive to complexing agents but similar otherwise; the lower sensitivity of mercury to complexing agents arises because polarization of the electrode enhances the formation of oxide.

9.4   MEASUREMENT OF pH OF AQUEOUS SOLUTIONS

The definition of pH as the negative logarithm of the hydrogen ion concentration or activity is rarely of direct use in practical measurements of pH. The glass electrode or other pH sensor is simply treated as a sensor having Nernstian response to hydrogen ion activity. For such an operational scale, the Nernst

TABLE 9.1   Values of pH for Five Standard Solutions[a]

t, (°C)	A	B	C	D	E
0	--	4.003	6.984	7.534	9.464
5	--	3.999	6.951	7.500	9.395
10	--	3.998	6.923	7.472	9.332
15	--	3.999	6.900	7.448	9.276
20	--	4.002	6.881	7.429	9.225
25	3.557	4.008	6.865	7.413	9.180
30	3.552	4.015	6.853	7.400	9.139
35	3.549	4.024	6.844	7.389	9.012
38	3.548	4.030	6.840	7.384	9.081
40	3.547	4.035	6.838	7.380	9.068
45	3.547	4.047	6.834	7.373	9.038
50	3.549	4.060	6.833	7.367	9.011
55	3.554	4.075	6.834	—	8.985
60	3.560	4.091	6.836	—	8.962
70	3.580	4.126	6.845	—	8.921
80	3.609	4.164	6.859	—	8.885
90	3.650	4.205	6.877	—	8.850
95	3.674	4.227	6.886	—	8.833

[a]The compositions of the five standard solutions are: A: KH tartrate (saturated at 25°C); B: KH phthalate, $c = 0.05$ mol/kg solvent; C: KH_2PO_4, $c = 0.025$ mol/kg solvent, and Na_2HPO_4, $c = 0.025$ mol/kg solvent; D: KH_2PO_4, $c = 0.08695$ mol/kg solvent, and $Na_2 HPO_4$, $c = 0.03043$ mol/kg solvent; E: $Na_2B_4O_7$, $c = 0.01$ mol/kg solvent.

equation becomes

$$pH(\text{unknown}) = pH(\text{standard}) + (E_u - E_s)F/2.303RT \tag{9.12}$$

where E_u and E_s are the potentials measured with the glass electrode in the unknown and standard solutions, respectively.

9.4.1   Standards of pH of Aqueous Solutions

All national and international scales of pH, being operational, depend upon the existence of standards. Much of the development of these has taken place at the U.S. National Bureau of Standards (USNBS) and is treated in Bates (G1). These data are accepted by most nations and by the International Union of Pure and Applied Chemistry (IUPAC) as the basis of their operational scales of pH, although the selection may vary between nations. The pH values of solutions prepared in appropriate ways can be taken as known to ±0.02 pH between sets of standards, ±0.001 within standards, and ±0.004 within sets of standards. That is, pH standardized using procedures of the British Standards Institution or the Japanese Standards Association will not vary by more than ±0.02 pH from pH standardized using the procedures of the USNBS. The USNBS supplies its seven primary standards in powdered form, and these can be used in the most precise work. The same, and other, standards can be prepared using reagent-grade chemicals in the laboratory and checked against values of USNBS samples. The preparation procedures of the USNBS should be carefully followed. The use of water free of acidic or basic substances is mandatory; distilled water should be boiled to remove carbon dioxide.

Of the available standard solutions, IUPAC has accepted five as defining an operational scale of pH. These are given in Table 9.1 and are identical to those used by the USNBS Additional standard solutions studied at the USNBS are given in Table 9.2. The values of solutions F and G in that table (S14) are accepted by USNBS as primary since they are known with no less accuracy than those given in Table 9.1. However, the pH values of F are very similar to those of A and B while the pH values of G are very similar to those of E and an essentially duplicate primary standard is redundant. The values for H and I, also determined at the USNBS (S15, S16, S17), are not redundant and constitute an extension of the pH standards to both high and low extremes. However, the consistency of the scale, as well as of the sensors, at these extremes is considerably less than in the intermediate pH range, and those standard solutions have therefore been considered secondary though their establishment follows the same procedure as establishment of the standards considered primary.

Several commercial supply houses provide standard buffer solutions at various pH values whose precision against primary standards is ± 0.01 pH unit. These buffers are usually made up as minor variants of the systems given in Table 9.1.

9.4.2   Single-Point Calibration Method

In this method, the potential of the glass electrode is measured in a solution that contains the standard buffer. The zero of the pH meter scale is adjusted until the pH meter reads the pH of the standard buffer. The electrode is then removed, washed, and placed in the unknown solution. The pH of the unknown solution, as read from the pH meter, is then the pH obtained assuming the Nernst equation, which incorporates the statement that the response of the electrode potential is not only linear with pH but also Nernstian. The British scale of operational pH, which uses only the $0.05m$ potassium hydrogen phthalate standard, is necessarily a single-point scale. The buffer used for calibration should be as close as possible to the pH of the solution being measured.

9.4.3   Double-Point Interpolation Method

In this method, one should employ a pH meter or voltmeter of high input impedance whose zero and gain are independently variable. Some pH meters possess only a zero control (often listed as a "calibrate" control) on the voltage reading of the meter. A pH meter that possesses only a zero control can still be used in this method by adjustment of the zero control to the best compromise

TABLE 9.2   Values of pH for Additional Standard Solutions[a]

t, (°C)	F	G	H	I
0	3.863	10.317	1.666	13.423
5	3.840	10.245	1.668	13.207
10	3.820	10.179	1.670	13.003
15	3.802	10.118	1.672	12.810
20	3.788	10.062	1.675	12.627
25	3.776	10.012	1.679	12.424
30	3.766	9.966	1.683	12.289
35	3.758	9.925	1.688	12.133
38	3.755	9.903	1.691	11.984
40	3.753	9.899	1.694	11.841
45	3.750	9.856	1.700	11.705
50	3.749	9.828	1.707	11.574
55	—	—	1.715	11.449
60	—	—	1.723	—
70	—	—	1.743	—
80	—	—	1.766	—
90	—	—	1.792	—
95	—	—	1.806	—

[a]The compositions of the standard solutions are: F: KH_2 citrate, $c = 0.05$ mol/kg solvent; G: $NaHCO_3$, $c = 0.025$ mol/kg solvent, and Na_2CO_3, $c = 0.025$ mol/kg solvent; H: $KH_3(C_2O_4)_2 \cdot 2H_2O$, $c = 0.05$ mol/kg solvent; I: $Ca(OH)_2$, saturated at 25°C.

point indicated by the two standard buffers, but compensation for lack of Nernstian response requires control of the internal amplifier gain of the meter.

To carry out a double-point calibration, the potential of the glass electrode is first measured in a solution that contains a standard buffer. The zero of the pH meter scale is adjusted until the meter reads the pH of the standard buffer. Then the electrode is removed, washed, and placed in a second buffer of different pH. Ideally, these two buffer solutions will bracket the expected pH of the unknown solution as closely as possible. The *gain* control of the amplifier is then adjusted until the meter reads the new pH correctly, then the electrode is removed, washed, and replaced in the first buffer solution. If necessary, the *zero* control is readjusted, and the process is continued until both pH values are read correctly. The electrode is then removed, washed, and placed in the unknown solution, whose pH is then read as before.

The advantage of the double-point interpolation is that it is no longer assumed that the response is Nernstian, but only that the response is linear, with pH. This is of particular importance in nonaqueous or partially aqueous systems and, in aqueous media, strongly acidic or strongly basic solutions. Modern pH meters such as that shown in Figure 9.2 generally have three controls rather than the zero

Figure 9.2   Modern pH meter. Beckman Zeromatic Model IV, courtesy of Beckman Instruments, Inc.

and gain controls mentioned above. One of these controls is the zero control; the remaining two, one of which is indicated as a temperature control, are both gain-adjustment controls. Under these circumstances the temperature control should be set to the actual temperature of the room and calibration carried out with the sensitivity and zero controls. Some pH meters are equipped with thermistors that sense the actual temperature and change the temperature setting automatically if the ambient temperature changes. In some meters only a zero and a temperature control are provided. In this case the temperature control should be used as a gain control, and no attention should be paid to the value of its setting.

9.5   MEASUREMENT OF pH OF NONAQUEOUS SOLUTIONS

Measurement of the pH of solutions that are wholly or partially nonaqueous, including mixed solvents, can in principle be done in either of two ways. These are the use of pH standards for the particular solvent or the use of aqueous pH standards. The former procedure, though in principle sound, is rarely used owing to the paucity of available nonaqueous pH standards. In practice, aqueous standards are used with a glass electrode filled with aqueous acid and an aqueous saturated calomel or silver chloride reference electrode. The potential difference measured includes not only the actual pH difference but a term representing the difference between solvents, which is unknown and generally not obtainable—except in cases where specific measurements have been made, such as methanol and methanol–water. Such measured potential differences ascribed to pH offer a sound operational pH scale for the nonaqueous or mixed solvents, but the pH so measured is not directly comparable to an aqueous pH measured by the same procedure.

9.6 REFERENCES

General

G1   R. G. Bates, *Determination of pH: Theory and Practice,* 2nd Ed., John Wiley, New York, 1973. Best work on the subject.

G2   M. Dole, *The Glass Electrode*, John Wiley, New York, 1941.

G3   G. Eisenman, Ed., *Glass Electrodes for Hydrogen and Other Cations, Principles and Practice*, Marcel Dekker, New York, 1967.

G4   C. C. Westcott, *pH Measurements*, Academic Press, New York, 1978, 172 pp. Concentrates upon the practical side of pH measurement.

G5   N. Linnet, *pH Measurements in Theory and Practice*, Radiometer, Copenhagen, 1970, 188 pp. A practical handbook.

Specific

S1   S. P. L. Sorensen, *Biochem. Z.* **21**, 131 (1909).

S2   S. P. L. Sorensen and K. Linderstrom-Lang, *Compt. Rend. Trav. Lab. (Carlsberg)* **15**, (6) (1924).

S3   G. J. Hills and D. J. G. Ives, in *Reference Electrodes*, D. J. G. Ives and G. J. Janz, Eds., Academic Press, New York, 1961, pp. 71 ff.

S4   M. Cremer, *Z. Biol.* **47**, 562 (1906).

S5   F. Haber and Z. Klemenciewicz, *Z. phys. Chem.* **67**, 385 (1919).

S6   G. Haugaard, *Nature* **140**, 66 (1937); *J. Phys. Chem.* **45**, 148 (1941).

S7   K. Schwabe and H. Dahms, *Monatsber. Deut. Akad. Wiss.* **1**, 279 (1959).

S8   R. G. Bates, in *Reference Electrodes*, D. J. G. Ives and G. J. Janz, Academic Press, New York, 1961, pp. 332ff.

S9   D. A. MacInnes and M. Dole, *Ind. Eng. Chem. Anal. Ed.* **1**, 57 (1929); *J. Amer. Chem. Soc.* **52**, 29 (1930).

S10   G. A. Perley, *Anal. Chem.* **21**, 391, 394, 559 (1949).

S11 L. Kratz, *Glastech. Ber.* **20**, 305 (1942).

S12 G. J. Janz and D. J. G. Ives in *Reference Electrodes*, D. J. G. Ives and G. J. Janz, Eds., Academic Press, New York, 1961, pp. 270ff.

S13 D. J. G. Ives, *ibid.*, pp. 332ff.

S14 B. R. Staples and R. G. Bates, *J. Res. Nat. Bur. Stand.* **73A**, 37 (1969).

S15 V. E. Bower, R. G. Bates, and E. R. Smith, *J. Res. Nat. Bur. Stand.* **51**, 189 (1953).

S16 R. G. Bates, V. E. Bower, and E. R. Smith, *J. Res. Nat. Bur. Stand.* **56**, 305 (1956).

S17 R. G. Bates, *J. Res. Nat. Bur. Stand.* **66A**, 179 (1962).

9.7 STUDY PROBLEMS

9.1 The pH of a certain aqueous solution is measured and found to be roughly 5.0. Give in detail an experimental procedure using the double-point interpolation method for measuring its pH. What would be the expected precision of this measurement?

9.2 An earlier chapter of this text stated that the Debye–Huckel limiting law is in excellent agreement with experiment up to an ionic strength of 0.01 in aqueous solution. Compute the activity of H^+ in a solution of ionic strength 0.01 in which the concentration of H^+ is 1.0×10^{-4} mol/dm^3. What is the difference between pH as measured on the activity and concentration scales?

CHAPTER
10
ELECTRODES SELECTIVE FOR
IONS OTHER THAN HYDROGEN

Electrodes of the first, second, or third kind, whether based on metal–metal ion couples or not, are often sensitive to the concentrations of ions other than hydrogen; so also are noble metal redox electrodes. These have already been discussed in Chapter 8, and the present chapter is therefore restricted to membrane potentials sensitive to ions other than hydrogen, popularly called ion-selective electrodes. The uses of these electrodes have been systematically reviewed by Buck (S1–S4). An applications bibliography for their product ion-selective electrodes is maintained by Orion Research, Inc., and is available on request.

In general, the term *ion-selective electrode* (Lakshminarayanaiah, G5) is applied only to electrodes that measure membrane potentials, which in turn are responsive to the concentrations (more correctly, activities, though the distinction is frequently of little importance) of ions other than hydrogen. The distinction of hydrogen is academic rather than real and will probably disappear in time.

The general theory of membrane potentials discussed in Chapter 8 (Hills, G6) applies to ion-selective electrodes. That is, the measured potential is the sum of the potentials of the electrodes, potentials of the liquid junctions, and the potential difference across the membrane:

$$E = \Delta E_{membrane} + E_1 - E_2 + \Sigma E_j \tag{10.1}$$

where E_1 is the potential of the electrode inside the membrane, E_2 is the potential of the electrode outside the membrane, and the E_j values are the liquid junction potentials of the cell. If the potentials of the liquid junctions and the electrodes are constant, as is generally the case, then

$$E = E' + \Delta E_{membrane} \tag{10.2}$$

where

$$\Delta E(membrane) = RT/zF \ \ln(a^+_{outside}/a^+_{inside}) \tag{10.3}$$

The notation a^+ is a summary notation for all of the ionic activities outside or inside the membrane. It is characteristic of membranes in general that they respond to a concentration difference in the activities of any one or more ions across them by establishing a potential difference. It is the purpose of the development of *ion-selective* membranes to ensure that the response in terms of potential is very much greater for some ions than for others so that variations in the activities of certain ions will be sensed while variations in the activities of others are not. In other words, if the value of $RT/zF \ln (a^+_{inside})$ is incorporated into the potential E'', which is legitimate because the composition of the solution on one side of (usually inside) the electrode does not change, then the summary notation of the Nernst equation for univalent ions is

$$E = E' + RT/F \ln[\Sigma_i K_i a_i] \tag{10.4}$$

For multivalent ions, the charge becomes important; IUPAC (S5) has recommended use of the form

$$E = E' + 2.303RT/Fz_A \log \Sigma_i K(A,i)a_i^{z_A/z_i} \tag{10.5}$$

where E, the measurable or experimentally observed potential of a cell, is given in terms of the usual constants and:

a_i is the activity of the ith ion (taken to the power z_A/z_i in order to compensate for the appearance of z_A in the prelogarithmic term)

A is the principal ion to which the electrode's response is related

$K(A,i)$ is the *selectivity coefficient* for the ith ion over the principal ion A. Since this constant is an equilibrium constant and is unrelated to rates, the upper case letter is used, following Koryta (G4).

z_A is an integer corresponding in sign and in magnitude to the charge of the principal ion A

z_i is an integer corresponding in sign and magnitude to the charge of the ith ion i.

This selectivity coefficient for the principal ion over itself is unity when the response of the electrode is Nernstian. When the response of the electrode is less than Nernstian, the selectivity coefficient of the principal ion over itself, $K(A,A)$, must become less than unity. Selectivity coefficients for interfering ions should be less than that for the principal ion over itself.

This terminology, which is adopted in the following discussion, replaces older terminology couched in terms of response factors. The constant E' includes the standard or zero potential of the indicator electrode, the potential of the reference electrode, and all of the junction potentials. The characteristics of useful ion-selective membrane electrodes, including the glass pH electrode, are as follows.

First, the selectivity coefficient for the principal ion, or ion of interest, is unity, or approximately so. In other words, the electrode response for the principal ion is Nernstian. In general, the response will be Nernstian only over a certain range of concentrations. Second, the selectivity coefficient for the principal ion, unity, is much greater than the selectivity coefficients for all other ions that may be present in the solution. In the above equation, this would be true when, for all $K(A,i)$ except $K(A,A)$, $K(A,i)$ is much less than 1.

The selectivity coefficients of an electrode can be measured using either of two methods, the *separate solution method* and the *fixed interference method*. In the separate solution method, the potential difference between the membrane electrode and the reference electrode is measured in each of two separate solutions. One contains the principal ion A at activity $a(A)$, but contains no ion B; the second contains the interfering ion B at this same activity $a(A)$, but contains no ion A. If the potential difference in the first case is E_1 and the second E_2, then by suitable substitution and rearrangement of the preceding equation a relationship giving explicitly log $K(A,B)$ is obtained:

$$\log K(A,B) = ((E_2 - E_1)/2.303\ RT/zF) + [1 + z(A)/z(B)] \log a(A) \quad (10.6)$$

The fixed interference method, which is more realistic since the solutions of interest are those in which the principal ion and the interfering ion appear together, is carried out using a more extensive series of solutions, each of which has the same activity of the principal ion $a(A)$, which must generally be varied over several orders of magnitude. A plot of the potential differences observed against the activity of the principal ion is made. Such a plot consists of a horizontal linear region in which the potential of the electrode is determined by the (fixed) activity of the interfering ion $a(B)$ and an ascending and generally Nernstian linear region in which the potential of the electrode is determined by the activity of the principal ion $a(A)$. Between these linear regions there is curvature, for here the electrode responds effectively to both ions. Extrapolation of both linear portions to intersection gives the value of $a(A)$ for which the electrode response to the two ions is equal, and at this point

$$a(A) = K(A,B)a(B)^{z(A)/z(B)} \quad (10.7)$$

or, more usually,

$$\log a(A) = \log K(A,B) + [z(A)/z(B)] \log a(B) \quad (10.8)$$

It should be noted that the selectivity coefficient is not in general constant, and that the term *selectivity constant* is therefore discouraged. The term *selectivity factor* has the same meaning as selectivity coefficient.

10.1   TYPES OF USEFUL ION-SELECTIVE ELECTRODES

All useful ion-selective electrodes contain an internal reference electrode and use an external reference electrode in the solution being analyzed. It is desirable, though not necessary, that they be the same.

The classification by type of membrane given here follows, with minor extensions, that recommended by IUPAC (S6), which divides ion-selective electrodes into three major categories: those in which the membrane is a solid; the liquid–liquid membrane electrodes; and special types of electrodes. Of these categories, the first is the largest, simplest, and of greatest analytical importance.

Solid-membrane electrodes may be either homogeneous or heterogeneous. Homogeneous solid membranes that have been used are of three types. The first is the *glass electrodes* in which the sensing membrane is a thin piece of glass. *Nonporous membrane electrodes* can also be constructed in which a membrane is a solid homogeneous mixture of an inert polymeric support such as polyvinyl chloride and some ionic or uncharged species. This group has received comparatively little study.

10.1.1   Glass-Membrane Electrodes

There are essentially three types of glass-membrane electrodes, which have different selectivity coefficients. The *pH-type glass membrane electrode* has selectivity coefficients of $H^+ >>> Na^+ > K^+, Rb^+, Cs^+ >> Ca^{2+}$ and other divalent ions. For most practical purposes, these electrodes respond to H^+ alone. The selectivity coefficients $K(H^+, Na^+)$ may be as small as 10^{-15} although values several orders of magnitude larger are more typical.

The *sodium-type glass membrane electrode* has selectivity coefficients $Ag^+ > H^+ > Na^+ >> K^+, Li^+, Rb^+, Cs^+ >> Ca^{2+}$ and other divalent ions (Table 10.1). The interference of Ag^+ can easily be removed by precipitation, so it is usually of no consequence. The pH must, however, be controlled if reasonable results are to be obtained. In general, the concentration of H^+ must less than 0.01 to 0.0001 times that of Ag^+ or Na^+; the selectivity of the electrode for any ion *against hydrogen ion* is poor. The selectivity coefficient $K(Na^+, K^+)$ is about 10^{-3} to 10^{-5} depending on the particular glass used and the manufacturer. Sodium electrodes are in general less sensitive than pH electrodes, slower in response, and less stable over a long term. Suppression of hydrogen ion by addition of $Ca(OH)_2$ or $Ba(OH)_2$ is normally sufficient to prevent its interference. The response to silver ion permits use of this electrode in following argentometric titrations. Similar glasses sensitive to lithium ions are used, and for these $K(Li^+, Na^+)$ is about 0.3, and $K(Li^+, K^+)$ is about 0.001.

The *cation-type glass membrane electrodes* have selectivity coefficients for $H^+ > K^+, Na^+ > NH_4^+, Li^+, Rb^+, Cs^+ > Ca^{2+}$, and so on. As with the sodium-type

TABLE 10.1   Selectivity Coefficients of Commercial Sodium Glass Electrodes[a]

Electrode	$K(Na^+,Ag^+)$	$K(Na^+,H^+)$	$K(Na^+,Na^+)$	$K(Na^+,K^+)$	$K(Na^+, NH_4^+)$
Beckman 39278	370	63	0.97	0.02	1.5×10^{-5}
EIL GEA 33C	200	22	0.98	0.002	5×10^{-5}
Orion 94-11	400	60	0.98	0.01	6×10^{-5}
Radiometer G502	100	10	1.00	0.001	3.5×10^{-5}

[a] Values are for pNa = 1 to pNa = 4 to 6 depending upon electrode; values are at 25°C and pH 7.8 ± 0.2 except for $K(Na^+,H^+)$, which is at pH 5.2. Adapted from Wilson, Haikola, and Kivalo (S29, S30).

electrode, the interference of hydrogen ion must be controlled. The selectivity coefficient $K(K^+,Na^+)$ is comparatively low, about 0.05 or perhaps slightly more after prolonged soaking, so that sodium ion is a serious interference unless its concentration can be held constant. The cation-type electrode will, in the absence of potassium ion, respond to the sodium ion about as well as the sodium-type electrode and can be used for such determinations. In the absence of *both* sodium ion *and* potassium ion (an unusual situation), the cation-type electrode responds to NH_4^+, Li^+, Tl^+, Cu^+, R_4N^+, Rb^+, Cs^+, and Ag^+ well enough to serve as the sensor for potentiometric titrations of these ions.

Glass-membrane electrodes can be used in partially aqueous solvents including ethanol, acetone, ethylene glycol, and formamide mixtures with water. In these media they retain their selectivity for particular ions, although a change in selectivity coefficients is usually found.

It can be seen from the selectivity coefficients that hydrogen ion is a serious interference for all glass ion-selective electrodes. Buffering of solutions is mandatory for precise ion-activity measurements to ensure that the potential due to hydrogen ion is constant. Removal of hydrogen ion by addition of $Ca(OH)_2$ or $Ba(OH)_2$ is an alternative. Use of ion-selective electrodes for following the course of a titration is often less sensitive to other ions than is a direct measurement, and their use in potentiometric titrations is significant.

In practice, glass solid-membrane electrodes are used as sensors for H^+, Na^+, K^+, and a few other monopositive cations both for direct activity measurements and in titrations.

10.1.2   Homogeneous Crystal-Membrane Electrodes

Crystal membrane electrodes are ion-selective electrodes in which the membrane is a solid crystalline material, either a single compound such as Ag_2S or a homogeneous mixture of compounds such as AgI/Ag_2S. They may be single crystals, cast disks, or pressed pellets of an ionic salt that is soluble only to a very slight extent in water. The construction of typical electrodes is shown in Figure 10.1. Examples are the silver halides, Ag_2S, PbS, CdS, $AgSCN$—Ag_2S, and LaF_3 (doped with EuO). The ions that can be determined using solid-state electrodes include F^-, Cl^-, Br^-, I^-, CN^-, SCN^-; S^{2-}; Ag^+; Cu^{2+}, Pb^{2+}, and Cd^{2+}.

Crystal-membrane electrodes can be classified as electrodes of the first, second, or third kind in the same manner in which electrode potentials were classified in Chapter 9. The response referred to is actually that of the membrane, since the internal and external reference electrodes maintain constant potential. Thus a crystal-membrane electrode of the first kind is one in which the ion sensed is the ion that is primarily responsible for transport of ionic current through the membrane; a crystal-membrane electrode of the second kind is one in which the ion sensed and the ion transported are linked by a single equilibrium; and a crystal-membrane electrode of the third kind is one in which the ion sensed and the ion transported are linked by two equilibria. The equilibrium constants are usually solubility products, since only materials of low solubility can be used for membranes.

The most important example of a *crystal-membrane electrode of the first kind* is the fluoride electrode developed by Frant and Ross (S22) in 1966. The electrode consists of a single crystal of LaF_3, whose structure is such that only the fluoride ion can migrate through the lattice. The ionic conductivity is increased by doping with EuO, which decreases the resistance from several megaohms down to less than about 100,000 Ω. The internal filling solution is about millimolar in fluoride ion as well as containing the ions necessary for the internal reference electrode.

The fluoride electrode responds rapidly in the Nernstian manner to fluoride ion activity from 1 to 10^{-5} mol/dm^3 fluoride; some response is obtained down to 10^{-6} mol/dm^3. The most serious interference is hydroxide ion, which is responsible for the concentration limit; $K(F^-, OH^-)$ is about 0.1. The only other significant interferences are the halides; $K(F^-, X^-)$ is about 0.001.

The most important example of a *crystal-membrane electrode of the second kind* is the silver sulfide membrane electrode. Below 176^0C, silver sulfide exists in a stable monoclinic structure that exhibits ionic conduction. Only the silver ion migrates in this compound. Either single crystals of silver sulfide or, more commonly, pressed polycrystalline silver sulfide can be used; the polycrystalline form has good mechanical properties. The internal filling solution must contain Ag+ in addition to the ions required for the reference electrode. Alternatively, the

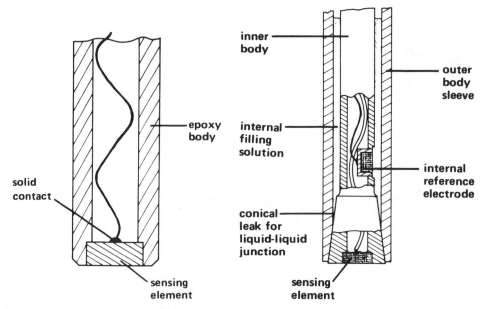

Figure 10.1   Homogeneous crystal-membrane electrodes. Design on left is ISE alone, while design on right includes the ISE and reference electrode in the same probe body. Courtesy of Orion Research, Inc.

internal filling solution and reference electrode can be replaced by a metallic silver contact; the mechanism by which this operates are not known, but the response of the electrode is still Nerstian.

The silver sulfide electrode responds both as an electrode of the first kind to silver ion activity and as an electrode of the second kind to sulfide ion activity. In the latter case the response is through the solubility product of Ag_2S, about 10^{-50} in molar units. Only cyanide ion is a significant interference to sulfide, owing to the formation of the very stable cyanide complex of silver ion, at high cyanide concentrations. Mercury(II) is a significant interference to Ag^+ because of the low solubility product of HgS (about 10^{-52} in molar units, as compared to 10^{-50} for Ag_2S). In the absence of silver ions, the electrode does respond to Hg^{2+}; it has occasionally been used as a sensor for this ion.

If the membrane is pressed from a mixture of Ag_2S and AgCl, a sensor for chloride ion is obtained. At the electrode interface, the activity of Ag^+ is determined almost entirely by the solubility product of AgCl (about 10^{-10}), which is much greater than the solubility product of Ag_2S, and by the concentration of chloride ion in the solution. The membrane thus acts as if it were permeable to chloride ion although it is really silver ion that moves. Membranes of single crystals of AgCl, AgBr, and AgI serve as similar sensors. These materials can also be cast (AgCl, AgBr) or pressed (AgI) from polycrystalline material. All of these types are available commercially. These electrodes are also crystal-membrane electrodes of the second kind.

A *crystal-membrane electrode of the third kind* can be prepared from a mixture of Ag_2S and CuS. The response of the electrode is still to Ag^+, but this is now linked to the activity of Cu^{2+} through the two solubility products, $Ag_2 S$ (10^{-50} and CuS (10^{-36}). The relationship is

$$a(Ag^+) = [a(Cu^{2+}) \ K_s(Ag_2S)/K_s(CuS)]^2 \qquad (10.9)$$

and incorporation of the constants into the standard potential gives the useful form of the Nernst equation at 25⁰C:

$$E = E' + 0.05915/2 \log a(Cu^{2+}) \qquad (10.10)$$

Similar types of electrodes can be prepared for other ions. The criterion is simply a solubility product for the metal sulfide which is larger than that of silver sulfide (so that dissolution of the metal sulfide rather than that of silver sulfide establishes sulfide activity in the solution at the electrode surface), as well as a solubility product low enough so that massive dissolution of the metal sulfide does not occur.

10.1.3   Heterogeneous Crystal-Membrane Electrodes

These electrodes have been developed primarily by E. Pungor of Hungary and his colleagues. They consist of silicone rubber, or some other hydrophobic binder, holding a precipitate of ionic material such as silver halide, which is the actual ion-selective membrane material. Heterogeneous crystal-membrane electrodes have been reported for F^-, Cl^-, Br^-, I^-, S^{2-}, Ag^+, SO_4^{2-}, and PO_4^{3-} ions, but the SO_4^2 and PO_4^{3-} ion electrodes may not be operable. There appears to be little advantage to heterogeneous membrane electrodes over solid-state membrane electrodes, although the former are reported to be somewhat less sensitive to oxidizing and reducing agents than are silver–silver halide-type electrodes, the other sensor commonly used for halides.

10.1.4   Homogeneous Noncrystal-Membrane Electrodes

Homogeneous noncrystal-membrane electrodes are very similar to liquid-membrane electrodes, except that the liquid ion exchanger is held in a homogeneous plastic matrix; in many cases the same ion exchanger can be used in both forms. The polyvinyl chloride (PVC) matrix-preparation procedure of Moody and co-workers (S7–S9) is to dissolve the liquid ion exchanger in tetrahydrofuran, dissolve polyvinyl chloride in the resulting solution, and slowly evaporate the tetrahydrofuran solvent from the solution held in a glass ring on a glass plate at room temperature. Coated wire electrodes can be prepared by dipping a platinum wire or graphite rod in similar solutions (followed by insulation of the uncoated portion of the wire or rod).

Commercial nitrate and perchlorate electrodes (Orion, Corning) can be of this type; response to other ions is also possible. Linear and approximately Nernstian response is obtained down to about 10^{-4} mol/dm^3 in the appropriate ion. The electrodes respond most satisfactorily in neutral or weakly acidic solutions; this response limitation is due to the ion exchanger rather than the matrix. Plastics other than polyvinylchloride have been used, but PVC appears to be the most convenient and widely used matrix.

10.1.5   Liquid-Membrane Electrodes

Liquid-membrane electrodes are electrodes in which a porous support separates an aqueous phase from a nonaqueous phase. The porous support is saturated with a cationic, anionic, or uncharged species. The response of these electrodes is due to the presence of this species in the membrane.

If cations are dissolved in a suitable organic solvent and held in an inert support, as can be done with bulky cations such as those of quaternary ammonium salts or of transition-metal complexes like 1,10.phenanthroline complexes, the membranes are sensitive to changes in the activity of anions. Conversely, if anionic complexing agents or bulky anions are held, the membranes are sensitive to changes in the activity of cations. The construction of an electrode of this type is shown in Figure 10-2. The ion-exchange liquid must not be appreciably soluble in the aqueous test and reference solutions. The two electrodes of this type that have seen the most use are those sensitive to Ca^{2+} and to Cu^{2+}. The former, available from Corning and Orion, use a calcium organophosphorus compound in the ion-exchange liquid membrane. The major interferences are hydrogen ion (useful only in the pH range 5.5 to 11); Sr^{2+}, for which $K(Ca^{2+},Sr^{2+})$ is 0.014; and magnesium and barium, for which the selectivity coefficients are somewhat lower. The selectivity coefficients $K(Ca^{2+},Na^+)$ and $K(Ca^{2+},K^+)$ are about 3×10^{-4}. For the Cu^{2+} electrode, hydrogen ion and Fe^{2+} interfere quite seriously while Ni^{2+} interferes only moderately.

It is also possible to base ion-selective electrodes on solutions of neutral molecular carriers of cations such as antibiotics, macrocyclic compounds, or other sequestering agents, which have a selectivity based upon preferential complexation of certain cations by the molecular carrier. The best-known example of this is the valinomycin-based electrode (S10, S11) which has a selectivity coefficient $K(K^+, Na^+)$ of 2×10^{-4}, almost two orders of magnitude more favorable than that of the best glass electrode.

10.1.6   Gas-Sensing Electrodes

Gas-sensing electrodes (S12), the first of the two important classes of special electrodes, are those in which a gas-permeable membrane or air gap is used to separate the solution being sampled from a thin film of an *intermediate solution*. This film is either held between the gas membrane and the ion-sensing membrane of the electrode or placed on the surface of the electrode using a wetting agent, as

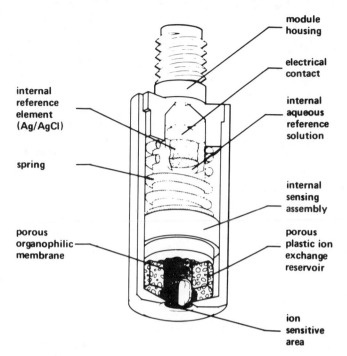

Figure 10.2 Liquid-membrane electrode. Courtesy of Orion Research, Inc.

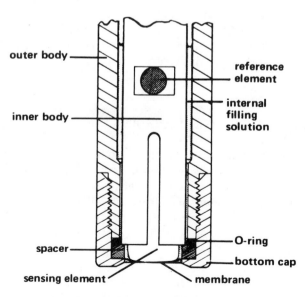

Figure 10.3 Gas-sensing electrode. Membrane separates internal filling solution from sensing element while permitting passage of gas. Courtesy of Orion Research, Inc.

in the air-gap electrode. This intermediate solution interacts with a gas species so as to produce a change in some property of the intermediate solution. This property is sensed by the ion-selective electrode, and is directly proportional to the partial pressure of the gas in the sample. (See Figure 10.3.)

One useful example (S13, S14) is the carbon dioxide-sensitive electrode, which consists of a normal pH-sensitive glass electrode. The intermediate solution, which is retained around the glass electrode by a gas-permeable membrane, is a solution of sodium hydrogen carbonate. Carbon dioxide diffusing into or out of the sodium hydrogen carbonate solution changes the pH of the intermediate solution, and the glass electrode responds to that pH change. Similar gas-sensing electrodes have been developed for ammonia, chlorine, and many other gases. Although IUPAC includes the oxygen electrode under this classification, this sensor operates on an amperometric principle rather than a potentiometric one and will be discussed in Chapter 18.

10.1.7   Enzyme-Substrate Electrodes

Enzyme-substrate electrodes, the second of the two important classes of special electrodes, are sensors in which an ion-selective electrode is covered with a coating that contains an enzyme. This enzyme catalyzes the reaction of an organic or inorganic species reaching the electrode to produce some species to which the ion-selective membrane responds. The first enzyme-substrate electrode was the glucose electrode developed by Clark and Lyons (S15) in 1962. Glucose oxidase was immobilized between membranes, and the hydrogen peroxide formed was determined amperometrically. Guilbault and co-workers (S16) devised electrodes of the potentiometric type for compounds such as urea, whose reaction to ammonium ion is catalyzed by a solution of the enzyme urease held in a polymer on the surface of a glass electrode responding to univalent cations. An interesting development of this work is use of living bacterial cells in place of the enzyme in the substrate.

10.2   USAGE OF ION-SELECTIVE ELECTRODES

Ion-selective electrodes can be used in any aqueous or partially aqueous medium; the significant junction potentials involved and the lesser interest in such media have led to little or no development of them in nonaqueous media. Typical aqueous media include seawater and biological fluids. Much of the interest in ion-selective electrodes has come from the latter area (S17), since physiological processes depend upon the activities of ions. Investigations within individual cells have been carried out (S18).

Ion-selective electrodes of all types must be employed with some attention to the properties of the sensor and the medium in which it is used. Since the measurement requires not only an ISE but also a reference electrode in the same solution, care must be taken to prevent contamination of the solution by leakage of

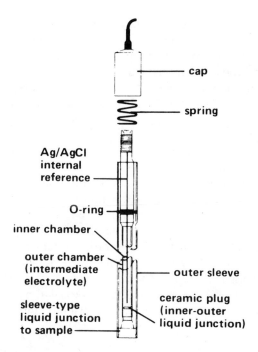

cap

spring

Ag/AgCl
internal
reference

O-ring

inner chamber

outer chamber
(intermediate
electrolyte)

outer sleeve

ceramic plug
(inner-outer
liquid junction)

sleeve-type
liquid junction
to sample

Figure 10.4 Reference electrode with isolation of reference solution. Used to prevent contamination of the solution being analyzed with the chloride ion of the inner Ag/AgCl reference solution; the intermediate solution anion is normally nitrate or sulfate. Courtesy of Orion Research, Inc.

a significant ion from the reference electrode junction. The electrolyte of the reference electrode normally contains chloride, and when this is a significant contaminant a reference electrode containing an intermediate electrolyte, often nitrate or sulfate, should be used, as shown in Figure 10.4.

The direct reading of an ion-selective and reference electrode pair in horse manure or horseradish is unlikely to produce an analytically meaningful result. To obtain useful results, three areas must be considered: general properties of ion-selective electrodes, interferences *specific to the particular ion-selective electrode*, and interferences *specific to the method*. Interferences specific to the method are those related to the properties of the ion being sensed or to the medium in which the ion-selective electrode is being used.

10.2.1   General Properties of Ion-Selective Electrodes

An ISE–reference-electrode pair responds to activity rather than to concentration. When activity is desired, the solution generally should not be altered prior to analysis, because the ionic strength, and thereby the activity coefficients of ions present, will be changed by addition of reagents. Standards of activity rather than concentration should be used. If it is necessary to add reagents to the solution prior to analysis to remove interferences specific to the particular ISE or for other reasons, the activity effects of these may be significant. It is more common for concentration to be desired, rather than activity. Adjustment of the ionic strength to a constant value is then mandatory to remove variation in activity coefficients. If the *total* concentration of a species rather than the concentration of the free ion is desired, additional reagents must be added to decomplex or dissociate other species to the free ion. These additional reagents will also contribute to the ionic strength.

The range of an ion-selective electrode is usually given as the range over which the concentration of principal ion can be measured with an essentially Nernstian response from the electrode. The high end of the range is due to the effect of activity coefficient changes at high concentrations and is not a serious limitation since it can be avoided by dilution with an inert electrolyte of fixed ionic strength. The low end of the range is imposed by the electrode seeing itself, as the loss of ions from the electrode controls the concentration of ions being sensed. This limit can be improved by lowering K_s, as by reducing the temperature or shifting to a wholly or partially organic solvent. The detection limit of the electrode is given by the point of intersection of the horizontal and Nernstian portions of the potential logarithm of concentration response curve as shown in Figure 10.5.

10.2.2   Interferences Specific to the Electrode

Interferences specific to the electrode arise from two sources: the material of the sensing element and the selectivity coefficients of the electrode for different interfering ions. For example, the sodium-type glass electrode has a serious hydrogen ion interference so the pH of a solution being analyzed for sodium ion is made basic with NH_3, $Ca(OH)_2$, or $Ba(OH)_2$. The fluoride ISE has interferences from both OH^- in basic solution and H^+ (from formation of HF) in acid solution; use of an intermediate pH buffer such as acetate avoids both trouble regions. The sulfide electrode responds to free sulfide, and in acidic solution HS^- and H_2S are formed; addition of NaOH to give a solution whose pH is greater than 11 eliminates this problem.

10.2.3   Interferences Specific to the Method

Handling method interferences requires knowledge of the chemistry of the medium and of the ion being sensed. For example, Al^{3+} forms a stable complex with F^- and if present is an interference; addition of a complexing agent for aluminum can release the fluoride. Atmospheric oxygen can oxidize sulfide ion; this interference is prevented by addition of ascorbic acid. On the other hand, sulfide is itself an interference to a chloride ISE, and addition of an oxidizing agent (H_2O_2, $NaNO_2$) is used to remove it. No general rules can be given for the handling of method interferences, and a general knowledge of the chemistry of the ions plus experience and the literature are the only useful guides.

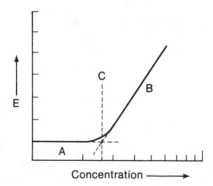

Figure 10.5   Detection limit of ion-selective electrode. A, horizontal region (electrode sees itself); B, Nernstian region; C, detection limit.

10.2.4   Advantages of Ion-Selective Electrodes

Ion-selective electrodes have many advantages over other methods of analysis. They are specific for certain ions or compounds, selecting for them against all others. They are comparatively low in cost. They are nondestructive sensors. They are reasonably portable and can be used in the field. Finally, they are useful for ions whose analysis by other procedures is difficult, such as Na^+, K^+, F^-, and Ca^{2+}, and for ion-ionic materials as well as by indirect methods.

10.2.5   Disadvantages of Ion-Selective Electrodes

Ion-selective electrodes have certain disadvantages that are inherent in the electrodes themselves. One of these is *drift*, the slow change of potential with time in a solution of constant composition and temperature. Drift is characteristic of the electrode itself and little effective remedial action can be taken. A more serious problem is *hysteresis*; some electrodes when immersed in a standard solution, placed in an unknown solution, and then restored to the standard will not give the same potential as originally observed in the standard solution. This error is systematic, being in the direction of the concentration of the solution in which the electrode was previously immersed, and cannot be effectively remedied. Other problems may arise from the specific electrode or medium being studied.

10.2.6   Procedures for Ion-Selective Electrodes

The methods of direct potentiometric measurement will be discussed here. Ion-selective electrodes can also be used to follow the course of a titration (Chapter 12).

Direct potentiometry using the Nernst equation is rarely used because of the high error involved; an error in potential of ± 1 mV leads to an error of $\pm 4\%$ in the direct determination of a univalent ion and $\pm 8\%$ in the direct potentiometric determination of a divalent ion at 298 K. It is more usual to prepare a full calibration curve for the particular electrode or to bracket the unknown with two calibration points as closely as possible.

The *standard addition method* is more commonly used. In this procedure, the concentration of a particular species in a solution is determined by adding small known amounts of that species to the solution and recording the change in potential of the ion-selective electrode as a function of the amount added; extrapolation to zero addition gives the original concentration. A similar *standard subtraction method*, in which the known additions are those of some species that quantitatively reacts with the species of interest, such as a complexing agent, is also known.

Standard addition and standard subtraction methods may also be carried out using a single addition of a known number of moles added to the unknown solution. This method is inferior, in both precision and accuracy, to that of multiple additions of standard, but is more convenient. The number of moles of the standard addition should be close to (ideally the same as) the number of moles of the sought-for species present; the number of moles of a standard subtraction should be enough to reduce the concentration to close to (ideally exactly) half of its original value.

The method of *null-point potentiometry* has been found useful with ion-selective electrodes. This method requires two electrodes, preferably identical, immersed in two solutions with identical reference electrodes. The standard solution contains the sought-for constituent at a known concentration or activity, the unknown solution is the solution to be analyzed. The addition of a different standard solution to either standard or unknown is carried out until the pairs of electrodes in the standard and unknown solutions read the same potential. Since this is the point at which their potential difference is null, a double-input amplifier chopping between the two inputs, such as is available on some laboratory oscilloscopes, serves as an effective output device. Null-point potentiometry does not require a stoichiometric, or indeed any, reaction between the standard being added and the sought-for constituent and thus is not a titration procedure.

10.3   STANDARDS FOR USE WITH ION-SELECTIVE ELECTRODES

Standards and procedures for their use are covered in the general references of this Chapter, in Bates and Alfenaar (S19), and in Toth and Pungor (S20).

10.4   REFERENCES

General

G1   H. Freiser, Ed., *Ion-Selective Electrodes in Analytical Chemistry,* Volume 1 Plenum Press, New York, 1978, 438 pp. A balanced collection of six excellent and authoritative review articles.

G2   R. A. Durst, Ed., *Ion-Selective Electrodes*, National Bureau of Standards (U.S.) Spec. Publ. 314, U.S. Government Printing Office, Washington, D.C., 196, 474 pp. Proceedings of a symposium held January, 1969.

G3   G. J. Moody and J. D. R. Thomas, *Selective Ion-Sensitive Electrodes*, Merrow, Watford, England, 1971, 140 pp. A brief, practical work with useful data on commercial electrodes.

G4   J. Koryta, *Ion-Selective Electrodes*, University Press, Cambridge, 207 pp.

G5   N. Lakshminarayanaiah, *Membrane Electrodes*, Academic Press, New York, 1976.

G6   G. J. Hills, "Membrane Potentials," *in Reference Electrodes*, D. J. G. Ives and G. J. Janz, Eds., Academic Press, New York, 1961, pp. 411f.

G7   G. Eisenman, Ed., *Glass Electrodes for Hydrogen and Other Cations: Principles and Practice*, Marcel Dekker, New York, 1967.

G8   P. L. Bailey, *Analysis With Ion-Selective Electrodes*, Heyden, London, 1976, 224 pp. Well-written, concentrating on the practical side.

G9   A. K. Covington, Ed., *Ion Selective Electrode Methodology*, CRC Press, Boca Raton, 1979, 2 vols., 416pp. Handbook of authoritative practical usage.

G10   G. E. Baiulescu and V. V. Cosofret, *Applications of Ion-Selective Membrane Electrodes in Organic Analysis*, Halsted Press, New York, 1977, 235 pp. A brief introduction followed by a functional-group organization of applications.

G11   K. Camman, *Working with Ion-Selective Electrodes*, H. Schroeder, trans., Springer-Verlag, Berlin, 1979, 225 pp.

Specific

S1   R. P. Buck, *Anal. Chem.* **44**, 270R (1972).

S2   R. P. Buck, *Anal. Chem.* **46**, 28R (1974).

S3   R. P. Buck, *Anal. Chem.* **48**, 23R (1976).

S4   R. P. Buck, *Anal. Chem.* **50**, 17R (1978).

S5   *Pure Appl. Chem.* **48**, 127 (1976).

S6   M. S. Frant and J. W. Ross, *Science* **154**, 1553 (1966).

S7   G. J. Moody, R. B. Oke, and J. D. R. Thomas, *Analyst* **95**, 910 (1970).

S8   J. E. W. Davies, G. J. Moody, and J. D. R. Thomas, *Analyst* **97**, 87 (1972).

S9   A. Craggs, G. J. Moody, and J. D. R. Thomas, *J. Chem. Educ.* **51**, 541 (1974).

S10   W. Simon, *Proc. Soc. Anal. Chem.* **9**, 250 (1972).

S11   L. A. R. Pioda, V. Stankova, and W. Simon, *Anal. Lett.* **2**, 665 (1969).

S12   J. W. Ross, J. H. Riseman, and J. A. Krueger, *Pure Appl. Chem.* **36**, 473 (1973).

S13   R. W. Stow, R. F. Baer, and B. F. Randall, *Arch. Phys. Med. Rehabil.* **38**, 646 (1957).

S14 S. R. Gambino, *Clin. Chem.* **7**, 336 (1961).

S15 L. C. Clark and C. Lyons, *Ann. N.Y. Acad. Sci.* **102**, 29 (1962).

S16 G. G. Guilbault, *Pure Appl. Chem.* **25**, 727 (1971).

S17 J. L. Walker, *Anal. Chem.* **43**, 89A (1971).

S18 B. Fleet, G. P. Bound, and D. R. Sandbach, *Bioelectrochem. Bioenerg.* **3**, 158 (1976).

S19 R. G. Bates and M. Alfenaar, in *Ion-Selective Electrodes*, R. A. Durst, Ed., National Bureau of Standards Special Publication 314, U.S. Government Printing Office, Washington, D.C., 1969, pp. 191ff.

S20 K. Toth and E. Pungor, *Amer. Lab.,* June, 9 (1976).

S21 T. M. Hseu and G. A. Rechnitz, *Anal. Chem.* **40**, 1054 (1968).

S22 M. S. Frant and J. W. Ross, *Anal. Chem.* **40**, 1169 (1968).

S23 G. G. Guilbault and J. Montalvo, *Anal. Lett.* **2**, 283 (1969).

S24 G. G. Guilbault and J. Montalvo, *J. Amer. Chem. Soc.* **91**, 2164 (1969).

S25 G. G. Guilbault and J. Montalvo, *J. Amer. Chem. Soc.* **92**, 2533 (1970).

S26 G. G. Guilbault and E. Hrabankova, *Anal. Chim. Acta* **52**, 287 (1970).

S27 G. G. Guilbault and G. Nagy, *Anal. Lett.* **6**, 301 (1973).

S28 G. G. Guilbault and G. Nagy, *Anal. Chem.* **45**, 417 (1973).

S29 M. F. Wilson, E. Haikola, and P. Kivalo, *Anal. Chim. Acta* **74**, 395 (1974).

S30 M. F. Wilson, E. Haikola, and P. Kivalo, *Anal. Chim. Acta* **74**, 411 (1974).

10.5 STUDY PROBLEMS

10.1 A sulfide-sensitive membrane electrode was used in a potentiometric study (S21) of the formation constant of thiostannate ion, which is given by

$$K_f = a(SnS_3^{2-})/a(S^{2-}) \tag{10.11}$$

Thiostannate ion is formed by reaction of solid stannic sulfide with sulfide ion in aqueous solution. Various concentrations of sodium sulfide were added to suspensions of stannic sulfide, and sulfide concentrations were determined potentiometrically. Some of the sulfide hydrolyzed to HS^- so that

$$c(S^{2-}) = c(S^{2-},total)(1 + c(H^+))/K_2 \tag{10.12}$$

$K_2 = 3.635 \times 10^{-15}$ at 25°C and ionic strength $= 0.1$

was necessary to correct for this. The concentration of thiostannate ion was obtained by considerations of mass balance. The concentration of thiostannate ion is equal to the original molar concentration of sodium sulfide less the total molar concentration of all forms of sulfide. The data shown in Table 10.2 were obtained.

TABLE 10.2

[Na$_2$S] original (mmol/liter)	$-E$ (V)	[S^{2-}] $\times 10^8$ (g ion/liter)	pH
8.000	0.6673	0.6310	8.298
12.00	0.6762	1.259	8.450
3.993	0.6732	1.000	8.998

Calculate the mean and standard deviation of K_f. Is the response of the electrode Nernstian or not?

10.2 Frant and Ross (S22) recommend the use of TISAB (total ionic strength adjustment buffer) with the lanthanum trifluoride ion-selective electrode. The components of TISAB are, in concentrations of mole per cubic decimeter: NaCl, 1.0; CH$_3$COOH, 0.25; CH$_3$COONa, 0.75; sodium citrate, 0.001. Explain the function or functions of each of these components.

10.3 The original urea-selective electrode of Guilbault and Montalvo (S23–S25) consisted of the enzyme urease immobilized on the surface of a cation-type glass electrode; the enzyme catalyzed the acid hydrolysis of urea to ammonium ion and carbon dioxide. This electrode could analyze for urea in blood and urine only after pretreatment of the sample with a suitable ion exchanger (S26). A urea-selective electrode using the same immobilized enzyme in which the glass electrode is replaced by a silicone rubber membrane containing nonactin (the internal filling solution being 1.0 mol/dm^3 ammonium chloride and the internal reference electrode AgCl/Ag) does not require sample pretreatment (S27–S28). Explain why.

10.4 The mercury–EDTA electrode of the third kind can be used as a potentiometric sensor for metal ion concentration. Write the explicit Nernst relationship relating electrode potential to the concentration of Ca^{2+} in aqueous solution containing such an electrode. The logarithm of the formation constant of the complex ion HgY^{2-} is 21.8, that of the complex ion CaY^{2-} is 10.7.

10.5 Your employer, Great Satanic Smelting Company, has just been served with a shutdown order by the Department of the Environment on the grounds that vegetation in the vicinity of its plant is being poisoned by its sulfur dioxide emissions. It is true that all plant life in the vicinity does look decidedly unhealthy, but you suspect that the real problem is fluoride poisoning caused by dust particles of cryolite (Na_3AlF_6) from the potlines of Empire Aluminum Company just upwind of your plant. Explain what standards and sampling procedures you might use in conjunction with a fluoride ion-selective electrode to prove that the fluoride content of the soil on and around your plant site is unusually high. Note that although you can extract the soil with anything you want, including digestion with perchloric acid, complete dissolution of the silica component of soil requires treatment with HF (and you can guess what that will do to a fluoride analysis).

10.6 An ion-selective electrode membrane sensitive to bromide is made from a pressed polycrystalline Ag_2S–$AgBr$ mixture. The response of such an electrode is found to be Nernstian to bromide, and in a solution of millimolar bromide concentration has a potential of -43 mV against a suitable reference electrode. As NaSCN is added to the solutions the potential remains constant, until the concentration of NaSCN equals 0.026 mol/dm^3; thereafter the potential becomes more negative by 59 mV for every order-of-magnitude increase in the NaSCN concentration. Explain these observations on the basis of solubility products.

10.7 Solutions suggested as standards for ion-selective electrodes that are sensitive to metal ions include solutions containing metal ion, complexing agent, and a pH buffer. Why should such solutions make better standards than solutions containing the metal ion and pH buffer alone? Draw an appropriate analogy with pH standards.

10.8 Ammonium ion can be determined potentiometrically using ion-selective electrodes of two types. One is the glass electrode. The other is a liquid-phase ion exchange membrane, incorporated into an ion-selective membrane electrode. The latter system employs a neutral carrier molecule that is the mixture of nonactin and monactin obtained directly from actinomycete cultures. This is incorporated in a liquid-phase–PVC membrane (by weight: 4.6% neutral carrier, 68.9% tris(2-ethylhexyl)phosphate, 26.5% polyvinyl chloride). Both can be used as sensors when the ammonium ion is produced by enzymatic action. Describe the reason that these two quite different materials can show ion-selective response to ammonium ion, showing in your answer that you understand both the individual reasons that they can serve as ion-selective electrodes and the difference in the fundamental mechanisms of their response.

CHAPTER
11
TITRATIONS USING EQUILIBRIUM TECHNIQUES

There are two distinguishing characteristics of titrations in general that must be recalled here. First, all titrations depend for their validity on the occurrence of a *stoichiometric chemical reaction* between the titrant species and the sought-for constituent species under the conditions in which the titration is carried out. Second, use of a titration technique requires a means of precisely detecting the *end point*, which should closely correspond to the *equivalence point* of the stoichiometric titration reaction. Only if both of these conditions are met can the titration procedure be used for quantitative analytical work.

11.1 POTENTIOMETRIC TITRATIONS WITH UNPOLARIZED ELECTRODES

Potentiometric titrations with unpolarized electrodes are those titrations in which the equilibrium potential of an electrode in the titration flask is measured during the course of a titration. The potential of the electrode is used to follow the course of the titration reaction.

The theory of potentiometric titrations is simplest when the indicating electrode responds in the Nernstian manner to either the species being titrated or to the titrant species and to these alone. The electrode is then serving as a sensor whose potential is proportional to the logarithm of the activity of the ion sensed, which in dilute solution is proportional to its concentration. The potential of the electrode changes rapidly in the immediate vicinity of the equivalence point because the concentrations change rapidly there, and the usual *S*-shaped form of a titration curve is obtained. The electrode potential at any point on the titration curve can be calculated by insertion of the activities or concentrations prevailing in the solution at that point directly into the Nernst equation appropriate to the indicating electrode. The same Nernst equation will apply at all points along the titration curve regardless of whether it is before, after, or even at the equivalence point.

11.1.1   Acid-Base Titrations

Potentiometric acid–base titrations are followed with a pH electrode, almost invariably a glass electrode. Such titrations are routine analytical practice.

11.1.2   Precipitation Titrations

Potentiometric precipitation titrations can be followed with electrodes that respond to the activities of anions or cations, although cation-sensitive electrodes are somewhat more common. The electrodes used can be of the first, second, or (much less often) third kind, or can be ion-selective electrodes. Since the activities of the ions sensed change very significantly over the course of the titration, care must be taken to make sure solution conditions are never such as to damage the electrode used. Again, these titrations are routinely used for analytical purposes, especially for reactions that form silver halide precipitates.

11.1.3   Complexation Titrations

Potentiometric complexation titrations are generally followed with an electrode sensitive to the activity of a particular cation, frequently an ion-selective electrode. Such titrations are well-suited to potentiometric sensors because the activity of the metal ion changes by orders of magnitude but generally remains very well-defined, and are routine analytical practice. As in the case of precipitation titrations, care must be taken that the sensor is not damaged, as may happen in the presence of a significant excess of complexing agent because material is extracted from the electrode itself.

11.1.4   Redox Titrations

Potentiometric titrations in which the stoichiometric chemical reaction is an oxidation–reduction reaction are normally followed with an inert metal electrode. Platinum or gold is used as the electrode material. Since an inert-metal electrode responds to all redox couples present in the solution, calculations along the course of a redox titration are less simple than for the other types. In the case of a titration of Fe(II) with Ce(IV) to yield Fe(III) and Ce(IV), the potential at any point along the titration curve could in principle be calculated from either the Nernst equation for the Fe(III)/Fe(II) couple or the Nernst equation for the Ce(IV)/Ce(III) couple, but in practice this is not possible. At the beginning of the titration no Ce(IV) has been added, and thus neither the cerium couple nor any determinable concentration of Fe(III) is present. An unstable potential of the electrode is established, by the very low concentration of Fe(III) produced by impurities in the solution. As the titration proceeds toward the equivalence point, an appreciable and determinable concentration of Fe(III) is rapidly established and the Nernst equation of the Fe(III)/Fe(II) couple can be used to calculate the electrode potential. Once beyond the equivalence point, the concentration of Fe(II) is very low, but those of both Ce(IV) and Ce(III) are appreciable and determinable; the electrode potential can therefore be more easily calculated from the Nernst equation for the Ce(IV)/Ce(III) couple.

Calculation of the equivalence-point potential for a redox titration is done from the Nernst equations for both of the couples involved and the physical solution that exists at the equivalence point. At any point in the titration the potential E calculated from either Nernst equation must be the physical potential of the electrode, so the two equations can be summed

$$E = +0.77 - 0.06 \log [a(Fe(II))/a(Fe(III))] \tag{11.1}$$

$$E = +1.61 - 0.06 \log [a(Ce(III))/a(Ce(IV))] \tag{11.2}$$

to give

$$2E = +2.38 - 0.06 \log [a(Fe(II))a(Ce(III))/a(Fe(III))a(Ce(IV))] \tag{11.3}$$

At the equivalence point, and only at the equivalence point, the moles of Ce(IV) added are equal to the moles of Fe(II) originally present. By the quantitative stoichiometry of the titration reaction, these have been quantitatively converted into Fe(III) and Ce(III) so that the concentrations of Fe(III) and Ce(III) are equal at the equivalence point. The quantitative stoichiometry of the reaction also requires the remaining small concentrations of the reactants Fe(II) and Ce(IV) to be equal. The logarithmic term of the summed Nernst equation above is therefore zero, and the equivalence-point potential is $+2.38$ V$/2 = +1.19$ V. For this iron–cerium titration, and in general for any 1:1 titration reaction, the equivalence-point potential is the mean of the standard or formal potentials of the two reacting couples obtained in the manner shown above.

If the titration reaction is not a 1:1 reaction, the equivalence-point potential is a weighted mean of the standard or formal potentials of the two reacting couples. For a general titration reaction between two couples 1 and 2:

$$aOx(1) + bRed(2) \longrightarrow aRed(1) + bOx(2)$$

the equivalence point potential is given by

$$E = [bE(1) + aE(2)]/[a + b] \tag{11.4}$$

Example
The titration whose reaction is

$$Sn^{2+} + 2Fe^{3+} \longrightarrow Sn^{4+} + 2Fe^{2+}$$

has an equivalence point potential given by

$$E = [2E^0(Sn^{4+}/Sn^{2+}) + E^0(Fe^{3+}/Fe^{2+})]/3$$

11.1.5 Problems in Potentiometric Titrations

The first problem that can arise in potentiometric titrations is a slow chemical titration reaction. Slow reactions, which are not unique to *potentiometric* titrations, will adversely affect both precision and accuracy if the analyst is not aware of them. In automated equipment, slow chemical titration reactions must be avoided completely, and in manual work they reduce the speed with which the analysis can be carried out. The general solution for this problem is to change the conditions of the titration so that the reaction becomes fast. If this cannot be done, an alternative method of analysis should be tried.

A second problem is potentiometric titrations is that of mixed potentials. Mixed potentials generally arise only at inert metal electrodes, although under unusual conditions they can be observed at electrodes of other types. A mixed potential arises when:

1. two more possible potential-determining couples are present in the solution at the same time, *and*

2. the conditions in the solution are such that these couples do not rapidly reach equilibrium with each other.

A mixed potential will generally, but not always, be observed to drift with time as the two couples move toward equilibrium. However, a steady state that is not an equilibrium may be established in the immediate vicinity of the electrode, and this may produce a stable potential, so that absence of drift does not guarantee the absence of mixed potentials.

A third problem that can arise in potentiometric titrations is that of electrode polarization. Potentiometric measurements must of necessity draw a finite but small amount of current from the cell whose potential is being measured. When this measurement current is of the same order of magnitude as the exchange current of the potential-determining couple, or larger, erroneous potentials will be observed caused by polarization of the electrode. These potentials often will be ill-defined. In principle, this problem could be avoided by the use of a measuring apparatus requiring less current, but in practice it is sometimes necessary to change the electrode reaction to which the (generally inert) electrode is responding.

11.2 MEASUREMENT OF END POINTS

End points of potentiometric titrations are often measured after the fact from data recorded over the course of the titration until well past the equivalence point. Such measurements can be carried out by computation from stored digital data as well as by hand from a recorded chart, and are performed equally well by man or microprocessor. For rapid work, especially in automated systems, real-time determination of the point of maximum slope or a preset equivalence point potential can be used.

11.2.1   Determination of the Point of Maximum Slope

This method relies upon the automatic or manual plotting of the titration curve, or sometimes its first or second derivative. The point of maximum curve slope is measured. This point is also the point of maximum in the dE/dV curve or the point at which the zero axis is crossed in the d^2E/dV^2 curve. For a symmetric case, which is observed when both couples are reversible and involve the same number of electrons, the three plots are well-defined as shown in Figure 11.1.

If these conditions are not met, the curve is asymmetric, as for the titration of iron with permanganate, which is also complicated by a slow equilibrium. The observed titration curve is shown in Figure 11.2. The reaction

$$MnO_4^- + 5Fe^{2+} + 8H^+ \rightarrow Mn^{2+} + 5Fe^{3+} + 4H_2O$$

is slow but is catalyzed by the manganese dioxide intermediate produced. Actually, in this titration the potentiometric response is probably to the highly reversible Fe^{3+}/Fe^{2+} couple well beyond the equivalence point as discussed by Laitinen (G3). Such behavior is found in other titrations as well.

11.2.2   Titration to (Preset) Equivalence-Point Potential

This approach will work well if the system is well-behaved; it depends upon previous knowledge of the actual equivalence-point potential. The equivalence-point potential can be calculated from the formal potentials of the couples involved as shown earlier in this chapter, or, alternatively, a manual measurement can be made from a stoichiometric standard identical to the equivalence-point solution or a previously titrated solution. Many equivalence-point solutions are not well poised and change potential with time owing to slow reactions or to air oxidation. This can be a problem if such solutions are used for standardization purposes.

11.2.3   Measurement of the Equivalence Point from the Titration Curve

Some electrodes, such as the glass pH electrode, are sensitive to ionic concentrations that are quite low, and accurate measurements can be made in the vicinity of the equivalence point. Many ion-selective electrodes, however, are not

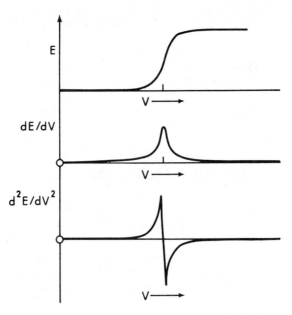

Figure 11.1 First and second derivatives of symmetric potentiometric titration curves.

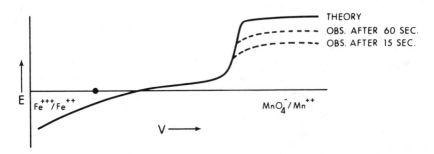

Figure 11.2 Potentiometric curve for titration of Fe(II) with permanganate.

sensitive to very low ionic concentrations such as those present near or at the equivalence point in complexometric or precipitation titrations. These titrations can still be accurately carried out by the method known as Gran's plot (S1) method.

In Gran's plot method, the potential is sensed and recorded after every small addition of titrant; a plot is made of the function antilog $(EF/2.303RT)$ on the vertical axis, and the volume of titrant is added on the horizontal axis. If the response of the electrode is Nernstian over some concentration range, Gran's plot will be linear over that concentration range, and extrapolation to the horizontal axis will yield an accurate value for the equivalence-point volume. Gran's plot will produce an accurate equivalence-point potential even when the titration curve is ill-defined. For greatest convenience, plots on semiantilog paper (available from Orion Research Inc.) are often used rather than calculations at each potential.

11.2.4   Automatic Measurements

The first step in the automation of potentiometric titrations is automatic plotting of the titration curve. This requires an automatic burette and a strip-chart recorder. Automatic burettes are simply hypodermic syringes driven by motor-actuated pistons; they are very precise but not capable of handling volumes over about 50 cm^3. There are two ways of using automatic burettes, of which the most common is a constant-rate burette synchronized with the strip-chart recorder. The only disadvantage of this method is that a slow rate of addition in the end-point region is not easily achieved unless the entire titration is carried out slowly. A second alternative is a variable-rate-drive burette used with an x–y recorder, but this is rarely used because it is expensive.

The next step in the automation of potentiometric titrations is the automatic cutoff of the burette. This requires consideration of many problems such as lead or lag, overshoot, and placement of electrodes. If automatic cutoff is to be used in the procedure of titration to the equivalence potential, the end-point potential must be precisely known. Moreover, overshoot of the end point must be prevented, which usually requires a rapid drive to initial preset point, followed by a slow approach to end point from preset point. This means that mechanical placement of the indicator electrode is critical, as is stirring, because the indicator electrode must lead, not follow, the bulk of the solution. One must also use an electronic circuit that compares the preset potential to the actual potential and shuts off the buret by opening an relay when they are equal.

There are some advantages in the alternative method of titration to the point of maximum slope (S2). These are that the end-point potential need not be known, the indicator electrode is somewhat less critical, and often fewer prior adjustments are necessary. The reaction rate must be reasonably rapid both in the solution and at the electrode, by which latter comment is meant that the exchange-current density at the indicating electrode must be high. However, the cut-off circuit involves taking the derivative of the titration curve, which amplifies high-frequency noise. Filtering circuits must be added to prevent the noise from cutting off the burette prematurely.

11.3   POTENTIOMETRIC TITRATIONS WITH POLARIZED ELECTRODES

Polarized electrodes are electrodes that are at some potential other than their equilibrium or zero-current potential. When the potential-determining couple or couples involved in a potentiometric titration are all rapid, reversible, and characterized by high exchange-current densities there is no advantage to the use of polarized electrodes. Acid-base titrations, precipitation titrations, and complexation titrations generally do not require them. In addition, these titrations

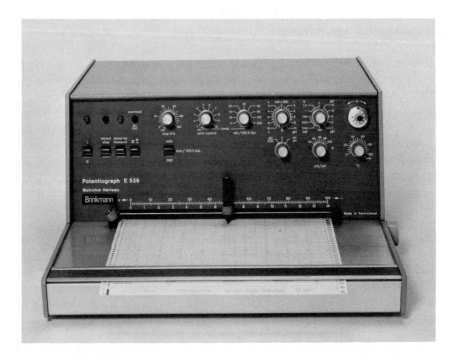

Figure 11.3   Modern Recording Potentiometric Titrator. Metrohm E536 Potentiograph/Recording Titrator, courtesy of Brinkmann Instruments Ltd. This unit plots the titration curve.

are normally followed with ion-selective electrodes, glass electrodes, or perhaps electrodes of the first or second kind; these do not lend themselves to polarization. If the rate of the stoichiometric chemical reaction of the titration itself is slow, the results obtained by polarized and unpolarized electrodes will be equally poor. If, however, one or more of the potential-determining couples are irreversible or have low exchange-current densities, use of polarized indicating electrodes may produce an analytically acceptable procedure when use of unpolarized indicating electrodes does not. Some of the couples frequently used in redox titrations suffer from this problem, such as thiosulfate/tetrathionate, MnO_4^-/Mn^{2+}, and $Cr_2O_7^{2-}/Cr^{3+}$.

Polarized electrodes originated around 1925 in the work of Fenwick (S3, S4). Noble metal electrodes are usually used to follow redox titrations and are the only polarized electrodes normally encountered.

Polarized electrodes can be used in either a single or a double configuration, that is, either one polarized indicator electrode in the solution being titrated or two of them. The first class is actually a normal potentiometric end point, but the double-indicator-electrode configuration is classifiable as a somewhat biased potentiometric system and will be considered here.

In a solution that contains a reversible redox couple, the potential of an inert electrode through which a small current is drawn will be nearly the equilibrium potential. Thus the imposition of a small current between two identical inert-indicator electrodes immersed in such a solution will give rise to only a small voltage difference between them, since the voltage drop IR_p will be small when R_p is low. If, however, the redox couple is either totally irreversible or slow, then R_p will be large, so will IR_p, and the potential difference between the electrodes will be larger.

In the titration of an irreversible couple by a reversible couple, the voltage difference due to a small imposed constant current prior to the equivalence point will be large because the reversible couple, as added, is removed by the stoichiometric titration reaction. The voltage difference will remain high until the reversible couple appears in solution, which will occur immediately following the equivalence point. Such a curve is shown in Figure 11.4A.

The converse, the titration of a reversible couple with an irreversible couple, normally produces an initial fall in voltage as the reversible couple is established at both electrodes, since either the oxidized or the reduced species of the reversible couple, rather than both, is present at the start of the titration. There is then a low voltage observed while the reversible couple is still present, but this voltage rises as the last of the reversible couple is removed just prior to the equivalence point by reaction with the irreversible couple, so that the polarizability of the electrodes is now characteristic of the irreversible couple; the resulting curve is shown as Figure 11.4B.

If a reversible couple is titrated with another reversible couple, the curve prior to the equivalence point is the same as that for the titration of a reversible couple with an irreversible one as discussed above. After the equivalence point, however, the voltage difference falls again as the second reversible couple establishes itself at both electrodes, as shown in Figure 11.4C. It is important to realize that the electrodes after the equivalence point are responding to a different redox couple than they were before the equivalence point. While their potential difference is small (as it was before), their absolute potentials, as could be measured against an unpolarized reference electrode in the same solution, may be significantly changed from their previous values.

11.4 PRACTICE OF POTENTIOMETRIC TITRIMETRY

Redox reactions and reagents have been discussed by Wawzonek (G1) and by Laitinen and Harris (G3) in considerable detail; these excellent treatments include much practical information. The most popular aqueous potentiometric titrations are those carried out in acid solution with strong oxidizing agents. The most commonly used strong oxidizing agents are dichromate, permanganate, cerium(IV), and iodine; the latter can also be used as a mild reducing agent in the form of I_3^-. Oxyhalogen compounds are used to a lesser extent. Titrations using reducing agents are less popular, possibly because of the ease of preliminary reduction of the solution to be titrated with amalgamated zinc, silver, or Sn(II) as well as the problem of preventing solutions of strong reducing agents from reducing atmospheric oxygen.

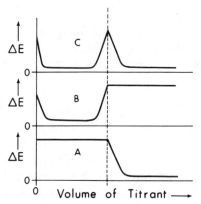

Figure 11.4   Titration curves observed with two polarized electrodes. (*A*) titration of irreversible couple with reversible couple; (*B*) titration of reversible couple with irreversible couple; (*C*) titration of reversible couple with reversible couple. Dashed vertical line indicates theoretical end point.

11.4.1   Nonaqueous Solvents

Potentiometric titrations can be carried out in nonaqueous or partially aqueous solvents; in general, there is no advantage in doing so unless the characteristics of the system being studied require it. Systems under study may require use of a solvent other than water for a variety of reasons, of which solubility is probably the most obvious. In acid–base titrations, an acid or base may be too weak for effective titration in water; then use of some other protic solvent such as acetic anhydride (S5) may change its strength sufficiently to permit a precise titration. In acetic anhydride, such a titration can be followed with a normal glass pH-sensing electrode and a normal AgCl/Ag reference electrode, provided that the glass electrode has been conditioned in acetic anhydride. Use of ion-sensitive electrodes to follow complexation or precipitation titrations in nonaqueous solvents is very similar to their use in water; again, the electrodes must have been conditioned in the medium being used.

11.4.2   Coulometric Titrants

Potentiometric titrations using coulometric titrant generation have the usual characteristics both of potentiometric titration and of coulometric titrant generation. They are, however, subject to unusual interferences because of the necessary presence of multiple electrode pairs with different functions in the same cell. Considerable care must therefore be taken to keep the electrodes of each pair out of the current path of the other. If either of the electrodes of the potentiometric sensor, polarized or not, is in the path between the generating electrodes, the *IR* drop from the generating system will appear in part as a potential at that electrode. If the indicating electrodes are polarized, the constant current flowing between them may adversely affect the coulometric calculations.

11.4.3   Miscellaneous Sensors

Before modern automatic recording potentiometers came into extensive use (about 1945), many ingenious techniques for manual titrations were developed (G2). The majority of these are no longer in use.

11.5   REFERENCES

General

G1   S. Wawzonek, *in Physical Methods of Chemistry,* Part IIA, A. Weissberger and B. W. Rossiter, Eds., Wiley-Interscience, 1971, pp. 1ff. Particularly good on organic redox systems.

G2   J. J. Lingane, *Electroanalytical Chemistry,* 2nd ed., Wiley-Interscience, 1958, pp. 91–167.

G3   H. A. Laitinen and W. E. Harris, *Chemical Analysis,* 2nd ed., McGraw-Hill, New York, 1975, pp. 282–381.

G4   G. Svehla, *Automatic Potentiometric Titrations,* Pergamon Press, Oxford, 1978, 219 pp.

Specific

S1   G. Gran, *Analyst* **77**, 661 (1952); *Orion Newslett.* **2**, 49 (1970).

S2   H. V. Malmstadt and E. R. Fett, *Anal. Chem.* **26**, 1348 (1954); *ibid.,* **27**, 1757 (1955).

S3   H. Willard and F. Fenwick, *J. Amer. Chem. Soc.* **44**, 2504 (1922).

S4   G. Van Name and F. Fenwick, *J. Amer. Chem. Soc.* **47**, 9 (1925).

S5   C. A. Streuli, *Anal. Chem.* **30**, 997 (1958).

11.6   STUDY PROBLEMS

11.1 An automatic titrator is found to designate an end point of an iron–dichromate titration at a potential that is 60 mV lower than the theoretical equivalence potential for the titration. Taking the formal potentials of iron(III)–iron(II) and chromium(VI)—chromium(III) to be, respectively, 0.67 and 1.06 V, estimate the percentage error in the titration. Explain qualitatively why it is valid to use the Nernst equation in the above calculation even though the chromium(VI)–chromium(III) couple does not exhibit reversible electrode behavior.

11.2 A certain automatic potentiometric titrator adds titrant at a constant rate of 0.1 cm³/s. It is designed to shut off when the second derivative of the titration curve is zero, but mechanical delays cause it to shut off 2 s later than this. If you are using this titrator to titrate 50 cm³ of 0.1 mol/dm³ Fe(II) with 0.1 mol/dm³ Ce(IV), what percentage error will this delay introduce? When the titrator finally shuts off, what will the potential be? What would it have been if the titrator had shut off at the equivalence point?

PART
3
NONEQUILIBRIUM
ELECTROANALYTICAL TECHNIQUES

CHAPTER
12
COULOMETRY AND COULOMETRIC TITRATIONS

The analytical uses of coulometry were reviewed by Bard (S1–S5) and Davis (S6, S7) for some years. The more recent titration reviews by Stock (S8, S9) include reviews of coulometric titrations. The principles of analytical coulometry are those that have already been presented in Chapter 6.

12.1 PRIMARY COULOMETRY

Primary coulometry is that branch of coulometry in which the material of interest reacts directly at the electrode surface. There are several significant analytical applications of primary coulometry. All forms of primary coulometry are somewhat slow because all the material of interest must be transported to the electrode surface; extensive stirring of the solution is required.

The calculations of primary coulometry are simple. The charge passed Q is obtained by an integrative method or (for constant-current coulometry) as the current–time product. Multiplication by the Faraday constant then gives the number of moles of electrons, which is directly related to the number of moles of other species through the balanced electrode reaction.

Example
One kilogram of metallic copper is to be deposited from a $CuSO_4$ solution. How many coulombs will be required and, if a constant current of 10 A is available, how long will the deposition take?

Answer
1000 g Cu/63.456 g Cu/mol Cu = mol Cu
mol Cu $\times$ 2 = mol e^- = 31.6 mol e^-
31.6 mol e^- $\times$ 96,487 C/mol e^- = 3.02 MC
3.02 MC/10 A = 3.02×10^5 s or 84.5 h

12.1.1 Neither Current- Nor Potential-Controlled

In the rare cases in which only one chemical reaction can occur in a cell virtually regardless of the current or voltage used, neither need be controlled. Although such methods are reported in older literature, the only useful area remaining for such primary coulometry is that of *electrogravimetric analysis* in which the current is not measured. The species being determined is plated in coherent form onto a previously weighed (generally platinum) electrode that is then weighed again after the deposition is complete; the weight difference gives directly the amount of the species of interest.

As a quantitative analytical technique, electrogravimetric analysis dates back to 1864 (S10). Modern use of electrogravimetric analysis is restricted to analyses of copper and copper-base alloys containing other metals such as bismuth, tin, lead, cadmium, and zinc. For such analyses, no method is superior in precision or accuracy, but the method is not convenient or rapid and must be considered obsolescent. Excellent references to these techniques are given by Lingane (G2).

12.1.2 Controlled Current

Coulometry at controlled current, which for analytical purposes normally means controlled at a constant value and thus is called *constant-current coulometry*, is one of the two significant coulometric alternatives. Either the current or the electrode potential can be the independent variable in almost any coulometric cell or analysis. The choice depends upon the nature of the analysis and upon the available instrumentation.

Constant-current coulometry is the simpler and less selective of the two alternatives. In constant-current coulometry, the current through the electrolytic cell is controlled, virtually always at a known constant value. If I is known and constant, the charge passed Q is the product of the constant current and the time interval during which it flows. Since measurement of time is simple and precise, as is measurement of a *constant* current, this method has long received wide use. The instrumentation required has been discussed in Chapter 2. Its major drawbacks are (as will be discussed in Chapter 13) that it requires comparatively long times for quantitative reaction, and that the current required, especially toward the end of electrolysis, may go to a reaction other than the one it is intended to be used for so that the current efficiency then becomes less than 100%. The *precision* of the technique is limited only by the constancy of the current and the precision of the time- and current-measuring devices, which in modern practice are never limiting. But the *accuracy* of the technique is severely limited by failure to achieve 100% efficiency in the electrolytic cell. It is difficult to achieve 100% conversion of any material present in solution since the material can only be oxidized or reduced at the electrode when primary coulometry is used.

12.1.3 Controlled Potential

Coulometry at controlled potential, which for analytical purposes normally means controlled at a constant value, is usually called *constant-potential coulometry*. The current is no longer constant over the time interval, and integration of the measured instantaneous current is required. This may be done by a chemical, mechanical, or electronic coulometer, or, alternatively, integration by calculation may be carried out. The methods of this type of measurement have already been discussed in Chapter 2. The precision of this technique is limited almost exclusively by the precision of the integrator, and precision of better than 0.1% is now attainable. The accuracy of this technique is much improved over the accuracy of coulometry at controlled current because control of the potential permits elimination of many undesirable side reactions that can occur in constant-current coulometry.

12.2 SECONDARY COULOMETRY

Secondary coulometry is that branch of coulometry in which the material of interest does not react directly at the electrode. The electrochemical reaction that occurs at the electrode generates some other chemical species, called an *intermediate reactant*. The intermediate reactant then reacts with the material of interest in the bulk solution. Secondary coulometry is used far more often than primary coulometry as the method of charge introduction in coulometric titrations. It is less often used in other analytical applications owing to the necessity of a suitable intermediate reactant.

The only disadvantage of secondary coulometry in comparison to primary coulometry is the necessity for quantitative reaction in both of the two steps. Generation of the intermediate reactant must take place with a Faradaic current efficiency of 100%, and reaction of that intermediate with the material of interest must be both rapid and quantitative. The reaction of the intermediate with the material of interest is usually a redox reaction, but hydrogen and hydroxide ions, coulometrically generated by electrolysis of an aqueous solution of sodium sulfate, permit the use of acid–base reactions as well. Some of the electrogenerated intermediate reactants found useful are Cu(I), Ti(III), and bromine.

In secondary coulometry, the generation of the intermediate reactant can be either external or internal. *External generation* means that the intermediate reactant is prepared by electrolysis of a separate (external) solution and that all or part of this solution is then added to a solution of the material of interest, reaction occurring on mixing. External generation is uncommon because of the inconvenience of physical addition; it is inherently inferior in precision to internal generation because quantitative transfer of the intermediate reactant is required. *Internal generation* means that the intermediate reactant is generated in the same solution that contains the material of interest, and is the preferred method.

Secondary coulometry has certain advantages over primary coulometry. It can be used for the analysis of certain species that do not react quantitatively at electrodes, but that do react quantitatively with oxidizing or reducing agents in solution. It is also more rapid, because the material from which the intermediate reactant is electrogenerated can be, and normally is, present in concentration greatly exceeding that of the material of interest; thus diffusion to the electrode does not limit the coulometric current that can be used to that which can be sustained by the material of interest. The intermediate reactant generated at the electrode then diffuses into the bulk solution, where it reacts with the material of interest. Finally, some materials that can be determined by primary coulometry can be more effectively determined by secondary coulometry. An example is thiosulfate, $S_2O_3^{2-}$, whose oxidation to tetrathionate, $S_4O_6^{2-}$, takes place only with a significant overvoltage. Secondary coulometry employing coulometrically generated iodine is a more effective analytical method for thiosulfate.

The most, and virtually only, significant analytical application of secondary coulometry is in coulometric titrations. Secondary coulometry in which neither the potential nor the current is deliberately controlled has no advantages. Secondary coulometry at controlled potential is less common than secondary coulometry at controlled current because the titration conditions can generally be adjusted to give the necessary selectivity.

12.2.1 Controlled Current

Secondary coulometry at controlled current normally means that the current is controlled at a constant value. However, especially in coulometric titrations or in routine analyses when the approximate values are known, it is useful to carry out constant-current coulometry at two different currents—to pass, say, 90% of the coulombs required at a higher current and then to switch discontinuously to a lower constant current for the final few coulombs required. Such a procedure can increase the speed of analysis considerably at no loss in accuracy.

Constant current is the normal method of secondary coulometric generation employed in coulometric titrations. The concentration of material from which the titrant is electrogenerated can be made high enough to sustain a constant current capable of completing the titration in a reasonable time. The product of current and time gives directly the coulombs used to the end point of the titration, which is then linked to the amount of electrogenerated reagent by the stoichiometric electrode reaction of the generation process. The amount of electrogenerated reagent is then related to the amount of the sought-for constituent by a stoichiometric redox reaction or other stoichiometric reaction.

Example
Marinenko and Taylor (S11) used Ag(I) generated by anodic coulometry of a silver electrode in the coulometric titration of halide ions. The stoichiometric electrogeneration reaction is:

$$Ag \rightarrow Ag^+ + e^-$$

and the stoichiometric titration (precipitation) reaction is:

$$Ag^+ + I^- \rightarrow AgI(s).$$

Example

Lingane (S12) used Sn^{2+} generated by coulometric reduction of a solution of Sn^{4+} in acidic bromide solution for the coulometric titration of Cu^{2+} ions. The stoichiometric electrogeneration reaction is

$$Sn^{4+} + 2e^- \rightarrow Sn^{2+}$$

and the stoichiometric titration (redox) reaction is

$$Sn^{2+} + Cu^{2+} \rightarrow Sn^{4+} + Cu$$

12.3   NONQUANTITATIVE APPLICATIONS OF COULOMETRY

Many organic and inorganic compounds can be prepared by using an electrode as an oxidizing or reducing agent of controllable strength. The idea of so using an electrode dates back to Haber (S13) in 1898. The use of controlled-potential electrolysis is normally required, since controlled-current electrolysis does not permit the required selectivity of oxidation or reduction. Preparative coulometry can be carried out in nonaqueous solutions as well as in aqueous, and, as a consequence, species and intermediates unstable in aqueous solutions may be prepared in aprotic media.

12.3.1   Preliminary Electrolysis

Controlled-potential coulometry can be used to separate certain major components of samples that may interfere in the analytical determination of other components. The usual components removed on platinum cathodes are copper and silver; on mercury cathodes these and other metals may be removed. The technique is effective but slow, and its modern use is uncommon. A significant modern use is in the preparation of pure supporting electrolytes by electrolysis of impurities into large mercury cathodes.

12.3.2   Species Identification and Mechanistic Studies

Especially in polarographic analysis of organic compounds, the species responsible for or formed in a particular polarographic wave may not be known. That is, either polarographic analysis yields multiple waves, with each half-wave potential corresponding to a different process, or perhaps only one wave is obtained, but the structure of the compound is such that either a one-electron process leading to a stable possible intermediate occurs or a two-electron process leading to a stable final product occurs. An example of this is the reduction of quinones to their corresponding hydroquinones, which may occur as the expected two-electron, two-proton process or a one-electron reduction to a stable semiquinone intermediate. In protic solvents, the reaction normally proceeds to the hydroquinone, while in aprotic solvents the semiquinone is more stable, but the course of the reaction depends also on the structure of the quinone itself.

Coulometry can be used to resolve questions such as this fairly directly. If a solution containing a known concentration of the compound plus the usual buffers and supporting electrolyte is exhaustively electrolyzed using a mercury-pool cathode, and the coulombs passed during this electrolysis are measured, the number of electrons required for this reaction of one molecule can be calculated from Faraday's laws. This requires use of controlled-potential coulometry, and therefore some integrative method, and is preferably done in a three-electrode cell configuration with the counterelectrode isolated by a porous frit or other means such that oxidized material cannot reach the mercury-pool cathode. Failure to use potential control and counterelectrode separation will lead to serious error. The potential of the mercury-pool cathode is controlled at a potential just on the diffusion plateau of the polarographic wave whose number of electrons is sought and well short of the succeeding wave, if any. The solution should be stirred during this procedure to facilitate mass transfer.

12.3.3 Analytical Uses of Simple Electrolysis

Simple electrolysis can be used for the direct analysis of only a few metals by *electrogravimetry*, as discussed above. Simple electrolysis finds considerably greater analytical use as a method of separation than as a direct analytical method. Simple electrolysis can be used in the removal of interfering ions before analysis by reduction or oxidation to a noninterfering form. Reduction or oxidation of the interfering ion must be easier than that of desired ion, and the potential difference must be significant. Since $E = E^0 + 0.059/z \, \log c(M^+)$, to separate ions of originally the same concentration by removing all impurities to the nearest part per thousand, or to obtain a situation where $[M^+]$ differs by a factor of 1000, the differences in E^0 for reversible couples must be 0.059 $\times$ 3 = 0.180 V if both are monovalent and 0.090 V if they are divalent. For irreversible couples, a separation of 0.2 V is generally adequate.

Simple electrolysis can also be used to effect the conversion of an ion being analyzed for by reduction or oxidation to a form whose analysis is easier. Reduction or oxidation of the sought-for ion must be easier than that of interfering ions, by the same voltages given above for the case of removal of an interfering ion.

12.4 COULOMETRIC TITRATIONS

Titrations are analytical procedures in which a material, the *titrant*, is added to a sample containing a sought-for constituent. The titrant must be selected so that it reacts rapidly and stoichiometrically with the constituent. When the amount of added titrant exceeds the amount required for the stoichiometric reaction (the *equivalence point*), the first excess of titrant is detected, and the addition of titrant is ended (the *end point*). Any type of chemical reaction that can be made rapid and stoichiometric can be used in a titration procedure, and any appropriate method can be used to detect the equivalence point.

Except in their method of addition of titrant, coulometric titrations do not differ from other types of titrations. The usual methods of detection for end points of coulometric titrations are electrochemical or colorimetric. Further details of electrochemical methods of end-point detection can be found in Chapters 4, 12, and 18.

Coulometric titrations date back to the 1938 work of Szebelledy and Somogyi (S14) in principle and the 1948 work of Shaffer, Briglio, and Brockman (S15) in extensive application. The use of constant current by Swift and co-workers (S16) in 1947 simplified the experimental technique required in coulometric titrations. Most coulometric titrations are now carried out at controlled current, while other forms of coulometry are usually carried out at controlled potential.

Coulometry has several advantages over other methods of addition of titrant. Standards derived from solutions of pure compounds that can be weighed out are not required, since the electron is its own primary standard. More correctly, the values of the charge on the electron and the Avogadro number, whose product is the Faraday constant, are so precisely known that they do not limit the precision of an analysis. A second advantage is that the ease of precise measurement of small quantities of charge is such that precise measurements of quantities of material as small as 1 μmol are possible coulometrically. Such small quantities are difficult to measure quantitatively by any other means. Additional advantages that arise from secondary coulometry are that reagents that are difficult to store or difficult to standardize may be prepared *in situ* and standardized by coulometry so that they can be used in quantitative work. Examples of such materials are Cu(I), Ti(II), Mo(V), and bromine.

12.4.1 Acid–Base Titrations

Acid–base titration reactions involve gain or loss of protons, and thus the coulometric generation of H^+ or OH^-, almost invariably from water or the aqueous component of a mixed solvent. Alternatively, the coulometric reaction may be the removal of H^+ or OH^- produced in the titration reaction. In any case large, inert (platinum) electrodes are generally used, as in the study of Kirowa-Eisner and Osteryoung (S17). The coulometric reactions can produce oxygen and hydrogen, which can be removed by continuous nitrogen bubbling if necessary.

12.4.2   Precipitation Titrations

The most common form of coulometric titration involving a precipitation reaction is that in which Ag^+ is anodically generated as a titrant for halide ions. The course of the titration reaction is easily followed with a potentiometric sensor responsive to either Ag^+ or halide ion. Such titrations cannot generally determine the different halide ions present in a mixture of them, but a total halide titration, oxidation, removal of Br^- and/or I^- as the respective halogens, and a titration of the remaining chloride can do so. Insoluble mercury compounds can be precipitated by coulometric anodization of metal-pool electrodes. An example of the anodization of silver can be found in the work of Nagy and co-workers (S18).

12.4.3   Redox Titrations

Titration reactions of the redox type often can employ coulometric generation of titrant to good effect. Oxidation of bromide produces bromine, perhaps the most widely used single redox titrant generated coulometrically. Iodine, Fe(II), and Sn(II) are also commonly used. An example of the use of bromine can be found in the method devised by Cheney and Fletcher (S19) for the coulometric titration of olefinic unsaturation in fatty acids in propylene carbonate.

12.4.4   Complexation Titrations

When complexation reactions are used in coulometric titrations, the complexing agent is not usually an electroactive species. The complexing agent is generated by reduction of a complex containing it, which thereby frees the complexing agent to react with the species of interest in the titration reaction. Such coulometric generation of complexing agents requires that the complex being coulometrically reduced is of greater stability than the complex formed in the titration reaction and is also more easily reduced than the complex formed in the titration reaction. Certain complexes of EDTA and of related compounds (S20) can be used in this way. Cases are also known in which the coulometric titration may involve more than one type of reaction; the determination of penicillins and penicillamine developed by Forsman (S21) may involve both formation of insoluble mercury sulfide bonds and chelation.

12.5   COULOMETRIC DETECTORS

Coulometric detectors are used in analysis of the output of a separation method, normally chromatography. They may be based upon either primary or secondary coulometry. In detectors employing primary coulometry, direct reduction or oxidation is carried out at an appropriate electrode as the effluent leaves the chromatographic column. The detector is coulometric if the conditions are such that complete oxidation or reduction of the species of interest occurs as it passes through the detector cell. If this is not the case, the detector may still give an acceptable response to the eluted species as an amperometric sensor, as discussed in Chapter 18. The same physical detector may be capable of operating in both modes under different conditions.

12.5.1 Primary Coulometric Detectors

A primary coulometric detector is one in which the total amount of the species of interest is reduced or oxidized as it passes through the detector cell. Such a detector is simply an amperometric detector (Chapter 18) operated under coulometric conditions. Thin-layer cells in which the column effluent passes between two closely spaced electrodes can be operated effectively in coulometric mode. In most cases there is no advantage to operating coulometrically rather than amperometrically, and primary coulometric detectors are not commonly used.

12.5.2 Secondary Coulometric Detectors

A secondary coulometric detector is essentially a continuous coulometric titration of the species of interest in the effluent of the chromatographic column. The effluent from the column flows into a coulometric titration cell in which the component of interest reacts stoichiometrically with an excess of titrant; the titrant is continuously regenerated coulometrically. The continual dilution of the solution in the titration cell is a problem with this approach. It is preferable to mix the effluent with a continuous stream of excess titrant, permit the titration reaction to occur, and pass the resulting solution through a primary coulometric detector in which the concentration *of the titrant* is continuously monitored.

12.6 REFERENCES

General

G1   S. D. Ross, M. Finkelstein, and E. J. Rudd, *Anodic Oxidation*, Academic Press, New York, 1975, 352 pp. Discusses anodic oxidations of organic compounds and the mechanisms of these oxidations.

G2   J. J. Lingane, *Electroanalytical Chemistry,* 2nd ed., Wiley-Interscience, 1958, pp. 91–167.

Specific

S1   A. J. Bard, *Anal. Chem.* **34**, 57R (1962).

S2   A. J. Bard, *Anal. Chem.* **36**, 70R (1964).

S3   A. J. Bard, *Anal. Chem.* **38**, 88R (1966).

S4   A. J. Bard, *Anal. Chem.* **40**, 64R (1968).

S5   A. J. Bard, *Anal. Chem.* **42**, 22R (1970).

S6   D. G. Davis, *Anal. Chem.* **44**, 270R (1972).

S7   D. G. Davis, *Anal. Chem.* **46**, 21R (1974).

S8   J. T. Stock, *Anal. Chem.* **48**, 1R (1976).

S9   J. T. Stock, *Anal. Chem.* **50**, 1R (1978).

S10   W. Cruikshank, *Ann. Phys.* **7**, 105 (1801).

S11   G. Marinenko and J. K. Taylor, *J. Nat. Bur. Standards* **67A**, 31 (1963).

S12   J. J. Lingane, *Anal. Chim. Acta* **21**, 227 (1959).

S13   F. Haber, *Z. Elektrochem.* **4**, 506 (1898).

S14   L. Szebelledy and Z. Somogyi, *Z. Anal. Chem.* **112**, 313 (1938).

S15   P. A. Shaffer, A. Briglio, and J. A. Brockman, *Anal. Chem.* **20**, 1008 (1948).

S16   J. W. Sease, C. Niemann, and E. H. Swift, *Anal. Chem.* **19**, 197 (1947).

S17   E. Kirowa-Eisner and J. Osteryoung, *Anal. Chem.* **50**, 1062 (1978).

S18   G. Nagy, E. Feher, K. Toth, and E. Pungor, *Anal. Chim. Acta* **91**, 97 (1977).

S19   M. C. Cheney and K. S. Fletcher, *Anal. Chem.* **51**, 807 (1979).

S20   M. Roynette and J-P. Sebwing, *Bull. Soc. Chim. France* **1978**, 195.

S21   U. Forsman, *Anal. Chim. Acta* **93**, 153 (1977).

S22   G. Marinenko and J. K. Taylor, *Anal. Chem.* **39**, 1569 (1967).

S23   G. J. Patriarche and J. J. Lingane, *Anal. Chem.* **39**, 168 (1967).

12.7 STUDY PROBLEMS

12.1 Analysis of NBS Standard Sample 83c Arsenic Trioxide (S22) by coulometric titration with triiodide ion and amperometric end-point detection gave assays of 99.9859, 99.9834, 99.9844, 99.9896, 99.9830, and 99.9904% As_2O_3. Compute the mean and standard deviation of these results. Is the sample 100% As_2O_3 or not?

12.2 Tetraphenylborate ion is used in gravimetric determinations of potassium and other ions. It has been suggested (S23) that a more accurate method would be to add an excess of tetraphenylborate and back-titrate the precipitate with coulometrically generated silver ion. The means of 2 to 4 analyses of aliquots of potassium by this method are:

mg K taken	mg K found
0.300	0.292(± 1)
0.500	0.497(± 1)
0.800	0.798(± 2)
1.000	0.995(± 4)
2.000	0.992(± 9)
3.000	2.991(± 8)
5.000	4.993(± 12)
10.000	10.00(± 30)

What is the lower limit of amount of potassium determinable, and what accuracy can be expected for this method above that limit?

12.3 In a certain analysis by spontaneous electrogravimetry of trace copper, a sample of zinc weighing 100 g was dissolved in nitric acid, diluted to 1.0 dm³, and the pH was adjusted to 6. Short-circuited electrodes of metallic platinum and zinc were then inserted. When the electrolysis had been allowed to continue to equilibrium, it was found that 21.6 mg of copper had been deposited on the platinum cathode. Compute the percentage of copper in the zinc. Compute also the percentage of the copper content of the zinc that was not recovered by this procedure.

12.4 A coulometric titration was used to standardize a solution of potassium dichromate, $K_2Cr_2O_7$, with Fe(II) electrolytically generated from Fe(III). A 25-cm³ aliquot of the potassium dichromate solution required, at a constant current of 200 mA, a total of 1800 s before the equivalence point was reached. Calculate the concentration of the potassium dichromate solution.

12.5 A certain organic compound containing a hydroquinone moiety as well as some side-chain carbonyl groups and other reducible functions was to be prepared by reduction of the naturally occurring quinone compound. When an organic chemist attempted to reduce the compound with sodium borohydride he found that the side-chain carbonyls were reduced as well as the quinone. He therefore thought to use controlled-potential reduction to obtain the desired compound. Classical polarography of the quinone starting material showed two waves, one due to the quinone at –0.54 V against SCE and one due to one or more of the side-chain carbonyls at –0.98 V. Approximately 0.45 mol of the hydroquinone is needed to complete the desired syntheses. The molecular weight of the quinone starting material, which is quite soluble in water, is 546.2. The quinone is stable in room-temperature air or aqueous solution indefinitely, while the hydroquinone is stable to air oxidation only when dry; in the presence of either water or a nonaqueous solvent it is rapidly reoxidized to the quinone by molecular oxygen. Give an experimental procedure that might produce the desired compound, the hydroquinone with the side-chain carbonyls unreduced, in reasonable yield.

CHAPTER
13

ELECTROLYSIS: PERTURBATION OF EQUILIBRIUM

When an electrode has no net passage of current, the interphase electrochemical system is at chemical equilibrium and the actual potential of the electrode is also the equilibrium potential of the electrode. When, however, a net current is flowing through the electrode–electrolyte interface, the electrochemical system is not at chemical equilibrium, and the actual potential of the electrode will not be the equilibrium potential of the electrode. The interrelation between the current and the potential is complex, for both depend not only on each other but also on the electrode material and the previous history of the electrode; on the nature of the electrolyte, including concentrations of all of its components; and on the structure of the specific electrochemical cell being used. All of these can, and do, vary in both space and time.

Consider the behavior of the electrode–electrolyte interface of the lead electrode, $//Pb^{2+}(aq)/Pb(s)$. The standard potential, against a standard hydrogen electrode, of such a lead electrode is -0.13 V. For an actual lead electrode at a practical concentration, say, 0.10 mol/dm^3, the equilibrium potential is given by the Nernst equation as

$$E = E^0 + 0.03 \log c(Pb^{2+}) \tag{13.1}$$

so that the actual equilibrium potential of the electrode is about -0.16 V. If this electrode, together with a saturated calomel electrode, constitutes an electrochemical cell whose potential is to be measured, the cell potential is the sum of the two actual electrode potentials. Both of these would, in fact, vary depending on the current flowing, but even assuming for the moment that they do not vary, the observed cell potential would vary with current owing to the ohmic drop in the solution resistance. Thus a cell composed of two actual electrodes, one a saturated calomel reference electrode that for all practical purposes does not change its potential when small currents are drawn and one the lead electrode described above, would exhibit the current–voltage behavior shown in Figure 13.1. In this figure the deviation from vertical behavior of the lead electrode is the ohmic or IR drop due to the resistance of the cell electrolyte and is shown considerably exaggerated. Curves such as those of Figures 13.1 and 13.2 are used throughout

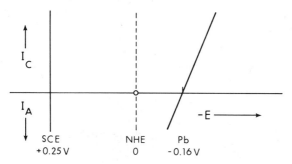

Figure 13.1 Potential curve of lead–calomel cell. Potentials of unpolarizable reference electrodes are shown as vertical lines, that of polarizable 0.1-mol/dm³ Pb^{2+}/Pb electrodes as the sloping line since the cell IR drop is assigned to that electrode. With increasing cell potential forcing the lead electrode cathodic, increasing anodic current is drawn from the SCE and increasing cathodic current from the lead electrode. The observed current–potential curve on the *inert* electrode would ascend anodic branch of SCE, traverse zero current line to Pb, then ascend the cathodic branch of the Pb electrode line.

the next several chapters. Each couple is represented by a more-or-less vertical line. The zero-current, or equilibrium, potential E is shifted, depending upon the actual activities of the species present in accordance with the Nernst equation. Cathodic (negative) currents are generally plotted upward, anodic (positive) currents downward, and increasing negative potential to the right as is traditional in electroanalytical practice (S1).

13.1 LIMITATION OF CURRENT

The IR drop of the solution of itself imposes no limit upon the current that may flow between the electrodes, but such *limiting currents* are frequently observed. An electroanalytical cell is normally constructed so that any limitation of the current is due to the indicating electrode rather than to either the reference or counterelectrodes, and for that reason only the limitations that arise at that electrode need be considered.

Limitations upon the current may arise from several different sources. For any reaction obeying Faraday's law, the instantaneous current is given by

$$I = zF \, dn/dt \qquad (13.2)$$

where n is the amount of the electrochemically active substance present. This change in moles is given by

$$dn/dt = AJ \qquad (13.3)$$

where J is the flux in moles per second per square meter, and A is the area in square meters. The instantaneous current is then

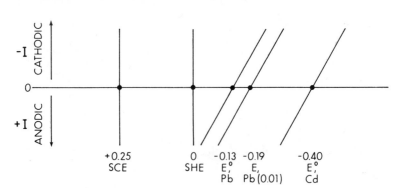

Figure 13.2   Lead-cadmium potential curve. Note that different lines are obtained for different couples and for the same couple at different concentrations–activities. The intercepts of all of these lines with the zero-current axis are the zero-current, or equilibrium, potentials of the electrodes.

$$I = zFAJ \tag{13.4}$$

It should be noted that in principle both A and J, or almost any of the terms on the right-hand side of any of the following equations, can be time-dependent. In practice only c^0 and dc^0/dx are so at electrodes of constant area. If the area A is also time-dependent, as it is at the polarographic dropping-mercury electrode, this will impose an *additional* time dependence upon the observed current.

13.1.1   Charge-Transfer Control

When the flux is controlled solely by the rate of charge transfer, the current is said to be *charge transfer limited* or under *activation control*. The limiting current is then given by the Butler–Volmer equation:

$$I_{lim} = Aj^0 \left[\exp(\alpha zF\eta/RT) - \exp(-(1-\alpha)zF\eta/RT)\right]$$

13.1.2   Kinetic Control

When the flux is controlled solely by the rate of a reaction that does not explicitly involve charge transfer, the current is said to be *kinetically limited* or under *kinetic control*. Frequently the step is heterogeneous, taking place within the double layer or on the electrode surface. A simple first-order heterogeneous kinetic step gives

$$I_{lim} = I_k = zFAk_h c^0$$

where k_h is the specific heterogeneous rate constant, and c^0 is the surface concentration of reacting species. The surface concentration c^0 is the concenetration of the reacting species at the electrode surface, the point at which the distance from the electrode out into the solution x is zero. The SI units for a first-order specific heterogeneous rate constant k_h are meters per second and for c^0 moles per cubic meter (mol/m^3 or mmol/dm^3), so the electrode area is then in square meters.

Adsorption is a heterogeneous process. If adsorption kinetics, rather than the equilibrium position of adsorption as given by an isotherm, need be considered it must involve a specific heterogeneous rate constant for the adsorption process.

13.1.3 Mass-Transfer Control

When the flux is controlled solely by the rate at which ions can reach the electrode, which is the rate of mass transfer, the current is said to be *mass-transfer-limited*. In electroanalytical practice ions reach the electrode primarily by diffusion through the solution because other forms of mass transport such as convection are prevented by cell design, so that the current is *diffusion-limited* or under *diffusion control*. In this case the time-dependent flux J is equal to a constant known as the *diffusion coefficient D* multiplied by the rate of change of concentration with distance from the electrode at the origin of distance, which is the electrode surface. This statement is known as *Fick's first law of diffusion*:

$$J = D(\mathrm{d}c^0/\mathrm{d}x) \tag{13.5}$$

Combination of the two equations above gives the time-dependent current:

$$I_d = zFAD(\mathrm{d}c^0/\mathrm{d}x) \tag{13.6}$$

The time-dependent concentration gradient, often called the *diffusion gradient*, is dc/dx. The SI units for D are square meters per second (cgs: cm^2/s).

13.2 NERNST DIFFUSION-LAYER THEORY

One of the most useful and simple ways to understand the time-dependent effects of current flow through an electrode is the concept of the *diffusion layer* introduced by Nernst (S2) in 1904. When an electrode has current drawn from it, this current must be carried by ions through the electrolyte solution to the electrode, where they are reduced or oxidized. This will lower the concentration of these ions in the immediate vicinity of the electrode surface. Additional such ions will then diffuse toward this region of lower concentration. For the lead electrolyte discussed earlier, the time-and distance-dependent concentration changes when cathodic current is drawn from this electrode would be as shown in Figure 13.3.

The Nernst diffusion-layer concept is that the concentration gradient is assumed to be linear. By taking at any instant of time the average slope of the actual gradient and assuming that this is the slope of the (nonexistent) linear

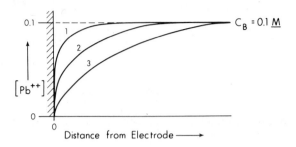

Figure 13.3   Concentration gradients for Pb(II) reduction. Concentration gradients for reduction of an initially uniform 0.1-mol/dm³ (c_B) solution of Pb(II). Curves 1, 2, and 3 correspond to different, and increasing, times after commencement of a constant cathodic current.

gradient, the assumption of linearity then defines a certain thickness of the diffusion layer δ as shown in Figure 13.4.

Application of the uniform concentration gradient of the Nernst diffusion layer implies that at any instant the concentration gradient is linear, and the slope is constant:

$$dc^0/dx = (c - c^0)/\delta \tag{13.7}$$

Combining the two preceding equations gives an expression for the diffusion-limited current:

$$I_{lim} = zFAD(c - c^0)/\delta \tag{13.8}$$

The SI units of I are amperes, A are square meters, D are square meters per second, c are moles per cubic meter, and δ are meters.

The thickness of a Nernst diffusion layer depends upon the solution into which it extends. For normal aqueous solutions that are not stirred, the thickness of the Nernst diffusion layer is 0.3 to 0.5 mm, but with stirring it can be reduced significantly, to 0.01 or 0.001 mm. In highly viscous media, even thicker Nernst diffusion layers can be achieved. Since the Nernst diffusion layer would, in the absence of influences other than diffusion, extend into the solution a distance that increases steadily with the passage of time, it follows that the thickness of a real diffusion layer must be controlled by another process. This process is generally convection, which achieves sufficient mass transport to keep the solution at some distance from the electrode effectively at the bulk concentration. It then follows that if a current is diffusion-limited and enough time has elapsed, a stable balance between current demand, diffusion gradient that meets current demand, and thickness of the Nernst diffusion layer will be established. The model for diffusion-limited current used later in this chapter will assume that this steady state is established instantly and maintained indefinitely. When the current is limited by diffusion in this way, the material reaching the surface is immediately

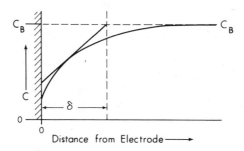

Figure 13.4   Nernst diffusion layer. Bulk concentration is c_B, thickness of Nernst layer is δ.

reduced or oxidized on arrival at the surface so that the surface concentration c^0 is held at zero. The preceding equation therefore simplifies to

$$I_d = -zFAD_{Ox}c_{Ox}/\delta \tag{13.9}$$

for a cathodic process and

$$I_d = zFAD_{Red}c_{Red}/\delta \tag{13.10}$$

for an anodic process. Since the thickness δ is limited by convection rather than diffusion, it may be considered time-independent.

Quantitative electrolysis studies are possible using the concept of the Nernst diffusion layer. This can most easily be seen in the treatment of an example. Starting with 100 cm³ of a solution that is 10 mmol/dm³ in Ag⁺ and 10 mmol/dm³ in Cu²⁺, consider what would happen if the solution were electrolyzed so as to achieve one of three possibilities: the maximum separation theoretically possible; the maximum separation possible at constant current; or the maximum separation possible at constant potential. As conditions of the electrolysis let us assume an electrode of real and geometric area A of 20 cm², a Nernst diffusion-layer thickness δ maintained by stirring of 0.001 mm, and an ability to control the potential E to ± 10 mV.

13.2.1   Maximum Separation Possible

If no current flowed in the system, the potentials of these couples would be their equilibrium potentials as given by the Nernst equation, the concentration of each species being its bulk concentration. For the conditions of this electrolysis, these are:

$$E(Ag^+/Ag) = E^0 + RT/F \ln c(Ag^+)$$

$$E(Ag^+/Ag) = +0.80 + 0.06 \log 10^{-2}$$

which is $+0.68$ V, and:

$$E(Cu^{2+}/Cu) = E^0 + RT/2F \ln c(Cu^{2+})$$

$$E(Cu^{2+}/Cu) = +0.34 + 0.03 \log(10^{-2}) = +0.28 \text{ V}$$

Upon electrolysis, silver will plate out first because it is reduced most easily; then copper will begin to plate out when E becomes equal to $+0.28$ V. At that point, the silver concentration remaining will be given by:

$$E = +0.28 = +0.80 + 0.06 \log c(Ag^+)$$

Since $\log c(Ag^+) = -0.52/0.06 = -8.7$, $c(Ag^+) = 2 \times 10^{-9}$ mol/dm^3. At this point $c(Cu^{2+})$ is 10^{-2} mol/dm^3, the original concentration of Cu^{2+}, since none of the copper has been removed. Thus the maximum separation theoretically possible is when the concentration of Ag^+ has become equal to 0.00002% that of Cu^{2+}, *regardless of the length of the electrolysis time.*

13.2.2   Constant-Current Electrolysis

There must be a maximum electrolytic current that can be supported by silver alone; any excess will have to go to some other electrolytic process, presumably first the unwanted deposition of copper. This maximum current will be given by the diffusion limit equation as

$$I_d = zFADc/\delta$$

To obtain its value it is necessary to assume a value for the diffusion coefficient D; the value taken, 1×10^{-9} m^2/s or 1×10^{-5} cm^2/s, is then approximately correct for metal ions diffusing through aqueous solutions. The maximum current is approximately 2 A. In other words, the diffusion of silver ion will support only 2 A; any greater current will plate out copper immediately.

This maximum current that can be sustained by silver deposition is the *absolute* maximum current. It is not a useful current for the electrolytic separation, because as soon as any silver is removed by the plating process the concentration will drop and the 2-A current will no longer be sustained. A considerably smaller constant current, perhaps 100 mA, must be used. The concentration of silver ion that will sustain that current is

$$0.1 = zFADc/\delta$$

A ratio of the currents is the ratio also of the concentrations since all other terms are constant; thus:

$$0.1 \text{ A}/2.0 \text{ A} = c(\text{final, 100mA})/c(0.01).$$

This concentration is 5×10^{-4} mol/dm^3, or 5% of the original concentration of silver ion; 95% of the silver ion originally present has been plated out.

Calculation of how long this will take requires Faraday's Law written in the form:

$$Q = It = zFcV$$

where V is the solution volume. The time for recovery of 95% of the silver is then given by:

$$t = zFcV/I$$

$$t = 96487 \times (0.01 \times 0.95) \times 0.1/0.1 = 917 \text{ s}$$

Similar calculations can be done for other electrolysis currents; for a current of 0.02 A, the calculation shows 99% recovery, which will require 9552 s (about 3 h). Clearly, recovery increases as current decreases, but so also does the required time.

13.2.3   Constant-Potential Electrolysis

Since the potential control is ± 0.01 V, an effective control potential would be $E = +0.28 + 0.01 = +0.29$ V so that the plating of copper does not occur. Then:

$$\log c(\text{Ag}^+, \text{final}) = (0.29 - 0.80)/0.06 = -8.5$$

Since the final concentration of silver ion is 3×10^{-9} mol/dm^3, recovery of silver is quantitative (99.99997%). To calculate the time required, let x be the fraction deposited per second. Initially $I_{\lim} = 2$ A at $c = 0.01$ mol/dm^3; the initial value of x is thus

$$2\text{A}/(96{,}487 \text{ A s/mol} \times 0.01 \text{ mol/dm}^3 \times 0.1 \text{ dm}^3) = 0.02(2\%)/\text{s}$$

At any instant the concentration will be given by the exponential decay form $c = c(\text{initial}) \exp(-xt)$;

$$c(\text{final})/c(\text{initial}) = (3 \times 10^{-9})/10^{-2} = \exp(-0.02t).$$

Thus $-xt = \ln(3 \times 10^{-7}) = 2.303 \log (3 \times 10^{-7})$, and $t = -2.303 \times (-6.52)/0.02 = 750$ s. During an electrolysis in which the potential is controlled at a constant value, the current varies, approximately 100% recovery is achieved, and 750 s are required. Clearly the controlled-potential method is not only more accurate but far more rapid.

This is a comparison of two types of electrolysis: constant-potential electrolysis, which is more selective, and constant-current electrolysis, which is simpler. In most cases constant-current electrolysis is used when potential control is not needed, when electrolysis is required as a complete conversion method but not as a selective conversion method—when interferences are absent. The apparatus for

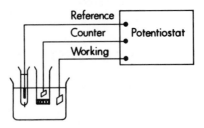

Figure 13.5   Controlled-potential electrolysis using potentiostat.

electrolysis in which the potential is controlled at a constant value is shown in Figure 13.5. The working electrode is normally a platinum or mercury cathode. Mercury cathodes are in the form of pools at the bottom of the electrolytic cell. Note that the counterelectrode is separated by a glass frit or membrane so that its products do not reach the working electrode.

13.3   VOLTAMMETRIC DIFFUSION-LAYER THEORY

Voltammetry involves consideration of the indicating electrode and the currents flowing through it as a function of the potential of this electrode. Henceforth this discussion will ignore the reference and counterelectrodes that are necessarily present. It will be assumed that they fulfill their functions ideally, that is, that the reference electrode will maintain a fixed potential and that the counterelectrode will serve as an infinite source or sink for electrons, regardless of the current or potential at the indicator electrode. If a solution containing a concentration of a reducible ion such as 10 mmol/dm³ Cu^{2+} is electrolyzed in a voltammetric arrangement at an inert electrode, the observed current–voltage curve will be as shown in Figure 13.6. The solid curve A of Figure 13.6 would be observed on any inert or noble metal electrode; its prolongation B in the anodic direction would be observed only at a copper electrode because it results from the anodic dissolution of copper. The cathodic continuation C would be observed if the concentration of Cu^{2+} were infinite in the sense of being capable of supplying any demand made for cathodic current at the electrode. Since the concentration is not infinite, the current is limited to I_{lim} by the amount of Cu^{2+} available at any instant. More correctly, the current is limited by the amount of Cu^{2+} available at the electrode surface at any instant, which is in turn limited by the rate at which Cu^{2+} can move from the bulk solution through the diffusion layer to the electrode surface. The deviation from the vertical may be greater (as shown in curve D) or lesser depending upon the ohmic drop of the solution. In a two-electrode configuration such an error can be quite significant; for example, in a cell of resistance 1000 Ω, a current of 1 one microampere will produce an ohmic drop of one millivolt. Henceforth this discussion will ignore IR drop and assume that it has been either corrected for or effectively eliminated by cell design.

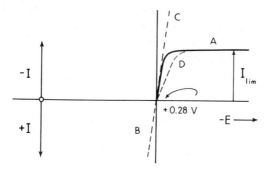

Figure 13.6   Voltammetry of Cu(II) solution. Cathodic voltammetric curve A (solid line) is obtained with a 10-mmol/dm³ solution of aqueous Cu(II). The anodic component B (dashed line) would be observed at a copper electrode, while the zero-current line would be followed at an inert electrode. Infinite rather than limited concentration of Cu(II) would give the cathodic curve C, and greater solution resistance the greater slope indicated by D (dashed lines).

The assumption that equilibrium or quasiequilibrium theory can quantitatively, account for the observed phenomena in voltammetry is equivalent to assuming that the exchange current I^0 for the equilibrium involved is relatively large. Quantitatively that will be true if the exchange current I^0 is much greater than the observed limiting current I_{lim} or, using within 1% to mean quantitative, $I^0 > 100I_{lim}$. When this is so the system is quantitatively reversible, and the Nernst equation can be used to give the potential observed at the electrode at any instant.

The simplest voltammetric curves are obtained when the current is limited only by the rate at which a species being oxidized or reduced reaches the surface of the electrode where it reacts immediately. If the sole mass-transport process is diffusion, the relationships between current and time derived previously hold. If the rate of change of the applied voltage is infinitely slow (in theory; in practice, sufficiently slow that further reduction in dE/dt does not alter the observed voltammetric curves), the observed potential will correspond to the potential that is given by the Nernst equation for the actual surface activities present. Moreover, if the current drawn is much less than the current required to produce significant overpotentials ($I^0 \gg I$), the current will actually be limited by mass transport to the electrode and not by charge transfer at the electrode.

For a general electrochemical reaction

Ox + $ze^- \rightleftharpoons$ Red

the Nernst equation is written

$$E = E^0 + (RT/zF) \ln (a^0_{Ox}/a^0_{Red}) \qquad (13.11)$$

If an electrochemical reaction is being carried out at the electrode, the surface activity of both components will be affected by the production and consumption of species in solution. This is more easily seen in terms of concentrations rather than activities, so that activity coefficients are incorporated into $E^{0'}$:

$$E = E^{0'} + (RT/zF) \ln(c^0{}_{Ox}/c^0{}_{Red}) \tag{13.12}$$

The maximum value of cathodic current, usually called the *cathodic diffusion-limited current* $I_{d,c}$, will be proportional to the bulk concentration c_{Ox}. This is most easily visualized from the Nernst diffusion-layer expression:

$$I = zFAD(c - c^0)/\delta \tag{13.13}$$

The maximum value of I is obtained when c^0 drops to zero, which corresponds to the physical situation just described, in which the species is reduced or oxidized as soon as it reaches the electrode. Since

$$I_{d,c} = zFADc_{Ox}/\delta$$

$$c^0{}_{Ox} = (I_{d,c} - I_c)\delta/zFAD_{Ox} \tag{13.14}$$

accounts for the $c^0{}_{Ox}$ due to the cathodic reaction. However, Ox is also produced by the anodic reaction. In the cathodic limiting region this is a totally insignificant contribution, but as E becomes closer to $E^{0'}$ the contribution increases. The cathodic current I_c should therefore be replaced by the actual current I:

$$c^0{}_{Ox} = (I_{d,c} - I)\delta/zFAD_{Ox} \tag{13.15}$$

To preserve the convention of negative cathodic current, the physical requirement that c^0 cannot be negative forces

$$c_{Ox} = (I - I_{d,c})\delta/zFAD_{Ox} \tag{13.16}$$

The anodic reaction is treated in the same way, but sign reversal is not required:

$$c^0{}_{Red} = (I_{d,a} - I)\delta/zFAD_{Red} \tag{13.17}$$

Insertion of these into the Nernst equation yields

$$E = E^{0'} + (RT/zF) \ln[(I-I_{d,c})D_{Red}/(I_{d,a}-I)D_{Ox}] \tag{13.18}$$

which simplifies to

$$E = E_{1/2} + (RT/zF) \ln[(I - I_{d,c})/(I_{d,a} - I)] \tag{13.19}$$

when

$$E_{1/2} = E^{0'} + (RT/zF) \ln(D_{Red}/D_{Ox}) \tag{13.20}$$

This equation gives the observed current as a function of potential, or vice versa, for a voltammetric process controlled by diffusion alone when the Nernst diffusion-layer concept is taken as the quantitative relationship governing diffusion. More complex relationships are obtained using more sophisticated diffusion relationships and/or if the process is not controlled by diffusion alone. In leaving this useful concept of the Nernst diffusion layer, the artificial nature of the Nernst model must be emphasized. The diffusion gradient is not linear and the boundary of the layer is not easily defined, but the simple model of Nernst is the last physical model easily visualized.

13.4 SEMI-INFINITE DIFFUSION THEORY

The next most simple theoretical treatment of diffusion is used when the diffusion can occur in only one direction from an infinite distance and is known as semi-infinite linear diffusion theory. This treatment was developed, and is applied, in many areas other than electrochemistry.

13.4.1 The Laws Of Diffusion

Consider a horizontal electrode of area A at the top of a vertical column of solution of length x, where x may extended to infinity, composed of distance elements dx. Such a diffusion column can be both constructed physically (Figure 13.7A) and described mathematically (Figure 13.7B).

The flux of material diffusing to the electrode from any distance x is denoted by the total flux $J'(x)$, which is the product of the flux per unit area $J(x)$ and the area A so that

$$J'(x) = AJ(x) \tag{13.21}$$

and the flux at distance $(x + dx)$ is

$$J'(x + dx) = AJ(x + dx) \tag{13.22}$$

which can be written as:

$$J'(x + dx) = A[J(x) + (dJ(x)/dx)dx] \tag{13.23}$$

The increase or decrease in the amount of substance n of a species in a volume element $A\ dx$ is the difference between the amount entering and the amount leaving the volume element, so

$$-dn = J(x) - J(x + dx) \tag{13.24}$$

which on substitution of the two preceding equations yields

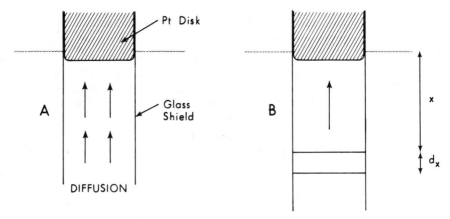

Figure 13.7 Linear diffusion to horizontal plane electrode. (*A*) physical electrode design; (*B*) mathematical model.

$$-dn = -A[dJ(x)/dx]dx \tag{13.25}$$

The amount of substance (moles) in any volume element n is given by the product of the concentration in that volume element $c_{x,t}$ and the volume of the element $A\,dx$ which yields

$$dn = d[c_{x,t}A\,dx]/dt \tag{13.26}$$

When x is taken as time-invariant and the area of the electrode is taken as constant the above equation becomes

$$dn = A\,dx[dc_{x,t}/dt] \tag{13.27}$$

Under a *steady-state* condition of diffusion as assumed earlier, the rate of loss of a substance must equal the rate of its gain in any volume element and therefore

$$-dJ(x)/dx = dc_{x,t}/dt \tag{13.28}$$

Fick's first law of diffusion (S3) used earlier is written as

$$-J(x) = Ddc_{x,t}/dx \tag{13.29}$$

where D is the diffusion coefficient. If the diffusion coefficient is invariant with time, then

$$dc_{x,t}/dt = d[Ddc_{x,t}/dx]/dx \tag{13.30}$$

If D is a true constant independent of distance x as well as time (which it will be if a large excess of supporting electrolyte is present so that the density and viscosity of the solution do not change appreciably with distance from the electrode because of any depletion or formation of the diffusing species), then

$$dc_{x,t}/dt = -D[d^2c_{x,t}/dx^2] \tag{13.31}$$

This equation is known as *Fick's second law of diffusion* (S3). Fick's second law of diffusion is a differential equation having many solutions under different conditions, known as *boundary conditions*. Solution of this equation requires use of the Laplace transform and is beyond the scope of this text. It has been discussed in monographs on diffusion (G4).

13.4.2 Diffusion to an Expanding Plane

When a planar electrode expands toward the incoming material, Fick's second law takes on an additional term to account for the additional transport. The equation is then

$$dc_{x,t}/dt = -D[d^2c_{x,t}/dx^2 + (2x/3t)\, dc_{x,t}/dx] \tag{13.32}$$

Again, solution of the equation under the appropriate boundary conditions is effected by means of the Laplace transform (G2, G3).

13.4.3 Diffusion to a Stationary Cylinder

Stationary cylinders are treated under the assumption that the area of the ends of the cylinder is negligible compared to that of the cylindrical surface. The volume elements used are now in the form of cylinders of varying diameter, and the distance is taken as radial distance r rather than linear distance x since the diffusion is radially rather than linearly symmetric. Effectively, there is linear symmetry along the cylindrical axis and radial symmetry with respect to the axis. Fick's second law again acquires an additional term, now in radius:

$$dc_{r,t}/dt = -D[d^2c_{r,t}/dr^2 + (1/4)\, dc_{r,t}/dr] \tag{13.33}$$

Solution of this equation under the appropriate boundary conditions generally requires either mathematical or numerical approximation.

13.4.4 Diffusion to a Stationary Sphere

Stationary spheres are more easily treated than stationary cylinders. The volume elements are now nesting spheres, and diffusion is radially symmetric in three rather than two dimensions. The form of Fick's second law is then

$$dc_{r,t}/dx = -D[d^2c_{r,t}/dr^2 + (2/4)\, dc_{r,t}/dr] \tag{13.34}$$

Solution of this equation, under the appropriate boundary conditions, is again effected by means of the Laplace transform (G2, G3).

13.4.5   Diffusion, Convection, and Migration

The effect of convection is to add terms to the form of Fick's second law. The effect of the expansion of a planar electrode can be visualized in that light. The only other electrochemically significant case is that of the rotating disk electrode in which the convection is hydrodynamically driven; in all other cases the effect of convection is difficult to express. The effect of the electric field upon the migration of ions, which would also be a complication in theoretical treatment, is experimentally eliminated by addition of enough supporting electrolyte to reduce it to insignificance in most nonequilbrium electroanalytical techniques.

13.5   CURRENT AT A STATIONARY PLANAR ELECTRODE

Application of the Laplace transform to obtain solutions of Fick's second law of diffusion for semi-infinite linear diffusion to a stationary planar electrode requires statement of the appropriate boundary conditions. The boundary conditions are that when $t = 0$, $x > 0$, then $c_{x,t} = c$; and when $t > 0$, $x = 0$, then $c_{x,t} = 0$. These are the conditions for a homogeneous solution at the start of electrolysis and diffusion control during electrolysis. The resulting solutions appear as the series of curves shown in Figure 13.8. The curves change continuously with time and no steady state is ever reached under these conditions. The same solutions can be expressed in explicit form:

$$c_{x,t} = c \; \mathrm{erf}(x/2D^{1/2}t^{1/2}) \tag{13.35}$$

where the error integral erf is

$$\mathrm{erf}\; a = 2\pi^{-1/2}\exp(-z^2)dz \tag{13.36}$$

in which the integral is a definite integral from zero to a, and z is an auxiliary variable. The equations defining erfc, the complement of the error function, and the derivative of the error function are

$$\mathrm{erfc}\; a = 1 - \mathrm{erf}\; a \tag{13.37}$$

$$d(\mathrm{erf}\; a(b))/db = 2\pi^{-1/2}\exp(-a(b)^2)da(b)/db \tag{13.38}$$

where b is a variable in terms of which the error function of a(b) is to be differentiated. The appearance of the error function is shown in Figure 13.9. The value of erf a is within 1% of unity when a is greater than 2. This gives the series of curves reaching steadily further into the solution with time shown previously in Figure 13.8.

The results thus far may be summarized mathematically by writing Fick's second law of diffusion:

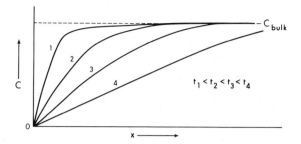

Figure 13.8   Linear diffusion to horizontal plane electrode. Curves 1 to 4 are concentration–distance profiles observed at different, and steadily increasing, times after commencement of electrolysis. The electrode is located at zero on the x (distance) axis.

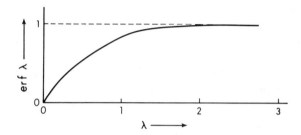

Figure 13.9   The error function.

$$dc_{x,t}/dt = -D[d^2c_{x,t}/dx^2]$$

and applying the appropriate boundary conditions, which are

1. $c_{0,t} = 0$ for $t > 0$ (diffusion limit)
2. $c_{x,0} = c$ for $t = 0$ (homogeneous start)
3. $c_{x,t} \rightarrow c$ as $x \rightarrow$ infinity (homogeneous bulk)

to obtain the solution, which is

$$c_{x,t} = c \ \text{erf}(x/2(Dt)^{1/2}) \qquad (13.39)$$

The current for an electrochemical reaction is a function of the flux at the electrode surface, located at $x = 0$, given by Faraday's laws as

$$I = -zFA[J(x)], x = 0$$

which when the current is limited by diffusion is

$$I_d = zFAD(dc^0/dx)$$

as derived in the previous section.

The time-dependent derivative of c^0 with respect to x is

$$dc^0/dx = c \, d(\mathrm{erf} \, (x/2D^{1/2}t^{1/2}))/dx \tag{13.40}$$

Since the exponential term in $-x^2/4Dt$ is unity whenever x is zero and t is not equal to zero, and these are the conditions at the electrode surface during electrolysis,

$$dc^0/dx = 2c\pi^{1/2} \, d(x/2D^{1/2}t^{1/2})/dx \tag{13.41}$$

$$dc^0/dx = 2c\pi^{-1/2}/2D^{1/2}t^{1/2} \tag{13.42}$$

$$dc^0/dx = c\pi^{-1/2}D^{-1/2}t^{-1/2} \tag{13.43}$$

Insertion of this expression into the general equation for diffusion-limited current yields the *Cottrell equation*:

$$I_d = zFAcD^{1/2}\pi^{-1/2}t^{-1/2} \tag{13.44}$$

The product of current and the square root of time is constant if the area and the diffusion coefficient are constant.

The Cottrell equation applies to *linear diffusion to a horizontal plane electrode*. According to it no stable time-independent current is ever achieved for diffusion to a horizontal plane electrode. Experimentally this is found to be true. It is sometimes possible to obtain a steady state or current in 2 to 3 min with a *vertical* plane electrode, but this does not arise by means of linear diffusion; the mass transport is convection-limited and arises from the density gradient produced by the electrolysis reaction. Steady, although poorly reproducible, states can also be obtained with diffusion to a *point* electrode, which is useful only for qualitative measurements.

13.6   CURRENT AT OTHER STATIONARY ELECTRODES

Solutions have been worked out for electrodes of other shapes by means of the Laplace transform. A useful summary is given in the monograph of Delahay (G1). No steady-state solution is obtained for cylindrical electrodes, but a steady-state solution is obtained for diffusion to spherical electrodes. The counterpart of the equation derived for semi-infinite linear diffusion, which can be written

$$I_d = zFADc/(\pi Dt)^{1/2} \tag{13.45}$$

in semi-infinite spherical diffusion is

$$I_d = zFADc/[r + (\pi Dt)^{1/2}] \tag{13.45}$$

The two equations are identical except for the term in reciprocal radius. The equation for semi-infinite spherical diffusion gives currents that approach a limit that is time-independent, as shown in Figure 13.10.

This equation was derived and tested by Lingane and Loveridge (S4). At a hanging-mercury-drop electrode, this equation is obeyed for up to 30 s in aqueous solution; after that the diffusion layer is destroyed by convection. Theoretical derivations have also been given by others (S7–S9). Experimental studies of current–time relations (S10–S14) and area–time relations (S15,S16) at a DME indicate that the theoretical one-sixth-power time dependence is not obtained because of minor variations in the mass flow rate of mercury, especially in the early part of drop life, which arise from inconstancy of the back pressure. Little practical advantage arises from use of more elaborate equations than that of Ilkovic. Static drops renewed by mercury extrusion are preferable to a gravity-driven DME when precision data are required, although they are less convenient to obtain. For further discussion, see the review of Markowitz and Elving (S15).

13.7   REFERENCES

General

G1   P. Delahay, *New Instrumental Methods in Electrochemistry*, Interscience, New York, 1954, 437 pp.

G2   Z. Galus, *Fundamentals of Electrochemical Analysis*, Ellis Horwood/John Wiley, New York, 1976, 520 pp.

G3   D. D. Macdonald, *Transient Techniques in Electrochemistry*, Plenum, New York, 1977, 329 pp. Unified theoretical presentation at a highly mathematical level.

G4   H. S. Carslaw and J. C. Jaeger, *Conduction of Heat in Solids*, Oxford University Press, London, 1947.

Specific

S1   *Pure and Appl. Chem.* **45**,133 (1976).

S2   W. Nernst, *Z. Phys. Chem.* **47**, 52 (1904).

S3   A. Fick, *Pogg. Ann.* **94**, 59 (1855).

S4   J. J. Lingane and B. A. Loveridge, *J. Amer. Chem. Soc.* **72**, 438 (1950).

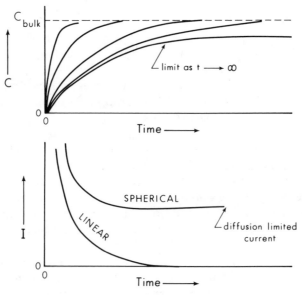

Figure 13.10   Spherical and linear diffusion. The upper curve is the set of spherical diffusion gradients, which, unlike those of linear diffusion, do achieve a stable level at infinite time. The lower curve shows the time dependence of the resulting diffusion-limited current for both linear and spherical diffusion.

S5   T. Kambara, M. Suzuki, and I. Tachi, *Bull. Chem. Soc. Japan* **23**, 220 (1950); *Proc. Intl. Polarographic Cong., Prague,* Part I, p. 126 (1951).

S6   H. Strehlow and M. von Stackelberg, *Z. Elektrochem.* **54**, 51 (1950).

S7   J. Koutecky and J. Cizek, *Chem. Listy* **50**, 196 (1956).

S8   L. Airey and A. A. Smales, *Analyst* **75**, 287 (1950).

S9   J. J. Lingane, *J. Amer. Chem. Soc.* **75**, 788 (1953).

S10 J. K. Taylor, R. E. Smith, and I. L. Cooter, *J. Research Natl. Bur. Standards* **42**, 387 (1949).

S11 M. von Stackelberg and V. Toome, *Coll. Czech. Chem. Commun.* **25**, 2959 (1960).

S12 J. J. MacDonald and F. E. Wetmore, *Trans. Faraday Soc.* **47**, 533 (1951).

S13 G. B. Smith, *Trans. Faraday Soc.* **47**, 63 (1951).

S14 R. J. Newcomb and R. Woods, *Trans. Faraday Soc.* **57**, 130 (1961).

S15 J. M. Markowitz and P. J. Elving, *Chem. Rev.* **58**, 1047 (1958).

13.8 STUDY PROBLEMS

13.1 Copper is to be analyzed by exhaustive electrodeposition on a platinum gauze electrode from a pH 2 solution of about 10 mol/m³ concentration. Given 100 cm³ of the solution, compute the constant current required to electroplate all of this copper within 1 h on an electrode of 30 cm² area. Given the ability to control a potential to ±5 mV, at what potential would you set a potentiostatic control to carry out this deposition in constant potential mode, assuming $D = 1.0 \times 10^{-9}$ m²/s and $\delta = 1 \times 10^{-6}$ m? How long would the potentiostatic deposition take?

13.2 A certain bottle of reagent grade KNO_3 was labelled as containing 0.0004% heavy metals (as lead). Controlled potential reduction of a 5.0-kmol/m³ solution of this with a mercury-pool cathode of 20 cm² area was employed to remove these impurities. For 1.0 dm³ of solution and a Nernst diffusion layer of 0.1 mm, how long would it take to reduce the impurities by two orders of magnitude? How would this time be affected if the solution were stirred?

13.3 An aqueous solution of 10 mol/m³ Sn^{4+} and 30 mol/m³ Sn^{2+} was studied voltammetrically in KCl supporting electrolyte. Sketch the overall voltammetric curve observable on an ideally inert indicator electrode (against SCE), assuming equal diffusion coefficients of Sn^{4+} and Sn^{2+}, if the cathodic limiting current was observed to be 16.3 μA.

13.4 Voltammetric curves for reduction of Cu^{2+} are affected by complexing agents in solution. Compare quantitatively the voltammetric curve obtained on an ideal electrode for the reduction of 1.0-mol/m³ Cu^{2+} aqueous solution and for the same solution that also contains 50 mol/m³ ammonia. Assume the supporting electrolyte functions ideally in both solutions.

CHAPTER
14
BASIC VOLTAMMETRY
AND POLAROGRAPHY

Voltammetry, or voltammetric analysis, includes polarography as a subclassification. A general scheme of organization of electroanalytical techniques proposed by IUPAC (S1) divides the class of techniques in which the electrode reaction is involved into two subclasses: techniques in which the excitation signal is constant or zero, such as potentiometry and its variants, amperometry and its variants, and electrogravimetry, and techniques in which the excitation signal varies with time. The second of these subclasses can be further divided into two groups. The first group is those techniques employing large variable excitation signals, where here "large" should be taken to mean greater than twice $2.3RT/F$ V. This includes all voltage or current-sweep methods such as voltammetry and its variants, polarography and most of its variants, and some chronopotentiometry. The second group consists of those techniques employing small variable excitation signals, where here "small" means less than $2.3RT/F$ V, which includes techniques such as AC polarography and square-wave polarography. This classification does not fit either historical or pedagogical organization and has not been used in this text. The following definitions are used:

Voltammetry is that class of electroanalytical measurements in which the controlled parameter, the potential of the indicator electrode, varies with time and in which the current flowing through the indicator electrode is the measured parameter.

Classical voltammetry is that subclass of voltammetry in which the rate of change of potential with time is sufficiently slow that the observed phenomena can be quantitatively described by equilibrium or quasi-equilibrium theories.

Polarography is that subclass of voltammetry in which the indicator electrode is a liquid-metal (not necessarily, but normally, mercury) electrode whose surface is continuously or periodically renewed such that long-term accumulation of the products of electrolysis at the electrode–solution interface is prevented.

Classical polarography is therefore a subclass of both classical voltammetry and polarography.

14.1   VOLTAMMETRIC LINEAR DIFFUSION THEORY

Heyrovsky and Ilkovic (S2) in 1935 used the result of linear diffusion theory in the same manner as the Nernst diffusion-layer concept was used in the previous chapter. They obtained the general equation known as the *Heyrovsky–Ilkovic equation,*

$$E = E^{0\prime} + (RT/zF) \ln(D_{\mathrm{Red}}^{1/2}/D_{\mathrm{Ox}}^{1/2}) + (RT/zF) \ln[(I - I_{\mathrm{d,c}})/(I_{\mathrm{d,a}}-I)] \quad (14.1)$$

which upon definition of

$$E_{1/2} = E^{0\prime} + (RT/zF) \ln(D_{\mathrm{Red}}^{1/2}/D_{\mathrm{Ox}}^{1/2}) \qquad (14.2)$$

simplifies to

$$E = E_{1/2} + (RT/zF) \ln[(I - I_{\mathrm{d,c}})/(I_{\mathrm{d,a}} - I)] \qquad (14.3)$$

The above equation, which is identical to Eq. (13.19), shows that the only change in going from Nernst diffusion-layer theory to semi-infinite linear diffusion theory, insofar as the time-independent relationships between current and potential are concerned, is a minor change in the definition of the half-wave potential. When the diffusion coefficients of the oxidized and reduced species are equal, there is no difference whatever.

Since the half-wave potential $E_{1/2}$ differs from the formal potential $E^{0\prime}$ only by the logarithm of the ratio of the square roots of the diffusion coefficients, the half-wave potential is very close to the formal potential $E^{0\prime}$ when these coefficients are similar, as they usually are. Thus the voltammetric or polarographic half-wave potential is an effective measure of the formal potential (and thus the free energy) of an electrochemical reaction.

It is sometimes useful to employ this equation in the form of its explicit solution for current, which can be more simply written if a parameter P is defined as

$$P = \exp(E - E_{1/2})zF/RT \qquad (14.4)$$

Rearrangement then gives

$$I = (I_{\mathrm{d,c}} + PI_{\mathrm{d,a}})/(1 + P) \qquad (14.5)$$

The complete form of the Heyrovsky–Ilkovic equation describes the composite anodic–cathodic wave obtained when both the oxidized and reduced forms of a single reversible redox couple are originally present in the bulk solution, diffuse to the electrode surface, react, and diffuse again into the bulk solution. Simpler forms are adequate to describe most processes.

14.1.1   Reduction of Soluble Species

If the bulk electrolyte initially contains none of the reduced form, the anodic limiting current due to its oxidation must be zero. The Heyrovsky–Ilkovic equation therefore simplifies to

$$E = E_{1/2} + (RT/zF) \ln[(I_{d,c} - I)/I] \qquad (14.6)$$

where the half-wave potential has the same definition as above.

The potential-determining redox couple described by the equation above for the reduction of a soluble species is one in which both the oxidized and the reduced form are soluble species, but the two species need not both be soluble in the same phase. Thus it could be used to describe the behavior of $Fe^{3+}(aq) + e^- \rightarrow Fe^{2+}$ (aq) or $Cd^{2+}(aq) + 2e^- \rightarrow Cd(Hg)$ equally well when the diffusion coefficients used are those of the species in the phase in which it is soluble. Use of a plot of $\log[(I_{d,c} - I)/I]$ to test diffusion control is a common voltammetric and polarographic practice; a straight line should be obtained.

The explicit solution for current in this case is

$$I = I_{d,c}/(1 + P) \qquad (14.7)$$

since the anodic diffusion-limited current is zero. Note that the current is always cathodic because $I_{d,c}$ is negative.

14.1.2   Oxidation of Soluble Species

If the bulk electrolyte initially contains none of the oxidized form, the cathodic limiting current due to its reduction must be zero. The Heyrovsky–Ilkovic equation therefore simplifies to

$$E = E_{1/2} - (RT/zF) \ln[(I_{d,a} - I)/I)] \qquad (14.8)$$

where again the half-wave potential has the definition given above. As in the case of reduction, the two forms need not be soluble in the same phase.

The explicit solution for current in this case is

$$I = PI_{d,a}/(1 + P) \qquad (14.9)$$

since the cathodic diffusion-limited current is zero, and is always anodic or positive.

14.1.3   Reduction of Soluble Species to Insoluble Species

Further simplification of the Heyrovsky–Ilkovic equation is obtained when reduction of Ox yields a species Red that is soluble in neither the electrolyte nor the electrode. The activity of Red is then approximately constant, often constant at unity, and its variation can be ignored. The resulting simplification is

$$E = E_{1/2} - (RT/zF) \ln[I_{d,c} - I] \tag{14.10}$$

However, the definition of the half-wave potential has now been changed for two reasons. Since the reduced species does not diffuse, its diffusion coefficient is zero and vanishes from the definition. However, the activity of the reduced species (even if it is unity) has been incorporated into $E^{0'}$ and thus the value of $E^{0'}$ may differ from that expected. With this second caveat in mind, the definition of the half-wave potential is

$$E_{1/2} = E^{0'} + (RT/zF) \ln[1/(D_{Ox})^{1/2}] \tag{14.11}$$

14.1.4   Oxidation of Soluble Species to Insoluble Species

This uncommon simplification follows the same pattern as reduction of a soluble species to an insoluble species. The simplified equation is

$$E = E_{1/2} - (RT/zF) \ln[I_{d,a} - I] \tag{14.12}$$

where

$$E_{1/2} = E^{0'} + (RT/zF) \ln[D_{Red}^{1/2}] \tag{14.13}$$

Again, any change in the activity of the insoluble material appears as a change in the value of $E^{0'}$.

14.2   ILKOVIC EQUATION

Consider linear diffusion to an electrode whose area increases as the area of a sphere increases. The equation for linear diffusion to a planar electrode of constant area was derived in Chapter 13, but the variation of area with time remains to be derived. The volume of a sphere is

$$V = (4/3)\pi r^3 \tag{14.14}$$

If the sphere is a mass of liquid then its volume is also given by

$$V = mt/\rho \tag{15.14}$$

where t is time, m is the mass flow rate of liquid, and ρ is the density. Equating these;

$$r^3 = (3/4)mt\pi^{-1}\rho^{-1} \tag{14.16}$$

Since the area of any sphere is $A = 4\pi r^2$, substituting into this area equation the value for r obtained above yields

$$A = (36\pi/\rho^2)^{1/3}m^{2/3}t^{2/3} \tag{14.17}$$

and thus

$$I = zFcD^{1/2}m^{2/3}t^{1/6}\pi^{-1/2}(36\pi/\rho^2)^{1/3} \tag{14.18}$$

Actually, the diffusion layer does not grow quite this fast, since it spreads out as it moves away from the sphere. This is a purely geometric factor that can be corrected for by multiplication by the square root of 7/3 (S3), giving the final result which is known as the *Ilkovic equation*.

$$I = (7/3)^{1/2}\pi^{-1/2}(36\pi/\rho^2)^{1/3}zFcD^{1/2}m^{2/3}t^{1/6} \tag{14.19}$$

The Ilkovic equation is usually seen in the form obtained by combining constants and using the density of mercury at 25°C as 13,534 kg/m³,

$$I = 7.339 \times 10^{-3}zFcD^{1/2}m^{2/3}t^{1/6} \tag{14.20}$$

which in SI units gives I in amperes with c in moles per cubic meter, D in square meters per second, m in kilograms per second, and t in seconds. This is the equation for the linear diffusion-limited current to an expanding sphere at any instant. The resulting current–time curve is shown in Figure 14.1.

At the instant the sphere is of maximum size, which is in a dropping-mercury electrode the instant at which the drop falls τ, the current is also given by this equation. Using the cgs metric units I (μA), c (mmol/dm³), D cm²/s), and m (mg/s), the equation for the maximum current becomes

$$I_{max} = 708.1zcD^{1/2}m^{2/3}\tau^{1/6} \tag{14.21}$$

The average current over drop life is obtained by integration of the instantaneous current over drop life, which gives the Ilkovic equation in the form

$$I_{avg} = 606.9zcD^{1/2}m^{2/3}\tau^{1/6} \tag{14.22}$$

with the cgs metric units as above.

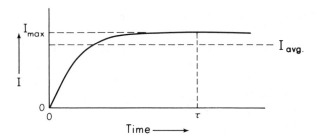

Figure 14.1 Current at an expanding spherical electrode. The maximum current I_{max} is achieved at τ, the instant the drop falls; it is equal to 7/6 of the average current I_{avg} during the entire drop life.

This equation was first derived in 1934 (S3), which is remarkably early considering that polarography began in 1922 (S4) and that the importance of the half-wave potential was not realized until 1935 (S2). The same equation is also obtained by the moving radial coordinate method of MacGillavry and Rideal (S5). In this case, the simplifying assumptions are mathematical, rather than the physical ones described below, but the result is the same.

14.2.1 Assumptions Inherent in the Ilkovic Equation

There are several assumptions that have implicitly been incorporated into this derivation. They are that:

1. The mass flow rate of mercury is constant during the lifetime of a drop. This can be shown to be only partially true in the first stages of drop life.

2. The drops are spherical. Photographs have shown that the drops are not perfect spheres, but the assumption is reasonably good for fine capillaries with no air bubbles.

3. The center of symmetry of each drop is fixed. The center of symmetry actually moves downward somewhat as the drop grows.

4. There is no shielding of the sphere by the capillary tube itself. This shielding can usually be ignored; capillaries with finely ground tips can be used to minimize it in the rare event it is necessary to do so, at the cost of a considerably more fragile capillary.

5. The concentration of reducible species is zero at the surface of the sphere and uniform throughout the bulk of the solution. This neglects depletion of the bulk concentration from the electrode reaction but is not a bad assumption for small electrodes.

6. There is no stirring. This is a reasonably good assumption unless the drop time becomes too fast, generally 0.1 s or less, because then the falling drops will stir the solution significantly.

7. Linear diffusion theory can be used instead of spherical diffusion theory.

14.2.2   Analysis of the Ilkovic Equation

The Ilkovic equation can be divided into two terms, the first of which is known as the *capillary characteristic*. This term $\tau^{1/6}m^{2/3}$ is characteristic of a particular capillary tube. It, or preferably m and s separately, should be reported in fundamental studies. The capillary characteristic depends upon the pressure driving the mercury flow, which is the difference between the pressure due to the height of the mercury column h and the back pressure due to the interfacial tension at the drop surface. The back pressure, in turn, depends on the surface tension, the potential, the solvent, and the composition of the solution. The components of the capillary characteristic are, first, the *drop time τ*. Unless mechanical drop detachment is used, the drop time for a given capillary depends upon the surface tension and thus upon the electrode potential, solvent, and solution composition. Although the drop time depends on potential, the second component, the *mass flow rate m*, does not vary significantly with potential. A plot of drop time against potential for a freely dropping mercury electrode is shown in Figure 14.2. The variation in drop time with potential is sometimes a problem with the dropping-mercury electrode; a drop time that is good at positive potentials often becomes too fast at extreme negative potentials.

The second term of the Ilkovic equation is called the *solution factor* because this term $zcD^{1/2}$ is characteristic of the particular electrochemical reaction taking place and the solution in which it occurs, and is independent of the capillary used. There are several factors included within this. The first is the *concentration c*. The Ilkovic relationship that current and concentration are directly proportional is true only if the residual current, which is not due to diffusion of the species, is subtracted out. Moreover, the direct proportionality between current and ionic concentration is a necessary but not a sufficient condition for diffusion control. The second component of the solution factor is the *diffusion coefficient D*. The rate of diffusion can be measured directly, as in a diffusion apparatus, or it can be calculated. For ions the diffusion coefficient can be calculated from conductance data at infinite dilution. For nonionic materials in solution the Stokes–Einstein equation works reasonably well for large, spherical molecules.

The third component of the solution factor is the number of electrons z. This is usually but not always known *a priori*. If it is not, then it must be determined polarographically, since the correct value of z is that for the polarographic reduction which may not correspond to the number of electrons determined under other conditions. This is usually done by means of the Ilkovic equation, because all the other terms are measurable. However, it is first necessary to prove that the process is diffusion-controlled. This is generally done by showing that I_d is directly proportional to $h^{1/2}$, which is a necessary but not a sufficient condition. If diffusion is controlling the current, then z must be an integer.

Having established diffusion control, there are two methods of obtaining the value of z. The first is an *absolute* method, in which the value of D is first estimated. Next, the values of I_d, m, c, and t are measured. Then z is successively

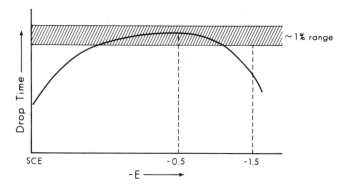

Figure 14.2   Variation of drop time with potential. The drop time shown is for a freely dropping mercury electrode. Electrodes with controlled drop time may go out of control at highly negative potentials if the spontaneous drop time decreases below the control setting.

assumed equal to 1, 2, 3, ..., and the value of D is calculated for each of these using the Ilkovic equation. The value of z giving the value of D closest to the original estimate is taken as correct. The second and better method is a *comparative* method. In this method it is necessary to measure, using the same capillary, the I_d values of two similar substances, whose reactions are both known to be diffusion controlled, and for one of which both z and D are known. Then:

$$I_d,1/I_d,2 = z(1)c(1)D(1)^{1/2}/z(2)c(2)D(2)^{1/2}$$

The ratio of electrons must be a small integer or the reciprocal of a small integer.

14.3   CURRENT AT THE DROPPING-MERCURY ELECTRODE

Currents flowing through dropping mercury electrodes may arise from, or be limited by, processes other than diffusion. A polarograph or other external sensor will record all current flowing as a sum and cannot distinguish between various processes that give rise to or limit it. It is therefore necessary to consider what other currents may arise and how they can be prevented or identified.

14.3.1   Diffusion-Limited Current

The diffusion-controlled current during the growth of a mercury drop at a dropping-mercury electrode as given by the Ilkovic equation increases as time to the one-sixth power. This is shown in Figure 14.3, which also shows actual current–time plot traced by a recording polarograph. This shape of the current–time curve during the lifetime of a single drop is a strong indication of diffusion control. The shape of the current-time curve can only be observed, however, with a recorder whose pen is fast enough to properly record the changes in current during drop life, nominally a 1-s pen recorder. With such a recorder it

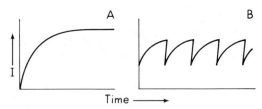

Figure 14.3   Diffusion-limited currents . (*A*) curve for one-sixth-power time dependence; (*B*) trace observed at a dropping-mercury electrode. Vertical drop does not reach zero owing to the finite response speed of recorder.

is best to read the maximum rather than the average current by avoiding damping.

14.3.2   Charging Current

The Ilkovic equation holds only for the current due solely to diffusion. Any residual current not due to diffusion must be subtracted from the total current observed. The major component of this residual current is due to the charging of the double layer and is called the *charging current*.

At reasonable polarographic concentrations of most supporting electrolytes, around 0.1 mol/dm³, the value of the double-layer capacitance of mercury changes comparatively little with potential, and a constant value of 200 to 250 mF/m² (20–25 μF/cm²) can be used. The equilibrium potential of the electrode is, as discussed in Chapter 5, the potential of the electrocapillary maximum E_{pzc} of the particular solution. To produce a different potential, charge must enter or leave the capacitor C_{dl} until the new potential E is achieved. This charge is given by

$$Q = AC_{dl}(E_{pzc} - E) \qquad (14.23)$$

The electrode area A is independent of potential, but in the case of a dropping-mercury electrode it is proportional to both the mass flow rate and time to the two-thirds power. The formula for polarographic charging current is then

$$I_{ch} = dQ/dt = C_{dl}(E_{pzc}-E)\, dA/dt \qquad (14.24)$$

Incorporation of the derivative of area with time yields:

$$I_{ch} = KC_{dl}[E_{pzc}- E]m^{2/3}t^{-1/3} \qquad (14.25)$$

with the constant K being 2/3 $(36\pi/\rho^2)^{1/3}$. This equation gives the decreasing curves shown in Figure 14.4.

Integration of the above equation shows that the average charging current over drop life is 3/2 that of the minimum charging current I_r, which is the observed

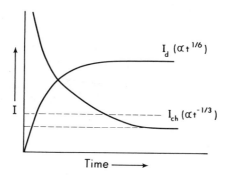

Figure 14.4 Charging currents. The lower dashed line gives the current observed at the end of drop life, $I(\tau)$, which is also the minimum charging current; the upper dashed line gives the average charging current, which is 3/2 the minimum. The time dependence of the current is to the $-\frac{1}{3}$ power. Vertical rise does not reach infinity owing to finite response speed of the recorder.

current at the end of drop life. Under normal polarographic conditions the charging current has a small potential-dependent value that passes through zero at the electrocapillary maximum potential, as shown in Figure 14.5. The residual current goes to zero at the electrocapillary maximum potential since the surface there possesses its equilibrium charge.

The significance of the charging current depends upon how large it is relative to the diffusion-limited current. For a 1-mmol/dm³ solution of a univalent ion, the diffusion current is about 5 μA and the charging current is 4% or less of this, so it can be ignored except in quite accurate work. If the concentration is 0.01 mmol/dm³ the charging current may now be four times the diffusion current, and a correction is obviously necessary. The charging current decreases through the life of a drop, while the diffusion current increases. The *relative* importance of the charging current therefore decreases during drop life; measurement of maximum rather than average current reduces its importance. Certain advanced polarographic techniques discussed in Chapter 16 take advantage of this by measuring current later in drop life.

14.3.3 Migration Current and Its Avoidance

In 1934 Heyrovsky (S6) noticed that a polarogram of a solution of a reducible ion, such as lead ion in a millimolar solution of lead nitrate, is affected by the addition of a large amount of a nonreducible ion such as potassium ion. The observed effect is shown in Figure 14.6.

The limiting current I_{lim} observed is given by

$$I_{lim} = I_d + I_m \tag{14.26}$$

that is, the observed limiting current is the sum of the current due to reduction of the lead ions moving under the influence of a diffusion gradient alone and the

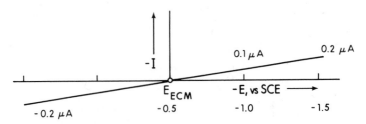

Figure 14.5 Charging component of residual current. Values shown are for a typical dilute aqueous solution using a DME and KCl supporting electrolyte.

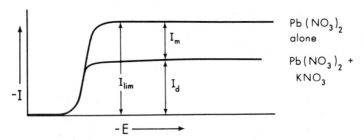

Figure 14.6 Migration current. The overall limiting current I_{lim} observed is the sum of the current for the reduction of lead ions moving under the influence of the diffusion gradient I_d and that for the reduction of lead ions moving under the influence of the electric field I_m. The latter is absent in the presence of a hundredfold excess of the nonreducible potassium ion.

current due to reduction of the lead ions moving under the influence of the electric field alone. In the presence of a large excess of supporting electrolyte thtat is not capable of oxidation or reduction at the potentials used, the value of the limiting current I_{lim} is decreased to that of I_d because the vast majority of the ions moving under the influence of the electric field are the ions of the supporting electrolyte. Since these are neither oxidized nor reduced there is no net current flowing due to their presence. In practice it is found that, within experimental error, the limiting current becomes equal to the diffusion-limited current when the concentration of the nonreducible supporting electrolyte concentration is more than 25 times the concentration of the reducible ions. Since the effect of migration current is decreased by addition of any inert electrolyte, a buffer solution may, if concentrated enough, remove the need for an additional supporting electrolyte.

Migration currents do not arise in the polarography of uncharged species. In the polarographic reduction of cations, the effect of migration is to increase the observed current above the normal diffusion-limited value since an additional accelerating influence on the ions is then present. In the polarographic reduction of anions, however, the effect of migration is to decrease the observed current below the normal diffusion-limited value since the effect of the potential gradient is to retard the motion of anions toward the negative electrode. The less common case of cation oxidation shows the same effect as anion reduction, while anion oxidation shows the same effect as cation reduction. The effect of migration can be quite significant, amounting to 50% of the observed current, and hence an inert electrolyte is absolutely vital for quantitative polarography.

The presence of certain types of electroactive species that are reduced prior to the reduction of the sought-for species can produce an exaltation of the migration current by increasing the IR drop of the solution. An uncharged species is preferred since the current can be thereby increased without simultaneously decreasing the solution resistance as more conducting ions are added. This effect, discovered by Kemula and Michalski (S7) and studied by Heyrovsky and Bures (S8) and by Lingane and Kolthoff (S9), is of no significant analytical utility. It is not a significant interference in the presence of a reasonable concentration of supporting electrolyte.

14.3.4 Current Maxima and Their Avoidance

Polarographic current maxima are due to additional mass transport due to convection, and therefore are observed as a current in addition to the diffusion-limited current. This convection is not the result of stirring or of vibration, but of the electrode reaction itself. Stirring or vibration can also lead to an increased current, but these effects are best avoided since the currents arising from them are not generally reproducible. Much of the study of polarographic maxima is due to Heyrovsky, and the best discussion is found in Heyrovsky and Kuta (G2). More practical information is given by Meites (G5).

Polarographic maxima are traditionally divided into two types.

Maxima of the First Kind

Maxima of the first kind are usually, but not always, of acute shape. They appear at the beginning of a reduction or oxidation wave in relatively dilute solutions. The height of the maximum may easily reach 10 times the height of the wave itself, and quantitative polarographic measurements may be made impossible by its presence. Maxima of the first kind are observed at potentials positive or negative with respect to the potential of the electrocapillary maximum, and are sometimes divided into positive and negative maxima on this basis. They are not observed at the potential of the electrocapillary maximum; thus the reduction of Cd(II), for which the half-wave potential is approximately equal to the potential of the electrocapillary maximum, shows no maximum of the first kind. The height of a maximum of the first kind can be reduced in three ways: by an increase in the drop time, by an increase in the concentration of reducible species, or by adding organic maximum suppressor. Only the third method is commonly used; the others are either impractical or do not decrease the maximum enough to be useful.

Maxima of the first kind are most easily identified by their acute appearance, as shown in Figure 14.8. The distinguishing characteristics of maxima of the first kind are the linear rise due to $E = IR$ followed by a discontinuous drop to the

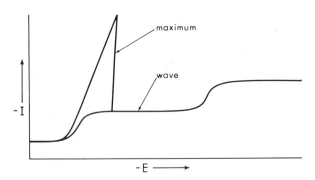

Figure 14.7   Maximum of the first kind. The lower curve, which is the diffusion-limited wave, would be obtained by the addition of sufficient organic maximum suppressor..

diffusion-limited plateau. They are observed at hanging- as well as dropping-mercury electrodes. These maxima arise from convective streaming in the solution around the drop under the influence of the potential gradient from the top to the bottom of the drop, which arises from the shielding of the capillary tip. Such streaming would appear as shown in Figure 14.9, and can be observed by study of the motion of small particles in the solution. The elimination of these maxima can be accomplished by the damping effect of a diffusion layer, once formed, which accounts for the disappearance of the maximum on the diffusion plateau, or by the damping effect of a layer of adsorbed capillary-active material. Addition of any amount of a capillary-active material, by which is meant a material that changes the potential of the electrocapillary maximum, decreases a maximum; enough such material will eliminate it completely. The capillary-active material must be in true solution; suspensions or colloids are ineffective. The usual maximum suppressors used in aqueous solution are gelatin, Triton X-100 (a detergent made by Rohm and Haas Company, Philadelphia, Penn.), or other organic materials of high molecular weight; in organic solvents, sulfur and sulfur-containing heterocyclic compounds are effective. Mixtures of several active substances may be more effective than the individual active substances.

The amount of maximum suppressor to be added must be determined empirically. Enough must be added to suppress the maximum, but an excess may produce adsorption effects, which will make measurement of the diffusion current impossible. So add as much as necessary but as little as possible. Typical amounts of Triton X-100 are 0.002% (0.1 cm³ of 0.2% stock solution in 10 cm³ of solution) and of gelatin are 0.05%. The latter should be freshly prepared, but Triton X-100 solutions can be stored. Some maximum suppressors are ineffective because distortion becomes significant before the maximum is completely suppressed.

Maxima of the Second Kind

Maxima of the second kind are rounded in shape. They appear in the middle of a

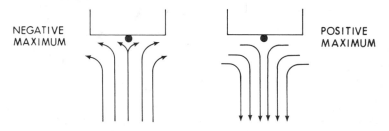

Figure 14.8   Streaming effects observed at hanging mercury drops.

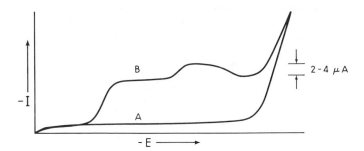

Figure 14.9   Maximum of the second kind. Curve A, supporting electrolyte alone; Curve B, supporting electrolyte plus reducible species.

diffusion-limited plateau, and in solutions whose concentration of supporting electrolyte is greater than 0.1 mol/dm³. The height of the maximum is comparatively low, as shown in Figure 14.9, but its presence is enough to cause serious error in measurement of the diffusion current. The height of the maximum is reduced when the flow rate of mercury is decreased, and no maximum is observed for stationary drops, but this fact is rarely of practical use. In practice, the maximum is eliminated by addition of capillary-active materials as are maxima of the first kind. Maxima of the second kind are believed to be due to convective processes occuring *within* the mercury drop rather than in the solution.

14.3.5   Catalytic Hydrogen Currents

Mercury has a sufficiently high hydrogen overvoltage that a hydrogen wave is usually negative of –1.4 V against SHE or –1.65 V against SCE except in quite acid solutions. Even at pH 0, a polarographic current of less than 10 µA is usually observed at potentials positive of –1.0 V against SHE on mercury, whereas on platinum a large current at 0.0 V is expected. This hydrogen overvoltage is reduced by either of two distinct classes of substance, and hydrogen ion reduction therefore appears as a wave. The first of these consists of the platinum metals, or complexes that on reduction will give platinum metals. The second class consists of organic substances containing groups capable of protonation that are adsorbed on the electrode, such as proteins, alkaloids, and quinine.

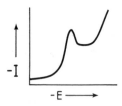

Figure 14.10   Catalytic hydrogen wave. Both peak and following plateau are part of the catalytic wave, while the following rise is due to a subsequent unrelated process..

Catalytic hydrogen waves are often in appearance somewhat similar to maxima of the first kind, as shown in Figure 14.10. They are not distinguishable solely by the presence of the peak, which is not always present. The useful diagnostics for catalytic hydrogen waves are, first, that the wave is diminished in the presence of oxygen, since the reduction product of oxygen is hydroxide and oxygen reduction precedes hydrogen reduction. Second, the height of the wave is decreased by increasing pH. Third, the limiting current either is independent of the mercury head or decreases with increasing mercury head, while for diffusion control, the limiting current is directly proportional to the square root of the height of the mercury column.

Catalytic hydrogen waves are of little use in analysis, and only the second group is occasionally a problem in routine work. Some use has been made of them in medical polarography (G2).

14.3.6   Currents Limited by Surface Processes

Currents that are limited by the surface are those for which each element of surface area can be used only once because it is blocked by a product of the electrode reaction; it is then not available for electrolysis of further molecules or ions diffusing in. The blocking can take place either by strong specific adsorption or by actual precipitation of insoluble material on the electrode.

Diagnosis of surface-limited currents is made fairly easily by observing the shape of the current–time behavior of individual drops on the current plateaus suspected of being due to surface-limited processes. For a surface-limited reaction, $I = Q \; dA/dt$, and the surface-limited current is obtained from the time-dependence of the area in the same manner as the charging current was previously:

$$I_{surf} = Km^{2/3}t^{-1/3} \tag{14.27}$$

the constant K being the same in both equations. The observed polarographic curves are shown in Figure 14.11.

Figure 14.11   Surface-limited cathodic process. Shapes of current–time curves for individual drops are shown; the rise time depends on the response speed of the recording device used.

This is one reason that a polarograph with a nominal pen response of one second or less should be used. With a polarograph with a slow pen response or heavy damping, the shape of the individual drop waveforms is not determinable.

Examples of surface-limited processes generally arise in the presence of strong specific adsorption, as discussed later, or of reactions whose products block the electrode surface. The reduction of $Co(NH_3)_6^{3+}$ gives polarograms as shown in Figure 14.12 when this complex is reduced (S10) at the DME to $Co(NH_3)_6^{2+}$; the increasing curves correspond to increasing concentrations of $Co(NH_3)_6^{3+}$.

The prewave is probably due to precipitation of $Co(OH)_2$ on the electrode. It is removed by higher concentrations of SCN^-, which complexes Co^{2+} and prevents precipitation, or by use of a NH_3/NH_4^+ buffer that prevents pH changes at the electrode sufficient to precipitate $Co(OH)_2$.

14.4   EFFECT OF REACTION KINETICS

The equation for a reversible voltammetric wave assumed that the current was limited only by the rate of diffusion of reacting species to and from the electrode surface. If other factors can also limit the current, the equation must be altered to take this into account. One of the most common such factors is the rate of a chemical reaction.

If the chemical step is infinitely fast relative to the rate of diffusion, then the shape of the voltammogram is not altered from that of a diffusion-limited process. It is only when the experimental conditions are such that the rate of the chemical step becomes comparable to or slower than the rate of diffusion that the voltammetric curve is affected. This is called *kinetic control* as opposed to *diffusion control*. As the rate of the preceding reaction decreases, the wave shifts from *pure diffusion control* through *partial kinetic control* to *pure kinetic control*. Kinetic control may arise from chemical reactions that precede, follow, or parallel the electrode process. *Activation control*, which arises from charge-transfer kinetics, may not always be experimentally distinguishable from kinetic control.

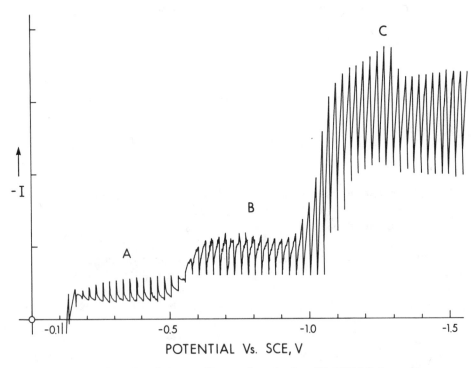

Figure 14.12   Film formation . Polarographic curve for reduction of $Co(NH_3)_6^{3+}$ in aqueous 1-mol/dm³ KSCN solution (author's laboratory).

14.4.1   CE Mechanism: Polarographic Effect

The mechanism for a reaction preceding the electrode process can be written as

$$A \rightleftarrows B, B + e^- \rightarrow C$$

The form A, which is electrochemically inactive at the potential being used, reacts to produce the polarographically active form B. The observed current is controlled by the rate of the preceding reaction and is affected by any parameter that might affect the rate of the preceding reaction. The ratio c_A/c_B, or the equilibrium constant, is of importance because, if [A]/[B] is very small, then the rate of the electrode reaction is totally controlled by the rate of the preceding reaction (pure kinetic control). If the ratio c_A/c_B is reasonably large, then diffusion of A from the bulk of the solution may also contribute to the observed current, leading to mixed diffusion-kinetic control.

Example
Formaldehyde in water exists in its hydrated form, methylene glycol, which is not reducible. The equilibrium lies far to the left:

$$H_2C(OH)_2 \rightleftarrows H_2C{=}O + H_2O$$

The polarographic reduction of formaldehyde is limited by the rate of dehydration of methylene glycol (S11, S12).

14.4.2   EC(R) Mechanism: Polarographic Effect

In this mechanism the electrochemically inactive form C reacts with the product B of the polarographic reduction of the active form A to regenerate A. The effect of this is that each molecule of A must be reduced more than once, and the height of the wave for reduction of A is therefore enhanced. The degree of enhancement depends upon the bulk concentration of C and the rate of the reaction $B + C \rightarrow$ A. This type of reaction is also called the *catalytic reduction* of the normally inert material C, the catalyst being the A/B couple. Under favorable circumstances, parallel reactions can be used to give quantitative information about the concentration of C, because if the rate of the $B + C \rightarrow$ A reaction is fast and first-order, the current controlled by that rate will be proportional to the concentration of C, [C].

Example

A catalytic wave for Fe(III) is observed in the presence of added H_2O_2 because the reduction product Fe(II) is reoxidized by the stronger oxidizing agent hydrogen peroxide.

$$Fe(II) + H_2O_2 \rightarrow Fe(III) + OH^- + OH\bullet$$

$$Fe(II) + OH\bullet \rightarrow Fe(III) + OH^-$$

The first step of this mechanism is rate-determining; the second is very rapid. Hydrogen peroxide cannot be electrolytically reduced until the potential is made much more negative than is necessary for reduction of Fe(III), and hence only the Fe(III) reduction is observed (S13).

14.4.3   EC Mechanism: Polarographic Effect

In this mechanism the product of a reversible reduction B is transformed into a polarographically inactive substance C. The effect of such reactions depend upon the rate of the $B \rightarrow$ C step and its overall reaction kinetics, but in general the effect will be to give a wave of normal reversible form at a half-wave potential that is significantly different from the reversible potential determined by other means.

Example

The anodic oxidation of ascorbic acid (S14, S15) gives a wave of reversible form whose half-wave potential is some 200 mV more positive than the potentiometric equilibrium potential. The limiting current is diffusion controlled. The irreversible reaction is to the electrochemically inactive dehydroascorbic acid. Similar behavior is observed in electrochemical reductions to unstable intermediates.

14.5   EFFECT OF CHARGE-TRANSFER KINETICS

In a reversible electrochemical reaction, the rate of electron transfer is considered infinitely fast relative to all other rates; thus the electron transfer does not give rise to any limitation on current. This means that $I^0 \gg I$, or in practice I^0 is greater than 10 times I, for any current observed. If the current drawn becomes greater than 10% of the exchange current, then the rate of electron transfer can limit the current observed. This can arise for a couple with a low exchange current density, a technique in which a high current density is required, or a combination of these two factors. A voltammetric wave in which the rate of electron transfer limits the observed current is said to be under partial or complete *activation control*.

14.5.1   Theory of Activation Control

At any potential, the net current I is the sum of its anodic and cathodic components (Chapter 5):

$$I = zFAc^0{}_{Red}k_{a,h} - zFAc^0{}_{Ox}k_{c,h}$$

The surface concentrations of both reduced and oxidized forms depend upon time, the rate constants for both anodic and cathodic reactions, and the diffusion coefficients of both the reduced and oxidized forms. The most appropriate method of dealing with such a complex dependence is to define the parameter ω;

$$\omega = (k_{c,h}t^{1/2}/D_{Red}{}^{1/2}) + (k_{a,h}t^{1/2}/D_{Ox}{}^{1/2}) \qquad (14.28)$$

which permits the expression of the dependences of both surface concentrations in the same form,

$$c^0 = c \exp(\omega^2) \operatorname{erfc}(\omega) \qquad (14.29)$$

so that

$$I = zFA \exp(\omega^2) \operatorname{erfc}(\omega) (c_{Red}k_{a,h} - c_{Ox}k_{c,h}) \qquad (14.30)$$

The general equation was developed independently by Smutek (S16), Delahay (S17), and Kambara and Tachi (S18). The more useful expressions for limiting currents are due to Delahay and Strassner (S19, S20) and Evans and Hush (S21); these relate to *totally irreversible* voltammetric waves, for which the backward reaction is neglected. It can be shown (G6, Chapter 4) that certain numerical values of the heterogeneous electrochemical rate constant $k^0{}_h$ apply in the case or normal polarography. When $k^0{}_h > 2 \times 10^{-4}$ m/s, the reaction is polarographically reversible, that is, the effect of the rate constant cannot be detected polarographically. It can be said to be *infinitely fast* on the polarographic time scale. At the other extreme, when $k^0{}_h < 3 \times 10^{-7}$ m/s, the reaction is polarographically totally irreversible, that is, the rate is *infinitely slow* on the polarographic time scale. In between, control is intermediate, and the process is usually called *quasi-reversible*.

Cathodic Limiting Current

Both the cathodic and anodic heterogeneous rate constants depend upon the electrode potential. This dependence is exponential in form and of opposite direction for the two rate constants as discussed in Chapter 5. When the potential is sufficiently cathodic (negative), the rate constant of the anodic reaction is so small that the expression for ω can be approximated as

$$\omega = k_{a,h} t^{1/2} / D_{Ox}^{1/2} \tag{14.31}$$

and the term $c_{Red} k_{c,h}$ becomes negligible with respect to $c_{Ox} k_{a,h}$. Therefore

$$I_{lim,c} = -zFAc_{Ox}k_{a,h} \exp(\omega^2)\, \text{erfc}(\omega) \tag{14.32}$$

Now the cathodic limiting current for a diffusion-controlled process at a planar electrode is:

$$I_{d,c} = -zFAD_{Ox}^{1/2} c_{Ox} \pi^{1/2} t^{-1/2} \tag{14.33}$$

Comparison of the two equations for limiting current requires the rewriting of the equation for activation control as

$$I_{lim,c} = -zFAD_{Ox}^{1/2} c_{Ox} t^{-1/2} \omega \exp(\omega^2)\, \text{erfc}(\omega) \tag{14.34}$$

The ratio of diffusion-limited to activation-limited currents is then

$$I_{lim,c} / I_{d,c} = \pi^{1/2} \omega \exp(\omega^2)\, \text{erfc}(\omega) \tag{14.35}$$

Anodic Limiting Current

The expression for ω is now approximated as

$$\omega = k_{a,h} t^{1/2} / D_{Ox} \tag{14.36}$$

hence

$$I_{\text{lim,a}} = zFAD_{\text{Red}}^{1/2}c_{\text{Red}}^{1/2}\omega \exp(\omega^2)\, \text{erfc}(\omega) \tag{14.37}$$

and

$$I_{\text{lim,a}}/I_{\text{d,a}} = \pi^{1/2}\omega \exp(\omega^2)\, \text{erfc}(\omega) \tag{14.38}$$

14.5.2   Polarographic Irreversibility

As the value of ω increases, the value of the function $\pi^{1/2}\omega \exp(\omega^2)\, \text{erfc}(\omega)$ increases toward unity. At sufficiently negative potentials, then, a cathodic current limited by both activation and diffusion approaches the diffusion limit alone. Likewise, at sufficiently positive potentials, an anodic current limited by both activation and diffusion approaches the diffusion limit alone. Thus for an irreversible wave in polarography, the *height* of the wave is not affected, since diffusion control still prevails there. The *shape* of the wave is altered, however; it becomes much more drawn out, as shown in Figure 14.13. The value measured for the half-wave potential of an irreversible reaction depends on the degree of irreversibility, or the magnitude of $k^0{}_h$, as well as on the free-energy change of the reaction, while for a reversible reaction the value of the half-wave potential depends on the free-energy change of the reaction alone. The half-wave potential for an irreversible reaction also varies with the drop time (by some 60 mV, over the range of polarographically reasonable drop times) as a consequence of its dependence upon $k^0{}_h$. Quantitative comparison of the half-wave potentials of different structures, as in organic polarography (G4), is a dangerous thing to do if the reactions are irreversible, since both the drop time and the value of the standard heterogeneous rate constant must be invariant across the series of compounds being studied.

In the derivation of the Ilkovic equation, it was assumed that the only process governing current at an expanding spherical electrode is diffusion. If activation also limits the current, solution of the differential equations is more complex; the result of the solution obtained by Koutecky (S22) is not expressible in simple analytic form. It is given numerically in Table 14.1, which gives the functional dependence of I/I_d on ω. The values of I/I_d range from zero at the foot of the wave ($\omega < 0.1$) to unity on the diffusion plateau ($\omega > 20$). Since for the limiting current that is the current at the end of drop life when $t = \tau$,

$$\omega = k_h\tau^{1/2}/D^{1/2} \tag{14.39}$$

and then since both τ and D are constant over the wave, it follows that ω depends only on the heterogeneous rate constant, which in turn depends upon the electrode potential. The physical consequence is that for potentials at the foot of the wave ω is small and the current is limited by activation; for potentials on the diffusion plateau ω is large and the current is limited by diffusion; and at intermediate

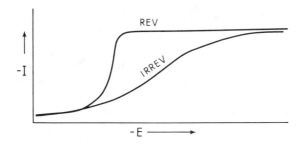

Figure 14.13   Reversible and irreversible polarographic reductions.

TABLE 14.1   Values of I/I_d for Irreversible Polarographic Wave

$(12/7)^{1/2}\omega$	I/I_d	$(12/7)^{1/2}\omega$	I/I_d
0.0	0.0	3.0	0.777
0.1	0.0828	3.5	0.803
0.2	0.155	4.0	0.827
0.3	0.219	4.5	0.844
0.4	0.275	5.0	0.858
0.5	0.325	6.0	0.880
0.6	0.369	7.0	0.895
0.7	0.409	8.0	0.909
0.8	0.444	9.0	0.919
0.9	0.476	10.	0.927
1.0	0.505	12.	0.939
1.2	0.555	14.	0.946
1.4	0.597	16.	0.953
1.6	0.632	18.	0.958
1.8	0.662	20.	0.963
2.0	0.690	∞	1.000
2.5	0.740		

points along the wave the current is limited by both diffusion and activation.

Diagnosis of irreversible behavior can be made by two methods. The first is that the shift of $E_{1/2}$ with drop time at 30°C should be 18 mV positive when the drop time is doubled, assuming $\alpha = 0.5$. Second, a plot of E against $\log[(I(d) - I)/I]$ gives a slope that corresponds to αz rather than z; the slope is therefore less by nearly a factor of 2, since α is usually about 0.5. Since the standard heterogeneous rate constant is a function of the electrode potential and the current depends upon that constant, the current is *not* independent of the electrode potential.

14.6   EFFECT OF ADSORPTION

Specific adsorption of material on the electrode surface is often blamed, frequently without justification, for all manner of deviations from expected polarographic behavior. Specific adsorption may be strong or weak and involve the substance being reduced, the reduction product, or other species present in the solution. The effects of adsorption of other species are complex, as are the combined effects of adsorption and kinetic control (G2). Only reversible reductions will show the simple effects described below.

14.6.1   Adsorption of the Product

If the substance being reduced is not adsorbed and the reduction product is sufficiently strongly adsorbed that it does not leave the electrode surface accessible to incoming material, elements of the surface are blocked to incoming reducible species and a surface-limited *adsorption prewave* is observed. The half-wave potential of the surface-limited wave will be positive of the half-wave potential of the diffusion-limited wave for the same diffusing species, and the difference between them is a measure of the free energy of adsorption of the product. It can be used for this purpose if both waves are observable.

The classic example of the influence of specific adsorption on polarographic curves was obtained in the study of the methylene blue (MB)–leucomethylene blue (LMB) system by Brdicka (S23). The electrochemical reaction is $MB + 2e^-$ $\rightarrow$ LMB; LMB is very strongly adsorbed on the surface of the electrode, while MB is not. The polarographic waveform obtained is shown in Figure 14.14. The height of the second plateau, for which the current-time curve over drop life is normal, is directly proportional to concentration as would be expected for a diffusion-limited plateau. The height of the first, or prewave, plateau, for which the current-time curve over drop life indicates a surface-limited process, is independent of concentration. The explanation for these two waves can be summarized in the following equations:

$$MB + 2e^- \rightarrow LMB(adsorbed) \qquad\qquad [prewave]$$

$$MB + 2e^- \rightarrow LMB(solution) \qquad\qquad [wave]$$

The process responsible for the prewave occurs more easily, that is, at lower applied voltage, because the adsorption energy of LMB favors it. The process responsible for the wave is not so favored and occurs only at higher applied potential.

Currents that are surface-limited are comparable in magnitude to diffusion-limited currents observed in dilute solutions. It can be calculated that if a monolayer with 0.5 nm^2 of surface area occupied by each molecule is formed then the current is, under normal polarographic conditions, 0.35 μA for I_{max} and $0.35 \times 3/2 = 0.54$ μA for I_{avg}.

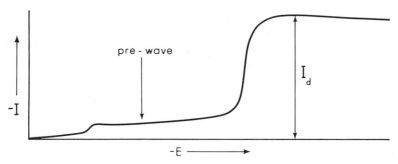

Figure 14.14   Adsorption prewave. Polarographic curves for the reduction of methylene blue; surface-limited current–time curves are obtained for drops on the prewave plateau but not for drops on the diffusion plateau.

14.6.2   Adsorption of the Reactant

If the reactant is sufficiently strongly adsorbed that it does not leave the electrode surface but the product is not adsorbed, and if the adsorption equilibrium is fast enough that equilibrium is achieved within the polarographic drop time, then in addition to the normal wave due to reduction of the nonadsorbed reactant there will be a wave due to the adsorbed reactant at more negative potentials, since the free energy of adsorption must now be overcome. The normal wave is at the expected half-wave potential, but its current is reduced since the adsorbed species diffuses but is not reduced at the potential of the normal wave. The height of the normal wave is concentration-dependent, but the height of the adsorption wave is not. Adsorption of reactant is much less common than adsorption of product in polarography.

14.7   REFERENCES

General

G1   I. M. Kolthoff and J. J. Lingane, *Polarography*, Interscience, New York, 1952, 2 volumes. Outdated classic.

G2   J. Heyrovsky and J. Kuta, *Principles of Polarography*, Academic Press, New York, 1966. Best current text for theory.

G3   S. G. Mairanovskii, *Catalytic and Kinetic Waves in Polarography*, Plenum Press, New York, 1968.

G4   P. Zuman, *Organic Polarographic Analysis*, Pergamon Press, Oxford, 1964.

G5   L. Meites, *Polarographic Techniques,* 2nd ed., Interscience, New York, 1964, 752 pp. Excellent discussion of polarography; also covers amperometric titration and, very briefly, other related techniques.

G6   P. Delahay, *New Instrumental Methods in Electrochemistry*, Wiley-Interscience, New York, 1954, 437 pp.

G7   G. W. C. Milner, *The Principles and Applications of Polarography and Other Electroanalytical Processes*, Longmans, London, 1957, 729 pp. The first 200 pages of this work are general; the remainder is a highly practical and well-organized compilation of specific useful methods. Covers the literature through 1955.

G8   O. H. Muller, *in Physical Methods of Chemistry, Part IIA: Electrochemical Methods*, A. Weissberger and B. W. Rossiter, Eds., Wiley-Interscience, New York, 1971, pp. 297ff.

Specific

S1   *Pure Appl. Chem.* **45**, 81 (1976).

S2   J. Heyrovsky and D. Ilkovic, *Coll. Czech. Chem. Commun.* **7**, 198 (1935).

S3   D. Ilkovic, *Coll. Czech. Chem. Commun.* **6**, 498 (1934); *J. Chim. Phys.* **35**, 129 (1938).

S4   J. Heyrovsky, *Chem. Listy* **16**, 256 (1922).

S5   D. MacGillavry and E. K. Rideal, *Rec. Trav. Chim.* **56**, 1013 (1937).

S6   J. Heyrovsky, *Ark. Hemiju Farmociju* **8**, 11 (1934).

S7   W. Kemula and M. Michalski, *Rocz. Chem.* **16**, 533 (1936).

S8   J. Heyrovsky and M. Bures, *Coll. Czech. Chem. Commun.* **8**, 446 (1946).

S9   J. J. Lingane and I. M. Kolthoff, *J. Amer. Chem. Soc.* **61**, 1045 (1939).

S10   H. A. Laitinen, A. J. Frank, and P. Kivalo, *J. Amer. Chem. Soc.* **75**, 2865 (1953).

S11   K. Vesely and R. Brdicka, *Coll. Czech. Chem. Commun.* **7**, 313 (1947).

S12   R. Bieber and G. Trumpler, *Helv. Chim. Acta* **30**, 706, 971, 1109, 1286, 1534, 2000 (1947).

S13   I. M. Kolthoff and E. P. Parry, *J. Amer. Chem. Soc.* **73**, 3718 (1951).

S14   Z. Vavrin, *Coll. Czech. Chem. Commun.* **14**, 367 (1949).

S15   C. Cattaneo and G. Sartori, *Gazz. Chim. Ital.* **72**, 351 (1942).

S16   M. Smutek, *Coll. Czech. Chem. Commun.* **18**, 171 (1953).

S17   P. Delahay, *J. Amer. Chem. Soc.* **75**, 1430 (1953).

S18   T. Kambara and I. Tachi, *Bull. Chem. Soc. Japan* **25**, 135 (1962).

S19   P. Delahay, *J. Amer. Chem. Soc.* **73**, 4944 (1951).

S20   P. Delahay and J. E. Strassner, *J. Amer. Chem. Soc* **73**, 5219 (1951).

S21   M. G. Evans and N. S. Hush, *J. Chim. Phys.* **49**C 159 (1952).

S22   J. Koutecky, *Chem. Listy* **47**, 323 (1953).

S23   R. Brdicka, *Z. Electrochem.* **47**, 721 (1941); **48**, 278 (1942).

14.8   STUDY PROBLEMS

14.1 The diffusion-limited current of a certain one-electron polarographic reduction is 8.0 μA in a 1.0 $mmol/dm^3$ solution of a certain organic compound. A similar compound was studied polarographically but was soluble only to the extent of 0.2 mol/m^3; the diffusion-limited current was 3.2 μA. What can you tell about the reduction process of the second compound?

14.2 Compute the (maximum) polarographic current observed with a DME at a drop time of 2 s if the polarographic diffusion current observed with a drop time of 5 s is 7.3 μA.

14.3 Explain qualitatively why the potential of a surface-limited process due to the formation of an insoluble film (such as $Co(OH)_2$) is positive of the potential of a reversible $Co(III)/Co(II)$ process while the potential of the reduction of $Co(NH_3)_6^+/Co(Hg)$ is negative of the reversible $Co(II)/Co(Hg)$ process.

14.4 Assuming that the charging current is the only significant component of the residual current and that the limiting sensitivity of a technique is the point at which the signal/noise ratio becomes unity, calculate the limiting sensitivity of polarography at drop times of 2 and 7 s.

14.5 One of the assumptions of the Ilkovic equation is the absence of depletion of the bulk solution. Comment on the validity of this assumption in the polarographic analysis of 10 cm^3 of 0.2 mol/m^3 Tl^+ if the analysis takes 10 min; choose reasonable values for any necessary parameters.

14.6 Polarographic analysis of a certain compound reduced irreversibly in a one-electron step gave a (maximum) diffusion current of 3.10 μA with a solution concentration of 0.88 mol/m^3. The electrode was dropping freely with a drop time of 3.53 s and the mass flow rate of mercury was 1.26 mg/s. Use the Ilkovic equation to calculate a numeric value for the diffusion coefficient of the species. Why is it proper to use the Ilkovic equation with an irreversible wave?

CHAPTER
15
PRACTICE OF CLASSICAL
VOLTAMMETRY AND POLAROGRAPHY

Classical voltammetry and its widely used subclass, classical polarography, differ only in that polarography uses a dropping-mercury electrode (DME) while voltammetry uses other types of electrodes. The other apparatus is virtually identical in voltammetry and polarography and is usually called a *polarograph*.

A polarograph consists of two circuits, a polarizing circuit that applies a slow DC ramp to the polarographic cell and a measuring circuit that monitors the cell current. Originally measurements were made using a hand-adjusted polarizing unit and an optical galvanometer, but this was soon replaced by more automatic instrumentation. For aqueous work a two-electrode cell and electromechanical polarograph based on a potentiometric recorder, shown in Figure 15.1, are perfectly adequate. Excellent commercial units have long been available (E. H. Sargent). With proper maintenance they, function well for decades.

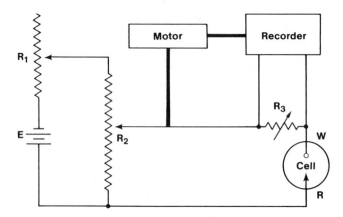

Figure 15.1   Two-electrode polarograph. Voltage source E, normally a battery, is adjusted so that desired full-scale voltage appears across potentiometer R_2. Motor drives potentiometer R_2 and also time axis of $x–t$ recorder. Cell current passes through calibration resistor R_3; voltage across R_3 is used as input to recorder. Recorded output, actually a voltage–position plot, thus appears as the desired current–voltage plot. Mechanical linkages shown in heavy lines. Standardization circuits and parameter selection circuits are optional and not shown.

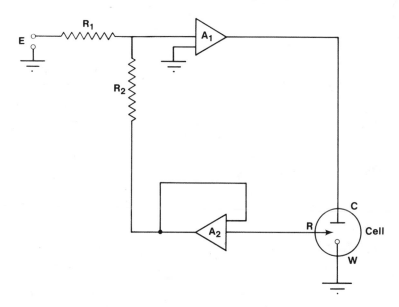

Figure 15.2   Three-electrode polarograph circuit. Voltage follower circuit A_2 is used to minimize current flow through the reference electrode R. The potentiostat circuit of A_1 puts the counterelectrode–working electrode (C–W) path in its output circuit. Potential between R and W gives a current via A_2 and R_2 which, summed with that of E via R_1, is the input of the amplifier A_1. The current output of A_1 is therefore controlled at the value required to hold W at the potential relative to R established by the input voltage E. This input voltage is normally the slow DC ramp potential. Actual circuits use additional components for ramp generation and parameter selection.

For nonaqueous work a three-electrode cell and electronic potentiostat circuit such as that shown in Figure 15.2 are usually necessary. A highly successful design is the Model 174A Polarograph (Princeton Applied Research) shown in Figure 15.3. This unit, in conjunction with a mechanical drop dislodging circuit sold as an accessory, is also capable of carrying out current-sampled DC polarography and both normal and differential pulse polarography. The modern Metrohm unit shown in Figure 15.4 has similar functions and uses similar cells.

15.1   CLASSICAL POLAROGRAPHY: THE DROPPING-MERCURY ELECTRODE

The construction of a practical dropping-mercury electrode is shown in Figure 15.5. The capillary tube is about 12 to 20 cm long. Suitable fine-bore tubing is sold as DME capillary (E. H. Sargent and Company, Chicago, Ill. or Princeton Applied Research) or as marine barometer tubing (Corning Glass Works). The standpipe is not necessary but is useful for measuring the height of the mercury column. The stopcock, that is also not necessary though useful, helps to save mercury and prevent plugging of the capillary. Suitable stands must be provided to hold the DME vertical and allow adjustment of the level of the dropping bulb.

Figure 15.3   P.A.R. Model 174A polarograph. (courtesy of Princeton Applied Research)

Figure 15.4   Polarograph with cells, DME, and HMDE. Metrohm E506 Polarographic Analyzer, courtesy of Brinkmann Instruments Ltd. The cells shown are for a DME (right) and HMDE (left), and are similar to those used by the P.A.R. unit. In this analyzer the recorder is an integral part of the unit, while the P.A.R. unit shown in the previous figure uses an external x–t or x–y recorder.

DROPPING MERCURY ELECTRODE

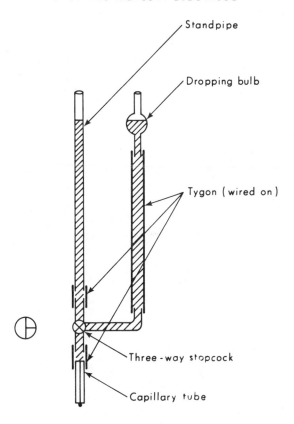

Figure 15.5 Simple dropping-mercury electrode. Detail at left indicates stopcock orientation when dropping. Teflon stopcock is preferable; use no grease.

Electrical contact to the DME is most easily made through the top of the dropping bulb. An efficient contact consists of a short platinum wire sealed through the closed end of a 10-cm piece of glass tubing. This is inserted into a split rubber stopper that is inserted into the top of the dropping bulb. Contact with the platinum inside the tube can be made with a copper wire and mercury; alternatively, the copper lead can be soldered to the platinum wire. This electrical contact will be lost if air enters the capillary thread. Loss of contact is not entirely disadvantageous as the air will prevent reliable polarographic measurements and must be removed by passage of additional mercury in any case. Normally the DME tip is held fixed and the polarographic cell held beneath it with an adjustable clamp so that the cell can be removed for washing the capillary or changing the solution.

Dropping-mercury electrodes will give very long service, over a period of months to years, when they are properly cared for. When abused, their lifetime is measurable in seconds. The key to successful maintenance of a dropping-mercury electrode is that solutions must *never* be allowed to enter the capillary. This means that the mercury flow must be started and the drops must be falling *before* the dry electrode is placed in solution, and that the mercury flow must continue until *after* the tip of the dropping mercury electrode is clean and dry. This can most easily be achieved by removing the solution, rinsing the tip with distilled water, draining most of the mercury from the standpipe, and letting a small amount of mercury continue to drop while the tip dries. If a capillary becomes dirty, attempts can be made to clean it with dilute nitric acid, but such attempts are rarely fully successful. The only cure is to break off the lower half-inch or so of the capillary, since the interfering material is normally inside the bore of the capillary just at the lower end. Repetition of this procedure will, of course, produce capillaries too short to maintain the desired drop time, which is of the order of 5 s for a freely dropping capillary. The capillary characteristic is changed by breaking off the tip and, in quantitative polarographic work, recalibration of the capillary is required.

Dirty capillaries should be suspected whenever the response of a polarograph becomes erratic. The drop times will generally vary randomly, and the currents observed will also. The effect of a dirty capillary is usually obvious when the capillary is freely dropping. If the drops are being mechanically detached, however, the effect is not as obvious and is more easily overlooked, though it may be no less deleterious.

The mercury used in the DME must be clean and free of impurities. Procedures for the proper cleaning of mercury are given in the general references cited at the end of this chapter. Mercury vapor is toxic, and precautions against spills are advisable. These include a tray beneath the DME assembly, careful wiring of all tubing, and a spare small beaker containing water that is placed under the DME while it is being washed or while solutions are being changed. Used mercury may be safely stored under water indefinitely while enough is being collected to justify repurification.

More elaborate electrodes that provide mechanical drop detachment or programmed dropping are commercially available. For classical polarography they have no advantages, although they are advantageous (and often required) for advanced polarographic and voltammetric techniques.

15.2   CLASSICAL VOLTAMMETRY: OTHER ELECTRODES

The apparatus used for classical voltammetry is identical to that used for classical polarography, and the technique is similar except for the electrode used. Since stable currents are not obtained by diffusion to a plane electrode of constant area, the electrodes often employ convection instead, as in the rotating platinum electrode or the rotating disk electrode, in which the disk may be fabricated of different metals. The electrodes employed in classical voltammetry are usually solids rather than liquids, although stationary mercury pools, hanging-mercury-drop electrodes (HMDE), and thin film mercury electrodes (TFMEs) are also used. Quantitative results are rarely obtained in classical voltammetry at any electrode other than the DME, although other voltammetric techniques are capable of providing quantitative information.

By comparison with the DME, other electrodes have several disadvantages. First, the surface of other electrodes is not continuously renewed, while that of the DME is. Second, electrode products may accumulate at the electrode surface; this is not usually a problem with the DME. Third, specific interactions of the solid electrode material or of mercury with the electrode process may occur.

There are various types of solid electrode materials, of which the most commonly used is probably platinum. With respect to the NHE, platinum has a useful voltage range of about 0.0 to $+0.75$ V, with perhaps an additional 250 mV available in each direction if circumstances are favorable. Use of platinum is complicated by surface oxide effects, and platinum also both absorbs and adsorbs hydrogen. Platinum can be sealed in Pyrex glass, which is useful in fabricating electrodes.

The second most commonly used electrode material is gold. With respect to the NHE, gold has a useful voltage range of about -0.75 to $+1.5$ V (except in acid chloride media where complex chlorides form) with the same type of extensions possible as for platinum; hence gold is much more useful for cathodic studies than is platinum. Gold is subject to surface oxide effects similar to those of platinum, but it does not absorb or adsorb hydrogen to nearly the same extent as does platinum. Unfortunately, however, metallic gold cannot be effectively sealed in glass.

A third class of electrodes now coming into wider use is carbon electrodes. There are three types of these. The first are simply carbon-rod electrodes. Generally, wax-impregnated graphite is now used. The potential range of the wax-impregnated graphite electrode (WIGE) is -1.0 to $+1.3$ V against NHE, making it suitable for many organic reductions and oxidations. Electrodes of the WIGE type have some problems with reproducibility, and generally also have high residual currents. The second type of carbon electrode is made out of pyrolytic graphite. These electrodes have a useful potential range from about -0.8 to $+1.0$ V against NHE, slightly more restricted, but are otherwise very similar to the WIGE. The third type of carbon electrode is the carbon-paste electrode. Like the carbon-rod electrode, it has a useful potential range of about -1.0 to $+1.3$ V. These electrodes are made of graphite powder mixed with an organic liquid such as Nujol; the mixture is then packed into an insulating holder. Mercury electrodes other than the DME are used in voltammetry both as the hanging-mercury drop and in the form of a mercury pool. Except for their nonrenewable surface, they are very similar to the DME and have a similar useful potential range. Details of their construction are given by Adams (S1).

15.3   CLASSICAL POLAROGRAPHIC MEASUREMENTS

The parameters that must be measured in classical polarography include the height of the diffusion-limited current plateau I_d, which is directly proportional to the bulk concentration of diffusing species c, and the half-wave potential $E_{1/2}$, which is characteristic of the particular electroactive species that is being either oxidized or reduced. Other useful parameters are the drop time τ, the height of the mercury column h, the mass flow rate of mercury m, and the residual current. These are either components of the Ilkovic equation and therefore necessary in quantitative polarography, or are useful diagnostic tools.

15.3.1   Residual Current

The current observed is the total current, which is the sum of the diffusion-controlled current and such other currents as may be present, which are collectively called the *residual current*. The largest component of the residual current is usually the charging current. The best procedure for obtaining the residual current is to run a complete polarographic curve on the *base electrolyte*, which is a solution containing the same concentrations of buffer, supporting electrolyte, maximum suppressor (if necessary), complexing agent (if used), and any other solute materials with the sole exception of the diffusing species whose behavior is to be studied. This polarographic curve will then give, at any potential desired, the residual current by direct measurement of either the maximum or the average currents. By "maximum currents" is here meant not the actual maximum current that exists during drop life but the current at the end of drop life. One of the major contributions, often the largest contribution, to the observed residual current is the charging current discussed in Chapter 14. For charging current, the current decreases during drop life and is a minimum rather than a maximum at the end of drop life. It is this *minimum* current that is to be taken as the residual current for measurements of the *maximum* diffusion-limited current. If the residual current contains other current components of magnitude similar to that of the charging current, the current during the life of an individual drop may be level or may decrease and then increase. In any event, the correct choice of residual current for measurement of the maximum *diffusion-limited* polarographic currents is the current observed at the end of drop life, while the correct residual current for measurement of the average *diffusion-limited* currents is the average current that flows during drop life.

A less sound but more convenient method of measuring the residual current is to use only the single polarogram of the base electrolyte plus the diffusing species whose behavior is to be studied. In some circumstances only this polarogram is available. The residual current is then estimated by drawing a straight line of slight ascending slope through the entire polarogram. The line should be coincident with the polarogram prior to the beginning of the rise due to reduction (or oxidation) of the most easily reducible (or oxidizable) species. The line may be constructed for either the average or the maximum diffusion-limited currents as discussed in the preceding paragraph. This technique assumes that charging current is the major component of the residual current.

If a polarogram contains more than one polarographic wave, the residual current for the second or later wave(s) is the diffusion-limited current for the preceding polarographic wave(s), extrapolated through the polarogram. This extrapolation will normally be a straight line of slight ascending slope as discussed in the preceding paragraph. If a maximum is present, it is ignored in the extrapolation. The residual current curve and the diffusion-limited current line(s) are normally expected to be straight lines of slight ascending slope due to the charging current, and they should be constructed so as to be parallel. That is, the height of the diffusion-limited plateau should be constant and independent, when corrected for residual current, of the location along the plateau.

15.3.2   Limiting Current

The limiting current is most easily measured with the help of a finely graduated transparent plastic ruler. Either a maximum or average line may be taken, but the line must be of the same type as the residual current line and should be parallel to it. In drawing the line, care must be taken to avoid being misled by the presence of maxima of the first kind. It is particularly difficult to measure heights of diffusion plateaus when the plateau itself is short or when the wave is irreversible and therefore drawn out and the start of the plateau is indistinct; in either of these cases, failure to construct a diffusion-limit line parallel to the residual current line may lead to serious error.

In taking the average current, care must be taken to ensure that the same damping factor is used in all measurements. Changes will make significant differences in the shape of the curves and in the results obtained.

Difficulties associated with measuring diffusion-limited currents of waves whose potential span is short can often be alleviated by use of a slower scan speed in order to obtain more drops. This does not help the problem of sloping curves due to irreversibility, although slower scan speeds do assist in defining such curves.

15.3.3   Drop Time

The drop time is most easily measured by stopping a polarogram at some potential and using a stopwatch to time the interval during which 10 or 20 drops fall. The drop time will generally vary significantly only with potential and height of the mercury column, so these parameters must be kept constant. The effect of changing the solution, though observable, is generally not significant. The drop times for a freely dropping electrode should be between 3 and 8 s. At longer times, convection can become a serious interference. At shorter times, the current is still changing very rapidly with time. With mechanical drop detachment, 2-s drop times are often used; 5-s times are perhaps preferable.

15.3.4   Mass Flow Rate Of Mercury

Approximate measurement of the mass flow rate of mercury is easily done simply by letting the DME drop in air and catching the mercury that flows out over some convenient period of time in a weighing bottle; this directly gives the flow rate in grams per unit time. Since the height of the mercury column is then the only experimental variable, and when dropping in solution is still the most significant experimental variable, the mass flow rate will be roughly the same dry or in the presence of solution. It is preferable to carry out the measurement by placing a small amount of mercury in a dry weighing bottle, weighing the capped bottle, inserting the DME capillary into the mercury, and weighing the recapped bottle after the flow has continued for some convenient period of time. In this way only the difference in back pressure between mercury and aqueous solution, rather than the entire back pressure, is ignored.

A more precise method (Lingane, S2) measures the time required for a fixed mass of mercury to flow into any solution at any desired height of mercury column, and is superior for the comparatively rare occasions when the mass flow rate must be known with greater accuracy.

15.3.5   Height of Mercury Column

The height of the mercury column is a useful diagnostic in classical polarography. This height is simply the vertical distance from the tip of the dropping capillary to the top of the meniscus at the upper end of the standpipe; if there is no standpipe, the height can be measured to the mercury level in the dropping bulb. A meter stick is usually sufficiently precise for this measurement, and DME stands often incorporate one for just this purpose.

The mass flow rate m of any liquid through a tubing of uniform bore diameter is given by the *Poiseuille equation*:

$$m = p\pi\rho r^4/8l\nu \tag{15.1}$$

where p is the pressure difference between the beginning and the end of the tubing, r is the radius of the tubing, l is the length of the tubing, ρ is the density of the liquid, and ν is the viscosity of the liquid. For any given capillary through which mercury is flowing, only the driving pressure p is variable. This driving pressure is provided by the height of the mercury column.

The physical height of mercury column measured as described above, however, is not the *effective* height of the mercury column, which governs the mass flow rate of mercury, because there is a back pressure. Actually, there are two back pressures, one resulting from the back hydrostatic pressure of the solution, which is perhaps 1 to 2 mm of mercury and therefore negligible, and another resulting from the interfacial tension at the drop surface, which is perhaps 1.5 to 2.0 cm of mercury and is not negligible, since the overall height of the column rarely exceeds 100 cm (G1, G6). A constant correction of 2 cm often suffices for all but the most precise work. The surface tension, and thus the back pressure, vary significantly at positive and particularly at negative extremes of potential.

In any polarographic observation when the same capillary drops into the same solution under different pressures, the weight of the falling drop $m\tau$ is constant. Thus the drop time τ is inversely proportional to m and also inversely proportional to the effective height of the mercury column. Insertion of the direct proportionality of m and the inverse proportionality of τ into the Ilkovic equation gives the useful diagnostic that *diffusion-limited currents are proportional to the square root of the effective height of the mercury column.* For surface limited processes, insertion of these dependences into the equation for surface-limited current of the preceding chapter shows that *surface-limited currents are directly proportional to the effective height of the mercury column,* not to its square root.

15.3.6 Half-Wave Potential

The half-wave potential of a polarographic wave can be easily measured by taking half the distance on the current axis between the residual current line and the parallel diffusion current line and at that distance constructing a line parallel to them. The half-wave potential is the potential at which that line intersects the actual current–potential curve measured. The current can be either the maximum or the average current. An alternative method of obtaining half-wave potentials is from the fundamental equations of the polarographic or voltammetric wave given in Chapters 13 and 14. Values of $E_{\frac{1}{2}}$ can be obtained from the intercept and values of z from the slope of these logarithm plots. It is important to realize that these equations, and hence this analysis, are of value only if the waves are in fact due to a reversible process. Such plots are often used as criteria of reversibility as well as for the calculation of z. If, however, the process is not reversible, the value of z obtained by this procedure is not reliable.

15.4   INORGANIC POLAROGRAPHY

Inorganic polarography involves ionic equilibria as well as the intrinsic ease of oxidation or reduction of particular inorganic species. The involvement of protonation or solubility equilibria is not of unusual importance in inorganic polarography (protonation is of greater concern in organic polarography), but complexation equilibria are of particular significance in inorganic polarography.

15.4.1   Polarography of Complex Ions

Most complex ions consist of a single metal atom or ion surrounded by a certain number of molecules or anions called *ligands*. The number of ligands associated with one central metal ion is called the *coordination number* of the central metal ion since the ligands are attached to it by coordinate covalent bonds. Metal M may form several different complex ions with different numbers of the same ligand L; these different complex ions will differ in stability.

The *stability constant* of a complex ion is simply the equilibrium constant for the reaction that forms the complex ion. If the equilibrium is

$$M + nL \rightleftarrows ML_n$$

then

$$\beta = a(ML_n)/a(M)a(L)^n \tag{15.2}$$

where β is the *cumulative formation constant* for the complex ion. Alternatively, the *stepwise formation constant(s)* may be preferred:

$$K_n = a(ML_n)/a(L)a(ML_{n-1}) \tag{15.3}$$

in which case one constant is needed for each step of the complexation. If only one stable complex ion is formed for a specific metal ion and specific ligand, the cumulative formation constant is often preferred (in the case of 1:1 complexes the cumulative and stepwise formation constants are the same).

Polarographic reduction of soluble complex ions to other soluble species, whether ions soluble in electrolyte or metals soluble in mercury, often gives waves that follow the Heyrovsky–Ilkovic equation. These waves differ from those of the uncomplexed ion only in half-wave potential and, usually only slightly, in the height of the diffusion-limited current plateau. The latter is due to the small difference in diffusion coefficient between the complexed and uncomplexed ion. The difference in half-wave potential is directly related to the free energy of formation of the complex ion through the Nernst equation, as pointed out by Heyrovsky and Ilkovic (S3) in 1935, and is a simple method of measuring the stability constant when a single stable complex ion is formed (von Stackelberg and von Freyhold, S4; Lingane, S5).

When the Heyrovsky–Ilkovic equation is not obeyed by a complex ion upon polarographic reduction, it may be due to a kinetic complication. Reduction of any complex ion must be a CE mechanism in which the decomplexation, deaquation, or rearrangement step precedes the electron transfer. If the preceding chemical step is slow relative to the electron transfer, kinetic distortion of the polarographic wave will be observed.

If the complexation equilibrium is rapid, as is usually the case, the effect of complexation is to shift the potential of reduction of the metal ion negative with respect to the potential of reduction of the uncomplexed metal ion. More correctly, if a stable complex (or complexes) form it is because the free energy of the complex ion is less than the free energy of the uncomplexed but hydrated ion. This is an example of the CE mechanism in which the first step is the chemical complexation equilibrium and the second step is the redox reaction. If both are fast, the effect is as before a reversible wave shifted in potential due to the preceding equilibrium. Alternatively, the complex ion may be directly electroactive (that is, reduction may take place *through* a bridging ligand such as chloride ion), and a preceding decomplexation step is not required. The result is the same if all of the reactions involved are rapid, since the quantity measured by half-wave potential is the free energy necessary to form a reduced species from the oxidation state of the complexed metal ion. Thus for a reversible wave, and only for a reversible wave, the mechanisms of the complexation and redox steps are irrelevant.

Extension of the theory to multiple stable complexes by DeFord and Hume (S6) in 1951 involved algebraic combination of the Nernst and Heyrovsky–Ilkovic equations under the condition that the ligand concentration is varied. If the concentration of ligand is large relative to the metal ion concentrations and only a single stable complex ion is formed, this extension simplifies to the discussion of Heyrovsky and Ilkovic.

It is convenient to define a ratio of activities R which is the ratio of the free metal ion activity to the total activity of the metal in all forms:

$$R = a(M)/a(\text{total},M) \qquad (15.4)$$

in which case the stepwise formation constants appear as a power series:

$$1/R = 1 + K_1 a(L) + K_2 a(L)^2 + \dots + K_n a(L)^n \qquad (15.5)$$

which, if K_0 is defined as unity, is:

$$1/R = \Sigma_i K_i a(L)^i \qquad (15.6)$$

If the activity of L is made much larger than the activity of the free metal ion M, which can be easily accomplished by making the concentration of L so large that the concentration of free L remaining after complexation greatly exceeds the *total* concentration of all forms of M whether free or complexed, then, if only one stable complex ML_n is formed,

$$1/R = 1 + K_n a(L)^n \tag{15.7}$$

or, approximately,

$$K_n a(L)^n a(M) = R a(\text{total},M) \tag{15.8}$$

From the Nernst equation:

$$E = E^{0\prime} + (RT/zF) \ln[a^0(M)/a^0(M,Hg)] \tag{15.9}$$

$$E = E^{0\prime} + (RT/zF) \ln R + (RT/zF) \ln[a^0(\text{total},M)/a^0(M,Hg)] \tag{15.10}$$

For the complex,

$$E = E_{1/2} + (RT/zF) \ln [(I_d - I)/I] \tag{15.11}$$

where

$$E_{1/2} = E^{0\prime} + (RT/zF) \ln (D(M,Hg)^{1/2}/D(ML_n)^{1/2} + (RT/zF) \ln R \tag{15.12}$$

but

$$E_{1/2} = E^{0\prime} + (RT/zF) \ln [D(M,Hg)^{1/2}/D(M)^{1/2}] \tag{15.13}$$

so that it is approximately correct to use

$$E_{1/2} = (RT/zF) \ln R \tag{15.14}$$

if the diffusion coefficients and currents of the complexed and uncomplexed M species are the same or nearly so.

The reduction of a complexed ion to the metal or metal amalgam is more difficult than is reduction of the uncomplexed ion because the formation constant of any reasonably stable complex ion is necessarily large. Reduction of one ion to another may be more or less difficult in the presence of a complexing agent because two stability constants are involved, and it is their ratio, rather than their individual magnitudes, that determines the ease of reduction. Thus reduction waves for a complex ion being reduced to a metal are negative of the reduction waves for the free ion, while the reduction waves for a complex ion being reduced to a complex ion of lower charge may be either positive or negative of the comparable reduction of the uncomplexed metal ion. Most of the latter reductions are found to be more negative because the stability of a complex ion generally increases with the charge on the central metal atom. Complexation is often used to alter the potential at which an inorganic species is reduced polarographically in order to improve its analytic determination.

15.4.2 Polarography of Inorganic Species

A brief summary table and comments are given (Table 15.1), but for analytical procedures and electrolytes the general references should be consulted.

Groups IA, IIA, IIIA

Polarographic reduction of alkali and alkaline earth metal ions is feasible in neutral or alkaline solutions because the free energies of formation of the metal amalgams shift the polarographic waves well positive of the reversible metal-ion–metal potentials. This shift, together with the almost 1 V of hydrogen overvoltage on mercury, is enough to make reduction of these metal ions polarographically feasible. In acid solution the waves are masked by hydrogen evolution, since the reversible hydrogen potential is itself 0.5 to 0.8 V more positive than in neutral or alkaline solutions.

Tetraalkylammonium salts must be used as supporting electrolytes. Removal of all other metal ions by controlled potential reduction on a mercury-pool cathode is required if they are present. The waves for the different metal ions are not sufficiently well separated for the analysis of components of mixtures of these ions, with the possible exception of Li^+. The rare earths are similar, with the exception of Sm, Eu, and Yb for which the M^{3+}/M^{2+} couple exists at less extreme potentials.

Group IVB

Polarography of Ti(IV) is possible in many media, reduction being to Ti(III), although the Ti(II) state is stable in some media; analysis of steel or minerals is feasible. Polarographic determination of zirconium, hafnium, and thorium is not feasible.

Vanadium, Niobium, Tantalum

Polarography of vanadium is possible in several oxidation states ($+5$, $+4$, $+3$, $+2$) and in different media; it has been used for analysis of vanadium in steel. Niobium is polarographically active in some acid media, but tantalum is not.

TABLE 15.1 Polarographic Characteristics of Inorganic Ion Reductions (1.0M Acetate buffer, pH = 5)[a]

Ion	$E_{1/2}$ (V vs. SCE)	Product	Remarks
Bi(III)	−0.307	metal	—
C(II)	−0.615	metal	—
Cr(III)	−1.18	Cr(II)	—
Cu(II)	−0.035	metal	—
Fe(II)	−1.378	metal	—
Fe(III)	−0.012	Fe(II)	—
In(III)	−1.09	metal	—
Mn(II)	−1.493	metal	—
Mo(VI)	−0.98	?	not reversible
Ni(II)	−1.29	metal	not reversible
Pb(II)	−0.455	metal	—
Sb(III)	−0.61	metal	—
Te(VI)	−1.29	Te^{2-}	—
Tl(I)	−0.458	metal	—
U(VI)	−0.438	U(IV)	—
V(IV)	−1.26	V(II)	not reversible
Zn(II)	−1.14	metal	—

[a]Waves well-defined unless otherwise indicated. Data from Meites (G1) and other sources. In this medium, reduction waves starting from the dissolution potential of mercury can be obtained for silver, gold, and the more noble metals of Group VIII as well as halogens. Addition of ammonia or chloride shifts many of these potentials negative by complex ion formation. For further data see Meites (G1), Meites and Zuman (G5), and Kolthoff and Lingane (G7).

Chromium

Chromium is polarographically active as Cr(III), Cr(II), and chromate ion, and polarography has been used for its analysis in steel. Molybdenum is likewise active and has been determined polarographically in soil, steel, and ores. Tungsten is also active and has been determined in steel; the tungstates have been the subject of mechanistic study.

Uranium

Uranium is reducible polarographically as UO_2^{2+}, UO_2^+, and U^{4+}; an anodic wave is obtained for U^{3+}. Hydrolysis of uranium species is media-dependent, and precipitation may complicate polarographic analysis.

Manganese, Rhenium

Well-defined waves of Mn^{2+} are obtained in most media and have been used for analysis of steels; permanganate does not give a useful wave. Rhenium catalyzes hydrogen evolution, and reduction of perrhenate is complicated by this.

Iron, Cobalt, Nickel

Reduction of Fe^{3+} occurs at all potentials negative of mercury oxidation unless complexing agents are present. Reduction of Fe^{2+} occurs near -1.4 V, more than one-half volt negative of the reversible potential; this is probably overvoltage due to the insolubility of iron in mercury. Reduction of Co^{2+} occurs near the same potential, while Ni^{2+} is reduced near -1.1 V. Polarography has been used to analyze alloys for these metals. Use of complexing agents can prevent polarographic reduction of these elements in mixtures.

Copper, Silver, Gold

Cuprous ion does not exist in simple form in aqueous solutions; reduction of Cu^{2+} occurs near $+0.04$ V. In complexing media the potential of Cu^{2+} reduction is shifted negative, and stepwise reduction is observed if more stable complexes of Cu^+ than Cu^{2+} are formed. Polarography of copper has employed varied media; analysis for copper in alloys, ores, salts, and milk has been practical. Silver is rarely studied polarographically since $Ag(I)/Ag$ is virtually coincident with $Hg(I)/Hg$ in almost all media; the complexes of the monopositive ions are similar. Polarographic determination of gold is possible only in complexing (cyanide) media.

Zinc, Cadmium, Mercury

Polarographic determinations of zinc and cadmium are direct in almost all media; the reduction of Cd^{2+} is the most well-characterized of reversible processes. The Hg_2^{2+} and Hg^{2+} ions both produce waves starting from the dissolution potential of mercury.

Aluminum, Gallium, Indium, Thallium

Aluminum(III) is reduced only at -1.75 V and is thus indistinguishable from alkali metals or alkaline earths; more reducible ions also interfere. Nevertheless, polarographic determinations in clay, cement, alloys, glass, ceramics, and wine have been made, and polarography is a useful method for such measurements. Polarographic measurements on gallium are difficult, but In^{3+} and Tl^+ can be determined.

Germanium, Tin, Lead

Germanium has many oxygenated species in aqueous solution, and it is not well characterized polarographically. Both Sn^{4+} and Sn^{2+} are reducible, but the wave of Sn^{2+} is much better defined and is analytically very useful. Tin has been determined in ores, steels, and biological material. Reduction of Pb^{2+} is likewise polarographically well-characterized, and lead is routinely determined by polarographic methods.

Nitrogen, Phosphorus, Arsenic, Antimony, Bismuth

Nitrate and nitrite are reducible at the dropping mercury electrode, as is nitric oxide, the respective products being hydroxylamine, nitrogen, and ammonium ion. Cyanide and thiocyanate give an anodic wave due to formation of the respective mercury compounds. Phosphorus compounds are not directly determinable polarographically, nor are compounds of As(V); the As(III) state does give poorly characterized waves. Antimony in the $+3$ state gives well-developed waves, and in some media so does Sb(V). Bismuth gives well-developed waves on reduction of the $+3$ ion.

Oxygen, Sulfur, Selenium, Tellurium

Oxygen is reducible at the DME in two waves, the products being H_2O_2 at -0.23 V (pH 7) and water at -1.0 V (pH 7) (in alkaline media, OH^-). This forces oxygen removal in polarography, but advantage can be taken of it in oxygen analysis in gas mixtures (equilibrated with electrolyte), water, sewage, and physiological or nutritional media. Hydrogen peroxide can also be determined, though less practically. Sulfate ion is not reducible at the DME, though sulfur dioxide is. Sulfite ion, thiosulfate ion, and sulfide ion give anodic waves due to the formation of stable complexes of mercury or, in the case of sulfide, HgS. Selenide ion gives an anodic wave similar to that of sulfide; selenium in the $+4$ oxidation state gives complex reduction waves. The $+6$ oxidation states of selenium and tellurium are not reducible polarographically; the Te(IV) state is similar to Se(IV), and telluride to selenide.

Halogens

Of the perhalates, only periodate is reducible at the DME, and of the halates only iodate and bromate. The elemental halogens all give waves starting from the anodic dissolution of mercury and are reduced to the respective halide ions. Solutions of halide ions yield anodic waves due to the formation of the respective Hg_2^{2+} halide salt that can be used for analytical purposes.

15.5   ORGANIC POLAROGRAPHY

Organic polarography could also be called functional group polarography. When the various different groups on organic molecules are reduced, the reductions occur more or less independently of each other. The oxidation reactions are virtually never used. In well-buffered media the observed wave height is generally dependent directly upon concentration even if the reaction is irreversible. The solutions are generally water–dioxane or water–alcohol mixtures, so that the solubility of the organic compounds is increased, while the solutions are still essentially aqueous. The reference electrode used is generally the aqueous SCE, to which the potentials given in this section refer. The classes of organic functional groups that are electrochemically active can be indicated briefly.

15.5.1   Hydrocarbons

The carbon—carbon single bond of hydrocarbons is electrochemically totally nonreactive. Carbon—carbon double bonds are not reducible unless activated by conjugation. The double bond in ethene is not reducible, while $CH_2 =CH—CH =CH_2$ gives a wave at $-2.6V$, and $(C_6H_5)CH=CH—CH= CH (C_6 H_5)$ gives a wave at -1.98 V, both against SCE. Carbon— carbon triple bonds react generally in the same manner as double bonds.

Among the aromatic systems (Wawzonek and Laitinen, S7), the benzene ring is not reducible. However, fused-ring aromatic systems sometimes are. Among the fused-ring systems, the three types shown in Figure 15.5 are the simplest. They correspond to the reductions of napthalene $(-2.47$ V), biphenyl $(-2.70$ V), and anthracene $(-1.95$ V). More complex types of fused-ring systems can be explained in terms of these three types of reductions. The reduction of phenanthrene, which consists of a two-electron step at -2.45 V followed by another two-electron step at -2.67 V, can be considered as one reduction of the napthalene type followed by one of the biphenyl type, while a more complex molecule, which also is reduced in a two-electron step at -2.03 V and a second two-electron step at -2.54 V, can be considered as one reduction of the anthracene type followed by one of the napthalene type.

All of these reductions are independent of pH and can be carried out without use of a buffer, although such practice is not generally recommended.

15.5.2   Halides

Organic halides are generally reducible. The reaction is a totally irreversible two-electron and one-proton reduction to the hydrocarbon and halide ion in one step that is pH independent. Multiple halogen substituents are reduced separately; this is useful for polarographic determination of insecticides. For example, the reduction of carbon tetrachloride proceeds in two-electron, one–proton steps to trichloromethane $(-0.78$ V), dichloromethane $(-1.67$ V), and chloromethane $(-2.3$ V). In general, dihalides are more easily reduced than are monohalides. The

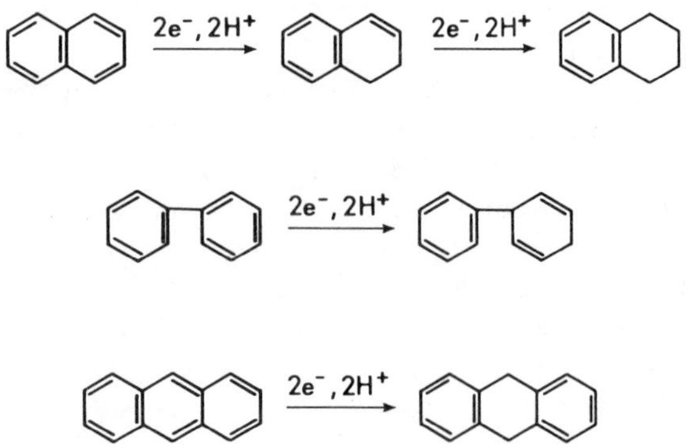

Figure 15.6   Reductions of aromatic fused rings. From top, reductions of napthalene, biphenyl, and anthracene.

ease of reduction is greater for iodine than for bromine than for chlorine, as would be expected. Short-chain aliphatic halides are more easily reduced than are long-chain aliphatic halides or aromatic halides.

15.5.3   Aldehydes

Aldehydes are generally easily reducible, the ultimate products being alcohols. The reduction waves of aldehydes generally show a pH-dependent height. The reduction of formaldehyde, HCHO, which exhibits strange kinetics, has half-wave potential of about -1.50 V. The larger aldehydes, RCHO, all reduce at -1.89 to -1.92 V if R is saturated. If it is not, then a typical reduction (Coulson and Crowell, S8) is

$$CH_2{=}CHCHO + 2e^- + 2H^+ \rightarrow CH_3CH_2CHO$$

$$CH_3CH_2CHO + 2e^- + 2H^+ \rightarrow CH_3CH_2CHOH$$

In general, sugars such as glucose give waves of the RCHO type (which are, however, controlled by the kinetics of the tautomerism reaction).

Aromatic aldehydes are reduced in acid solutions in two steps, probably to the free radical and from the radical to the alcohol since both steps involve one proton and one electron. However, the second wave is affected by coupling reactions, as in the case of benzaldehyde, so that the second wave may not be of the expected height. The half-wave potential for reduction of benzaldehyde (extrapolated to pH 7) is -1.262 V (Laitinen and Kneip, S9).

15.5.4   Ketones

Ketones are generally somewhat more difficult to reduce than are aldehydes. The ultimate reduction product is again alcohols. Saturated aliphatic ketones are particularly hard to reduce; acetone has a half-wave potential of -2.46 V. These waves are rarely used in analysis. In unsaturated aliphatic ketones an adjacent double bond may make the keto group easier to reduce and a pH-dependent wave similar to that for aldehydes is then observed; one or two waves may be seen, depending upon the pH. Thus the compound $CH_3CH{=}CHCOCH{=}CH_2$ has a half-wave potential of -1.42 V. Aromatic ketones behave like unsaturated aliphatic ketones; they do have a tendency to couple through the free radical produced in the first one-proton, one-electron reduction step.

Quinones are easily reducible in a reversible two-electron, two-proton step to the corresponding hydroquinones. In nonaqueous solvents, reversible reduction to the semiquinone and from the semiquinone to the hydroquinone is observed, while in aqueous solution the semiquinone is generally unstable. In aqueous phosphate buffers of pH 7, parabenzoquinone is reduced at $+0.04$ V, 1,4-napthoquinone at -0.13 V, and anthraquinone at potentials of about -0.55 V.

15.5.5   Acids and Esters

Acids and esters are rarely reducible; their oxidation is possible at platinum electrodes (Kolbe process), but not on mercury. Saturated aliphatic acids are not reducible at all, and even in unsaturated aliphatic acids the acid group is not reducible. In aromatic acids, however, reduction to alcohols may be possible at very negative potentials, as in the case:

$$C_6H_5COOC_2H_5 + 4H^+ + 4e^- \rightarrow C_6H_5CH_2OH + C_2H_5OH$$

15.5.6   Nitrogen Compounds

Nitro compounds are generally reducible; half-wave potentials are low, perhaps -0.25 V, but they are highly dependent on pH. Aliphatic nitro compounds generally exhibit two irreversible waves in acid solution, a four-electron, four-proton reduction to water and the hydroxylamine followed by a two-electron, two-proton reduction of the hydroxylamine to the amine.

In aqueous solution aromatic nitro compounds are generally reducible through the corresponding phenylhydroxylamine to the aniline:

$$ArNO_2 + 4e^- + 2H^+ \rightarrow ArNHOH$$

$$ArNHOH + 2e^- + 2H^+ \rightarrow ArNH_2$$

The reduction to the phenylhydroxylamine is suitable for quantitative analysis, and stepwise reductions are observed when multiple nitro groups are present.

Reduction of the phenylhydroxylamine whether as the second step in the reduction of a nitro compound or as the hydroxylamine itself is observed only in acid solution, where two-electron reduction occurs to the amine. In basic solution phenylhydroxylamine couples to give azoxybenzene, which is itself reducible to hydrazobenzene and water at the dropping-mercury electrode. The reduction of nitrobenzene occurs at -0.59 V against SCE. The reduction of p-nitrophenol involves an overall six-electron reduction to the amine; the polarographic and coulometric z values are not equal.

Aromatic nitroso groups are generally reversibly reduced to the corresponding aromatic hydroxylamines, which in turn are oxidizable to the nitroso compounds. These reversible redox couples have found some use in potentiometric titrations; the potentials are fairly positive, that of C_6H_5NNO being -0.10 V against SCE.

Azo linkages are also reducible; a typical two-electron, two-proton reduction is found for azobenzene:

$$C_6H_5N=NC_6H_5 + 2H^+ + 2e^- \rightarrow C_6H_5NNHC_6H_5$$

The usual reduction of $(CH_3)_2N(C_6H_4)N=NC_6H_5$ (Kolthoff and Barnum, S10) is a four-electron step:

$$(CH_3)_2N(C_6H_4)N=N(C_6H_5) \rightarrow (CH_3)_2N(C_6H_4)NH_2 + H_2N(C_6H_5)$$

A disproportionation reaction that is pH-dependent also occurs for the intermediates:

$$2H_2NC_6H_5 \rightarrow (C_6H_5)NHNH(C_6H_5) \rightarrow (C_6H_5)N=N(C_6H_5)$$

Thus at pH 9, for example, the polarographic number of electrons z is 2 while the coulometric number of electrons z is 4.

Amines are difficult if not impossible to reduce except in liquid ammonia. Coupling is obtained in water upon reduction of quaternary ammonium ions:

$$(CH_3)_4N^+ + e^- \rightarrow (CH_3)_3N: + CH_3\bullet$$

$$CH_3\bullet + CH_3\bullet \rightarrow C_2H_6,$$

and/or

$$CH_3\bullet + Hg \rightarrow CH_3Hg$$

$$(CH_3)_3N: + H_2O \rightarrow (CH_3)_3NH^+ + OH^-$$

followed by further alkylammonium reduction.

Heterocyclic nitrogen compounds in acid solution often give rise to hydrogen discharge, as in the reaction catalyzed by pyridine. In alkaline solution, free radicals can be formed; a radical may couple with itself or other species in solution. All nitrogen compounds, including those that are heterocyclic, have a tendency to catalyze the discharge of hydrogen ion.

15.5.7   Peroxides

Peroxides can be reduced polarographically, although the waves are usually quite long and drawn out. Reduction is to alcohols; thus ROOR′ yields ROH and R′OH, ROOH yields ROH and water, while formic acid HOOH is reduced to water alone.

15.5.8   Sulfur Compounds

Sulfur compounds are reducible; a study of the cystine—cysteine system by Kolthoff and Barnum (S10) found overlapping waves for the reduction:

$$RSSR + 2H^+ + 2e^- \rightarrow 2RSH$$

$$2RSH + Hg \rightarrow (RSHg)_2 + 2H^+ + 2e^-$$

15.6   REFERENCES

General

G1   L. Meites, *Polarographic Techniques,* 2nd ed., Interscience, New York, 1964, 752 pp. Excellent discussion of polarography; also covers amperometric titration and, very briefly, other related techniques; particularly useful for practical classical polarography.

G2   P. Zuman and S. Wawzonek, in *Progress in Polarography, Vol. I*, P. Zuman, Ed., Wiley-Interscience, New York, 1962, pp. 303, 319.

G3   P. Zuman, in *Advances in Analytical Chemistry and Intrumentation, Vol. 2*, Wiley-Interscience, New York, 1963, p. 219.

G4   R. N. Adams, *Electrochemistry at Solid Electrodes*, Marcel Dekker, New York, 1969, 401 pp.

G5   L. Meites and P. Zuman, eds., *Handbook Series in Organic Electrochemistry*, CRC Press, West Palm Beach; Vol. 1, 1977, 896 p.; Vol. 2, 1977, 640 pp.; Vol. 3, 1978, 672 pp. Tabulation of electroanalytical data on organic compounds.

G6   O. H. Muller, "Polarography," in *Physical Methods of Chemistry, Part IIA*, A. Weissberger and P. W. Rossiter, Wiley-Interscience, New York, 1971, p. 297ff.

G7   I. M. Kolthoff and J. J. Lingane, *Polarography*, 2nd ed., Wiley-Interscience, New York, 1952, 2 volumes, 990 pp. Somewhat dated now, but still well worth study. Many application references, critical presentation.

Specific

S1 R. N. Adams, "Voltammetry at Electrodes with Fixed Surfaces," *in Treatise on Analytical Chemistry*, I. M. Kolthoff and P. J. Elving, Eds., Part I, Volume 4, Chapter 47, pp. 2381ff.

S2 J. J. Lingane, *Anal. Chem.* **16**, 329 (1944).

S3 J. Heyrovsky and D. Ilkovic, *Coll. Czech. Chem. Common.* **7**, 198 (1935).

S4 M. von Stackelberg and H. von Freyhold, *Z. Elektrochem.* **46**, 120 (1940).

S5 J. J. Lingane, *Chem. Rev.* **29**, 1 (1941).

S6 D. D. DeFord and D. N. Hume, *J. Amer. Chem. Soc.* **73**, 5321 (1951).

S7 S. Wawzonek and H. A. Laitinen, *J. Amer. Chem. Soc.* **64**, 2365 (1942).

S8 D. M. Coulson and W. R. Crowell, *J. Amer. Chem. Soc.* **74**, 1290, 1294 (1952).

S9 H. A. Laitinen and T. J. Kneip, *J. Amer. Chem. Soc.* **78**, 736 (1956).

S10 I. M. Kolthoff and C. Barnum, *J. Amer. Chem. Soc.* **62**, 3061 (1940); **63**, 520 (1941).

S11 E. Casassas and L. Eek, *J. Chim. Phys.* **64**, 971 (1967).

S12 A. A. Vlcek, *Coll. Czech. Chem. Commun.* **22**, 948, 1736 (1957); *Chem. Listy* **50**, 1072, 1416 (1956).

15.7 STUDY PROBLEMS

15.1 A certain metal ion was polarographically reduced in aqueous solution. The current (on the diffusion plateau) at –0.600 V was 10.0 μA. The currents (in microamperes) measured manually at various potentials were: –0.464 V, 1.0; –0.485 V, 2.0; –0.509 V, 4.0; –0.531 V, 6.0; –0.555 V, 8.0; –0.576 V, 9.0. Calculate the value of the half-wave potential and the apparent number of electrons involved in the reduction.

15.2 The diffusion-limited current for a certain Pb(II) solution was found to be 15.0 μA. To 5.0 cm^3 of this solution was added 10.0 cm^3 of a standard 1.0-mmol/dm^3 solution of Pb(II). The diffusion-limited current was then found to be 22.5 μA on a second polarogram. Compute the concentration of Pb(II) in the original solution.

15.3 The following data were obtained for reduction of 1.0-mmol/dm^3 Pb(II) in aqueous 0.1-mmol/dm^3 KNO$_3$ buffered with acetate using an SCE reference electrode: -0.450 V, -3.15 μA; -0.470 V, -11.35 μA; -0.485 V, -19.25 μA; -0.500 V, -26.06 μA; -0.650 V, -32.25 μA. The final value given above for the current is clearly on a diffusion plateau. Use the Heyrovsky–Ilkovic equation to obtain the half-wave potential for Pb(II) reduction and the number of electrons involved in the reduction. Estimate the error in the half-wave potential obtained from these data.

15.4 Casassas and Eek (S11) reported values for the stability constants of Co(II) and Ni(II) complexed by 1-nitroso-2-napthol. The measurement was made using the polarographic half-wave potential for reduction of 1-nitroso-2-napthol, which is much less negative than the half-wave potential for the reduction of the metal ions; the half-wave potentials of the metal ion reductions when complexed were not accessible. Assume, as they did, that all activity coefficients are unity and that the reduced form of 1-nitroso-2-naphthol forms no stable complexes with these metal ions. Use a large excess of metal ion so that only the 1:1 complex need be considered. Now sketch the experimental polarographic results that should be obtained upon addition of these metal ions to an aqueous solution of 1-nitroso-2-napthol, and explain how these results can be used to measure the stability constant of the complex ion. The nitroso group on the molecule is both the most reducible group and the group essential for formation of a stable complex ion.

15.5 The polarographic reduction of the complex ion $Ni(CN)_4^{2-}$ has been studied by Vlcek (S12). The reversible reduction of the complex ion is followed by decomplexation reaction(s) that produce species that cannot be reoxidized to the original complex ion. Explain, in qualitative terms, what the formation of these species might do to the form of the reversible wave observable in classical polarography.

15.6 Ionic forms of platinum act as a catalyst for the reduction of aqueous Ni(II) by the hypophosphite ion. This catalysis can be used as the basis of an electroanalytical kinetic method for the determination of platinum ions in the following way. A fixed aliquot of a standard solution of Ni(II) is added to a fixed aliquot of a standard solution of hypophosphite ion, a variable aliquot of a standard solution of platinum ions is added to that, and the whole is immediately mixed and diluted to a volume of 10 cm³. Classical polarography is used to measure the height of the reduction wave for the irreversible reduction of Ni(II) 10 min after mixing and dilution. Neither platinum ions nor hypophosphite is reducible at this potential. A calibration curve of the logarithm of the observed wave is found to be linear. Explain how you would use this procedure and this calibration curve to obtain the concentration of platinum ions in a solution in which their concentration is not known. Then explain, in as much detail as you can, why this analytical method works.

15.7 An analytical chemist determined the exchange capacity of an ion-exchange resin by the following procedure. Classical polarography gave a wave of height 63 μA for a 10-cm^3 aliquot of a certain solution of Cu(II). A 1.0-cm^3 aliquot of a 0.001 mol/dm^3 standard solution of Cu(II) was added to the aliquot of the original solution; the height of the reduction wave was then increased to 67.2 μA. An aliquot of 100 cm^3 of the original Cu(II) solution was equilibrated with 2.50 g of the resin. A 10-cm^3 aliquot of the equilibrated solution was separated from the resin; the wave height observed in this solution was 6.3 μA. A 1.0-cm^3 aliquot of the 0.001-mol/dm^3 standard solution was added to the aliquot of the equilibrated solution; the wave height then increased to 34.6 μA. Calculate the exchange capacity of the resin in terms of moles Cu(II) per gram of resin.

CHAPTER
16
ADVANCED VOLTAMMETRIC
AND POLAROGRAPHIC TECHNIQUES

Advanced voltammetric and polarographic techniques are derived from the basic techniques already described. Those of analytical significance are classified in four groups: rapid-scan techniques, pulse techniques, AC techniques, and stripping techniques. Useful biennial reviews of these techniques are published in *Analytical Chemistry*, the emphasis varying with the authors (S1–S15).

In advanced polarographic techniques, a current sampling method must generally be employed; when current sampling is the only modification of classical polarography, the technique is referred to as *tast polarography*. Tast polarography, named from the German for "touch," is not really a polarographic technique at all, but rather the technique of measuring, or sampling, the current shortly before the drop falls. This technique is also known as *current-sampled DC polarography*. For example, in the P.A.R. Model 174A polarograph used in this mode, the sampling interval is 16.7 ms; the current measured during this interval is held by the sample-and-hold circuit and displayed until shortly before the *subsequent* drop falls. Provision must be made so that the sample is taken on the same portion of each drop. This can be achieved if the drop time is mechanically controlled by a "drop knocker" or if synchronization is achieved by a circuit that senses the fall of the previous drop and samples at a preset time thereafter.

The major advantage of current-sampled DC polarography (S16–S17) is that the effect of charging current, which decays as the drop grows as time to the $-\frac{1}{3}$ power, is reduced, and the baseline is more horizontal than if the average current is measured. A minor advantage is that the rapid pen fluctuations are reduced, because the stored sample of current is displayed until the next sample is taken. For the same reason, the polarogram is more suitable for digitization and subsequent computer analysis. The effect of surface-limited waves is reduced, because their waveshape is like that of charging current; however, since drop curves for the current flowing throughout drop life are not plotted, their diagnosis is not possible as it is in classical polarography. Waves are obtained just as in classical polarography, although the output signal appears in staircase form rather than a smooth curve. The sensitivity is not improved over that of classical polarography, neither as to diffusion current nor as to resolution of $E_{1/2}$, by sampled current measurement rather than measurement of the current at the end of drop life from an ordinary recorder tracing. On a diffusion-limited plateau the sampled current will be very close to the maximum current, but it will not be so for a residual current whose major component is charging current.

16.1   RAPID-VOLTAGE-SCAN TECHNIQUES

When the rate of voltammetric scan is increased above a few millivolts per second, the interphase region that is the electrode does not have sufficient time to achieve stable quasi-equilibrium gradients with the bulk solution before the voltage is significantly changed. The theory presentation in Chapters 13 and 14 can no longer be applied, since it was based upon the equilibrium state at the electrode necessary for the use of the Nernst equation.

The nomenclature of these techniques has varied widely although fundamentally all of them are simply voltammetry at high rates of voltage scan. Rapid voltage scan voltammetry is most correctly called *chronoamperometry with linear sweep*. It is also known by the obsolete names of single-sweep peak voltammetry, fast linear-sweep voltammetry, and stationary-electrode voltammetry. The simplest chronoamperometic excitation signal, a potential step, is as shown in Figure 16.1, along with the resultant current output. Chronoamperometry without the linear sweep component is discussed further in Chapter 17.

16.1.1   Rapid-Voltage-Scan Voltammetry

Rapid-voltage-scan voltammetry, or chronoamperometry with linear sweep, employs the applied waveform shown in Figure 16.2. The assumption that the electrode potential is always an equilibrium potential, as assumed in the previous chapters, is no longer valid when the rate of voltage sweep becomes sufficiently high. The sweep rate at which this effect becomes observable depends upon the electrolysis conditions and upon the electrode. At a DME under normal polarographic conditions, scan rates above 10 mV/s can be expected to show effects of the scan rate. At an electrode of constant area such as an HMDE, even lower voltage-scan rates may show observable effects. The author has used scan rates between 20 mV/s and 50 V/s in rapid-voltage-scan voltammetry.

The usual scan speeds at an HMDE in chronoamperometry with linear sweep are 15 to 50 mV/s, as compared to the 2 to 3 mV/s of classical polarography, and a 1-s potentiometric recorder is then adequate for response; a full voltammogram would be completed in about 1 min. At speeds of 100 mV/s and upwards, an oscilloscope or electronic digital or analog recorder is required to follow the trace.

When a potential step is applied as shown in Figure 16.1, the decay of the current is due to the decrease of the amount of electroactive species in the immediate vicinity of the electrode as the diffusion gradient establishes itself. In the absence of an electroactive species, only the exponential decay of charging

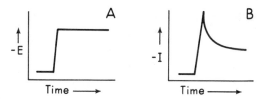

Figure 16.1 Chronoamperometry without linear sweep. (*A*) Excitation step waveform for reduction; (*B*) resultant current–time curve for a single reversible reduction. Slope of rise exaggerated. Dashed line in (*B*) shows decay of charging current.

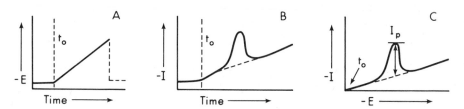

Figure 16.2 Chronoamperometry with linear sweep. (*A*) Excitation signal; the ramp starts at t^0 and the drop at the end of the ramp is not used. The curve in (*B*) is the resultant signal for a single reversible process.

current is observed; this is usually negligible in rapid-scan techniques and will not be considered until the following chapter. If the electrode area is not constant, its dependence upon time must also be considered, but modern rapid-scan polarographic techniques employ current sampling in the latter part of drop life so that the electrode area of the DME does remain approximately constant. Other electrodes, such as the HMDE, actually do have constant area.

The effect of the rapid-voltage-scan waveform shown in Figure 16.2 is to produce the resultant peak current waveform as shown in the same figure. Measurement of the peak current is made from the ascending baseline of charging current, which should be measured separately before the electroactive species is added to the base electrolyte, as shown in the figure. Qualitatively, the initial rapid increase is exponential. For processes limited by charge transfer, this rise should follow the Tafel equation, and values of $z\alpha$ can be obtained from it. Beyond about 10% of the peak current the exponential rise will begin to slow as the electrode reaction reduces the amount of electroactive species in the immediate vicinity of the electrode. The current then falls toward the diffusion-limited current as the diffusion layer establishes itself. Since the diffusion-limited current at an electrode of constant area decreases with time, no stable time-independent current is ever achieved.

When two or more electroactive species are reduced or oxidized independently in the same solution, the baseline for the subsequent peaks is taken as the extension of the decay curve for the preceding peaks. This extension can be measured experimentally by halting the voltage scan prior to the onset of the subsequent peak and following the decay of the current with time at constant potential. If the halt occurs at a potential sufficiently beyond the potential of the preceding peak, the current will have decayed to the diffusion-limited value, and the extension will include only the decay of the diffusion-limited current. Subsequent peaks can then, when measured from that baseline, be treated as if the preceding peaks were not present. When the potential separation of the processes is less than about 100 mV, or, when the processes are not independent, this baseline measurement will not fully correct for the preceding peak process.

The sole advantage of any rapid-voltage-scan method *as an analytical technique* is that one can quantitatively determine material at lower concentrations than in classical polarography, down to about 1 μmol/dm^3. The methods have several significant fundamental disadvantages in addition to those imposed by the recording technique, which must be more elaborate. First, there will always be a sloping baseline, especially on the DME because of the increase in both Faradaic and capacitive current during the continuing growth of the drop. Second, either calibration curves with known standards or the method of standard addition must be used. The theoretical equations linking peak height to concentration are not of great practical use. And third, unless a DME with its attendant more sloping baseline is used, the electrode surface is not renewed, and electrolytic products will accumulate. The proper role of rapid-voltage-scan methods is in the study of rates and mechanisms of electrode processes; pulse methods are superior for analytical determination of species in solution.

16.1.2   Cyclic Voltammetry

Rapid-voltage-scan techniques in which the direction of voltage scan is reversed are called cyclic techniques. In these techniques a ramp is applied over the full voltage-scan range and then reversed so that a descending ramp returns, almost invariably to the original potential. The scan rate in the forward and reverse directions is normally the same, so that the excitation waveform is actually an isosceles triangle. Cyclic voltammetry is that voltammetric technique in which the current that flows in a system is measured as a function of time and in which the excitation signal is a triangular potential. (See Figure 16.3.) The potential is normally that of an isosceles triangle, although this is not strictly necessary. Cyclic voltammetry can be used in single-cycle (Figure 16.4) or multicycle modes, depending upon the electrode, the reaction in question, and the information sought. In most cases the first and later scans are not identical. Multicyclic techniques have not been used for quantitative measurements because quantitative theory is not available for them. They are, however, useful diagnostic tools when complex mechanisms are involved. The theoretical equations for cyclic voltammetry have been developed by Nicholson and Shain (S26–S28), and an excellent discussion, both theoretical and practical has been given by Piekarski and Adams (S36).

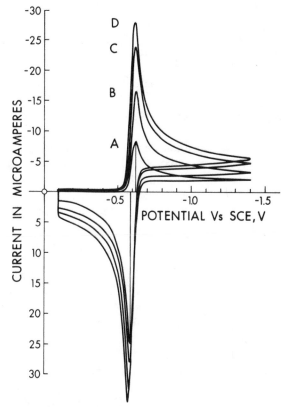

Figure 16.3   Cyclic voltammetry of a reversible process. Aqueous solution, 1.0 mol/m³ Cd(II) in 100 mol/m³ KCl, HMDE. Curve A, 20 mV/s; curve B, 50 mV/s; curve C, 100 mV/s; curve D, 200 mV/s. Initial scan cathodic (author's laboratory).

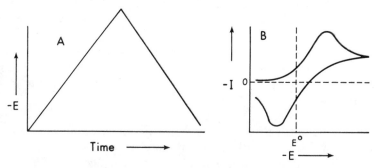

Figure 16.4   Cyclic voltammetry. (*A*) Excitation signal; (*B*) response is obtained for a single, simple reversible reduction when the voltage–time excitation signal extends considerably on both sides of the E^0 for the reversible process.

It is usual to plot the cyclic voltammogram on an *x–y* potentiometric recorder, although an ordinary *x–t* recorder can be used. On an *x–y* recorder, the response for a reversible system would be as shown in Figure 16.5. The measured parameters in cyclic voltammetry are the anodic and cathodic peak potentials $E_{p,c}$ and $E_{p,a}$; the anodic and cathodic peak currents $I_{p,c}$ and $I_{p,a}$; and the half-peak potentials, which are the potentials $E_{p/2,c}$ and $E_{p/2,a}$ at which the cathodic and anodic currents reach half of their peak values. The independent variables are the voltage scan rate and the range of potential over which the scan is made. The former is the most important parameter in a diagnostic sense, although proper selection of scan range often can eliminate interference from other processes.

16.1.3   Single-Sweep Polarography

Single-sweep polarography, also known as *fast linear-sweep polarography* (S24, S25), is simply chronoamperometry with linear sweep carried out at a DME. Typical sweep rates range upward from 100 mV/s. In this technique, a scan over the full potential range desired is applied to the drop over a short time interval late in drop life. It differs from pulse polarography in the range of the voltage used and in the fact that the pulse is a linear ramp rather than a square wave.

The excitation signal applied and the response curve obtained for a single reversible process are shown in Figure 16.6. Over the range of time employed in the sweep, the area of the electrode may be considered to remain effectively constant. The decay in current follows as the diffusion layer extends further into the solution. The peak potential differs from the half-wave potential of classical polarography in the same manner as does the peak potential in chronoamperometry with linear sweep. The height of the peak is proportional to the bulk concentration of electroactive species and to the square root of the scan rate dE/dt. Calibration by known solutions or the method of standard addition is used for standardization. Although the height of the peak is increased by increasing the scan rate, voltage resolution is not. Moreover, sloping baselines are obtained that are due to the increase in both Faradaic and capacitive currents with drop life, so that the precision of single-sweep polarography is worse than that of classical polarography or chronoamperometry with linear sweep. The sole advantage of single-sweep polarography over classical polarography is the ability to measure lower concentrations of electroactive species, down to a few micromoles per cubic decimeter. This can be done because the diffusion layer does not extend as far into the solution, since it has not been depleted by electrolysis during the first part of drop life. In this way single-sweep polarography is similar to pulse polarography.

An older variant of single-sweep polarography employed no drop timer; the curves were watched on an oscilloscope with only the highest peak used. In this method, successive fast ramps were applied continuously, and the drops were not synchronized with them. This method, called *cyclic fast-linear-sweep polarography* or *multisweep polarography*, is inferior in precision and is now obsolete.

Single-sweep polarography has sometimes been referred to as *oscillographic polarography*. Such terminology is misleading, because oscillographic polarography is actually an AC technique.

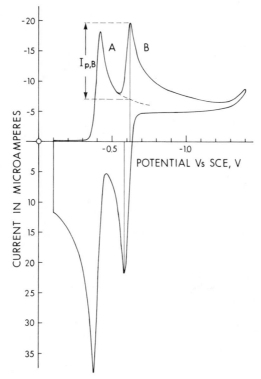

Figure 16.5   Cyclic voltammetry of multiple reversible processes . Aqueous solution, 10.0 mol/m³ Cd(II) and Pb(II) in 100 mol/m³ KCl, HMDE. Initial scan cathodic, scan rate 50 mV/s; A, Pb(II) reduction; B, Cd(II) reduction; C, potential range –0.20 to –0.55 V (author's laboratory).

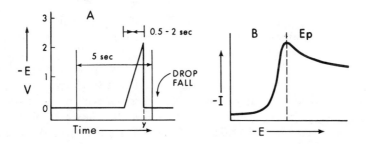

Figure 16.6   Single-sweep polarography. (*A*) Excitation signal; range of sweep potential is the full voltage sweep of the polarogram. (*B*) Response curve obtained for a single reversible reduction. Multiple processes will give rise to an ascending series of peaks.

16.1.4   Triangular-Sweep Polarography

This technique (S26) is identical to cyclic voltammetry except for the use of a dropping-mercury electrode; the excitation signal and responses are shown in Figure 16.7. Triangular-sweep polarography gives a waveform identical to that of

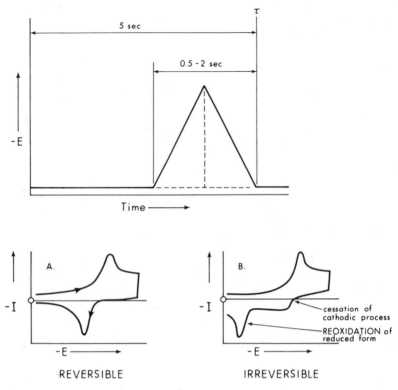

Figure 16.7 Triangular-sweep polarography. Upper curve, excitation waveform. Lower curves, response signals; (*A*) reversible process, (*B*) irreversible process. Drop must fall at, or at a fixed time after, cessation of the excitation waveform.

single-sweep polarography on the rising side of the triangle and a reversal waveform on the falling side. Triangular-sweep polarography has a major advantage over single-sweep polarography in that the reversibility of the process is immediately apparent. The time scale of triangular-sweep polarography, like that of single-sweep polarography, is much faster than that of classical polarography. Thus a process that is clearly irreversible when studied by triangular-sweep polarography may appear reversible when studied by classical polarography; a process that is reversible to triangular-sweep polarography will be reversible to classical polarography as well.

The older variant of triangular-sweep polarography, *cyclic triangular-sweep polarography*, in which a continuous series of triangular pulses is applied during drop life, is not obsolete, although it is the counterpart of multisweep polarography. It remains useful because, while normal triangular-sweep voltammetry will show on the return (usually oxidation) half-cycle any electroactive intermediate created on the forward (usually reduction) half-cycle, an electroactive intermediate created on the return (usually oxidation) half-cycle will not be observed unless additional cycles occur.

The use of *oscillographic polarography* to describe triangular-sweep polarography should be avoided.

16.2 RAPID-VOLTAGE-SCAN THEORY

In the theoretical treatment of rapid-voltage-scan methods it is necessary to consider the effect of the voltage-scan rate dE/dt, usually symbolized by χ, on the curves obtained. In the limit as $\chi \rightarrow 0$, the theory of each of these methods simplifies to that of its counterpart at infinitely slow rates of scan.

16.2.1 Reversible Processes

The differential equations and boundary conditions for reversible electrode processes studied by rapid voltage scan were solved by Randles (S20) and Sevcik (S21) around 1950. Under the conditions of a single reversible electrode process and semi-infinite linear diffusion to a planar electrode of constant area, the instantaneous current is given by the *Randles–Sevcik equation*:

$$I = PzFAcD^{½}(zF\chi/RT)^{½} \tag{16.1}$$

where the parameter P is a function of $zF\chi t/2RT$ that varies as shown in Figure 16.8, and χ is the voltage scan rate dE/dt. At 25°C the constant of the Randles–Sevcik equation is $0.4463F^{3/2}/(RT)^{½}$ and

$$I_p = 2.69 \times 10^5\, z^{3/2}\, AcD^{½}\chi^{½} \tag{16.2}$$

when the SI units are current in amperes, area in square meters, concentration in moles per cubic meter, diffusion coefficient in meters squared per second, and scan rate in volts per second. Since the peak current is directly proportional to the bulk concentration of diffusing species, this technique is useful for quantitative analysis.

Revisions to the Randles–Sevcik equation (S29, S30) have been made, as have extensions to totally irreversible (S31) and quasireversible (S32) processes and processes that are governed by chemical kinetic steps (S33–S35). Cylindrical (S36) and stationary spherical (S26,S37–S43) electrodes have also been considered. The best theoretical summary is that of Nicholson and Shain (S44), who give 0.4463 as the maximum value of P. This maximum occurs at a potential related to the polarographic half-wave potential for the reaction,

$$E_p = E_{½} \pm 1.109RT/zF \tag{16.3}$$

the sign being positive for anodic peaks and negative for cathodic ones. The potential at half the peak maximum current is also related to the reversible polarographic half-wave potential,

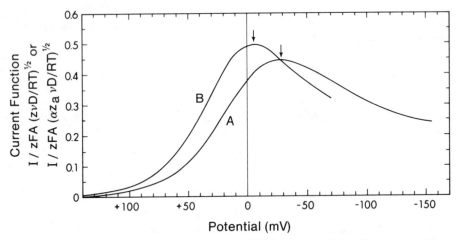

Figure 16.8 Theoretical curves for reversible and irreversible processes. Data of Nicholson and Shain; upper curve for reversible process, lower for totally irreversible process. Vertical axis differs for processes.

$$E_{p/2} = E_{1/2} \pm 1.109RT/zF \tag{16.4}$$

the sign now being positive for *cathodic* peaks and negative for anodic ones. For a reversible process, then:

$$E_{p/2} = E_p \pm 2.218RT/zF \tag{16.5}$$

The potentials at 25°C are related by

$$E_{p/2} \pm 28 \text{ mV} = E_{1/2} \tag{16.6}$$

$$E_{1/2} \pm 28 \text{ mV} = E_p \tag{16.7}$$

the positive sign applying to anodic processes and the negative to cathodic processes. For reversible processes, the peak and half-peak potentials are independent of scan rate—a useful diagnostic.

The Randles–Sevcik equation also shows that the peak current of a reversible process is directly proportional to the square root of the voltage-scan rate. Since the voltage-scan rate is easily varied over orders of magnitude, this is a useful diagnostic—especially in questions of possible influence of reaction kinetics.

16.2.2   Effect of Charge-Transfer Kinetics

If the process is totally irreversible, the corresponding relationship (S24, S27) is

$$I = \pi^{1/2}P'zFAcD^{1/2}\alpha z_aF\chi/RT)^{1/2} \tag{16.8}$$

where the parameter P' is a function of $\alpha z_a F \chi t / RT$. The shape of $\pi^{1/2} P'$ is similar to that of P and its maximum value is 0.4958. The parameters are those of the Randles–Sevcik equation save that z_a is the *apparent* number of electrons involved in the charge-transfer step. For an irreversible process at 25°C:

$$I_p = 3.00 \times 10^5 \, z(\alpha z_a)^{1/2} A c D^{1/2} \chi^{1/2} \tag{16.9}$$

No relationship between the irreversible peak potential and a reversible half-wave potential exists, but the peak and half-peak potentials are related:

$$E_{p/2} = E_p \pm 1.857 RT / \alpha z_a F \tag{16.10}$$

At 25°C this equation is

$$E_{p/2} = E_p \pm 48 \text{ mV} \tag{16.11}$$

Both peak and half-peak potentials shift with changes in scan rate, by some $30/\alpha z$ mV per 10-fold change if the process is totally irreversible. Potentials of cathodic processes shift negative and those of anodic processes shift positive.

If the process is irreversible, the peak current is still proportional to bulk concentration of diffusing species, but its value is lower. For a one-electron process where α is 0.5 the peak current for an irreversible process is about 80% of the peak current for a reversible process. Use of this difference to determine the reversibility of the electrode process has been suggested in the literature; the writer does not recommend it save as a supporting technique, because cyclic voltammetry is more versatile for this purpose.

For a reversible electrode process at 25°C, the peak potential and the half-peak potential will differ by $0.057/z$ V. For an irreversible system the peak potential and the half-peak potential will differ by $0.048/\alpha z$ V, so that with α 0.5 the peak will be more spread out for an irreversible reaction, in addition to being lower in height.

If a system is observed not to be reversible, this irreversibility can arise from preceding, parallel, or following slow chemical steps as well as from the electrode process itself. Moreover, chronoamperometry with linear sweep is a much faster technique than classical polarography so that processes that are observed to be reversible in classical polarography may not be observed to be reversible when this technique is used. The formal equations for the complete slope of the peak are known but rarely used; the situation is similar with the equations of electrode processes that are not reversible.

16.2.3   Cycling Diagnostics

Additional information can be obtained in rapid-voltage-scan methods when the direction of voltage scan is reversed. If the process is reversible, a peak in the opposite (normally anodic) current direction should be observed. The peak voltage $E_{p,a}$ should be positive of $E_{1/2}$ in exactly the same way that $E_{p,c}$ is negative of it so that

$$E_{p,a} - E_{p,c} = 57/z \text{ mV} \tag{16.12}$$

If the voltage-scan reversal occurs at a potential sufficiently beyond that of the process being studied, then the peak current of the anodic process will be independent of that potential and will be equal to the peak current observed for the cathodic process if the oxidized and reduced forms of the reversible couple have equal concentrations and diffusion coefficients. For a single simple reversible process for which this is true, then regardless of scan rate,

$$I_{p,a}/I_{p,c} = 1 \tag{16.13}$$

The peak current of the reversal (normally anodic) peak should, like the cathodic peaks of subsequent processes, be taken using as a baseline the extension of the curve of the cathodic peak rather than the absolute or recorder zero of current.

Cyclic voltammetry is of considerable assistance in establishing the influence of charge-transfer kinetics. For a slow charge-transfer step it is found that $E_{p,c}$, $E_{p,a}$, $E_{p/2,c}$, and $E_{p/2,a}$ are no longer independent of dE/dt. Instead, $E_{p,c}$ and $E_{p/2,c}$ shift negative, while $E_{p/a}$ and $E_{p/2,a}$ shift positive, as dE/dt increases. As the charge-transfer step moves through quasireversible to totally irreversible, the magnitude of this shift increases until, for charge-transfer steps that are totally, irreversible, the shift at 25°C is $30/\alpha z$ mV for a 10-fold increase in the scan rate. At 25°C, $E_{p,a} - E_{p,c}$ may approach $57/z$ mV at lower scan rates, but as the scan rate increases so does the potential separation of the peaks. The peak currents $I_{p,a}$ and $I_{p,c}$ remain almost proportional to the square root of the scan rate, although as the process becomes less reversible the values of $I_{p,a}$ and $I_{p,c}$ decrease. And finally the ratio $I_{p,a}/I_{p,c}$ remains equal to unity only when α equals 0.5; even then, the ratio goes to zero as the charge transfer becomes irreversible, because the anodic peak on the return cycle disappears. The effect of slow charge transfer is shown in Figure 16.8 and 16.9.

16.2.4   Effect of Reaction Kinetics

If the charge-transfer step is fast and the kinetics of the accompanying kinetic step are slow, the cyclic voltammetric response will differ depending upon whether the slow reaction precedes, follows, or parallels the charge-transfer step.

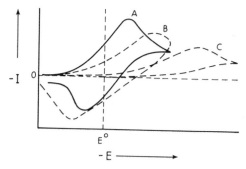

Figure 16.9 Cyclic voltammetry: effect of charge transfer kinetics. Curve A, reversible system; curve B, quasi-reversible system ($\alpha = 0.5$, $k^o_h = 3 \times 10^{-4}$ m/s, or 0.03 cm/s); curve C, irreversible system ($\alpha = 0.5$, $k^o_h = 10^{-6}$ cm/s).

CE Mechanism

If the slow chemical reaction step precedes the charge-transfer step:

1. The potential of the preceding (normally cathodic) peak shifts (positive) as the scan rate increases, while that of the subsequent (normally anodic) peak is independent of scan rate.

2. The preceding (normally cathodic) peak current is no longer proportional to the square root of scan rate; the ratio of peak current to square root of scan rate decreases as the scan rate increases.

3. The ratio $I_{p,a}/I_{p,c}$ is generally greater than unity and approaches unity as the scan rate decreases.

EC Mechanism

If the reaction follows the charge transfer step:

1. The potential of the subsequent (normally anodic) current peak shifts (negative) as the scan rate increases while that of the preceding (normally cathodic) current peak is independent of scan rate.

2. The ratio of the preceding (normally cathodic) peak current to square root of scan rate is virtually constant with the scan rate if the reaction is reversible, while the ratio of the subsequent (normally anodic) peak current to square root of scan rate is virtually constant with the scan rate if the reaction is totally irreversible.

3. The peak current ratio $I_{p,a}/I_{p,c}$ decreases from unity as the scan rate increases if the reaction is reversible, and increases toward unity as the scan rate increases if the reaction is totally irreversible, when the cathodic peak is the preceding peak (as it usually is).

EC(R) Mechanism

If the reaction parallels the charge-transfer step, then the case is actually the catalytic regeneration of the reactant for which the diagnostic criteria are:

1. The potential of the preceding (normally cathodic) peak shifts (positive) with increasing scan rate though with a maximum positive shift of $60/z$ mV per 10-fold increase in the scan rate.

2. The ratio of the preceding (normally cathodic) peak current to the square root of scan rate increases as the scan rate decreases, finally becoming independent of the scan rate.

3. The peak current ratio $I_{p,a}/I_{p,c}$ is unity at all scan rates.

The situation here is complex, and the original literature must be consulted (G6).

ECE Mechanism

If the ECE mechanism is involved, separate peaks are observed for the two charge-transfer steps only if they differ significantly in potential and the second step occurs at greater potential than the first. Otherwise the situation is complex. The original literature must be consulted (G6).

16.2.5   Effect of Adsorption

Strong specific adsorption of product or reactant will in general produce additional peaks; a cathodic and anodic peak centered on the same potential will normally be encountered, as shown in Figure 16.10. If the adsorption is weak, no additional peak will appear, but the anodic and cathodic peaks of the process of charge transfer itself will be enhanced. These adsorption effects on cyclic voltammetric curves have been studied theoretically and experimentally by Wopschall and Shain (S37–S39).

16.3   PULSE TECHNIQUES

Pulse applications to voltammetry do not appear to have received any significant attention, so *pulse polarography* may be a more accurate title for this section. In pulse techniques, a square-wave-type waveform is superimposed upon the normal slow DC polarographic sweep either by electronic superimposition of a square wave or by a mechanical or electronic switch that alternates the electrode potential between two levels at a rate rapid compared to that of the polarographic voltage sweep. Pulse polarography was originally developed by Barker (S42) around 1960 and became widespread with the development of practical instrumentation in about 1964; commercial instrumentation dates from about 1970.

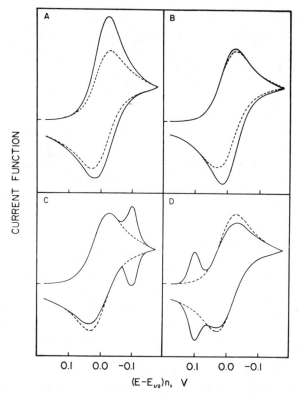

Figure 16.10 Cyclic voltammetry: theoretical adsorption peaks. Curve A, reactant adsorbed weakly; curve B, product adsorbed weakly; curve C, reactant adsorbed strongly; D, product adsorbed strongly. Dashed lines indicate behavior of the uncomplicated Nernstian charge transfer. From (S32), reprinted by permission of the American Chemical Society.

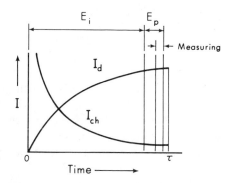

Figure 16.11 Normal pulse polarography: time scales. The times when the electrode is held at the initial potential E_1 and the pulse potential E_{pulse} are shown. Note the short interval of actual measurement time and the relative magnitudes of the diffusion and charging currents. Time values for the P.A.R. Model 174A courtesy of Princeton Applied Research.

All forms of pulse polarography have the same advantage of greater sensitivity over classical polarography; a usable signal can be obtained from significantly lower concentrations of electroactive species. There are no other advantages inherent in pulse techniques. Ordinary polarography can be used at concentrations of 1 mmol/dm³ without difficulty; the usual values of diffusion-limited currents are 3 to 6 μA for such solutions. Since current decreases linearly with concentration, at 1×10^{-5} mol/dm³ it is 0.03 to 0.06 μA, approximately the same as the charging current. Since the charging current is a nonremovable part of the residual current, or "noise," the signal-to-noise ratio approaches unity. This is the fundamental reason that concentrations in classical polarography must be above 1×10^{-5} mol/dm³. To improve the signal-to-noise ratio, the charging current must be reduced and/or the Faradaic current must be increased. Both occur in the pulse polarographic methods.

16.3.1   Normal Pulse Polarography

In the technique of normal pulse polarography, the potential of the drop is held at some nominal initial value E_I during most of the drop life. Near the end of drop life, a potential step (pulse) moves the potential of the drop to a new potential E_{pulse} and holds it there. The duration of this pulse is usually some 50 to 100 ms, and the current is measured during the last few ms of the pulse. In the pulse mode of the P.A.R. Model 174A, for example, the pulse is 57 ms long, and the current is measured (averaged) over the final 16.7 ms of the pulse. The pulses are of slowly increasing height, such that E_{pulse} changes slowly just as the potential ramp does in classical polarography. The pulse application and sampling times are shown in Figure 16.11.

It is not possible to measure the current during the initial portion of the pulse, because there is a charging current due to application of the pulse itself that must be allowed to decay; this is shown in Figure 16.12. In normal pulse polarography the charging current will be effectively reduced, although it will not be eliminated completely, and hence a slightly sloping baseline will still be noticed. The current observed will be larger than in classical polarography because the material at the electrode surface has not yet been depleted by reaction in the early part of drop life, as is the case in classical polarography, and hence the flux, and the current, are much larger. The increase has been calculated by Parry and Osteryoung (S43) as about 5 to 7 times (ratio of wave height observed in classical polarography to that observed in normal pulse polarography). The procedure gives useful polarograms at concentrations as low as 1 μmol/dm³ in aqueous solutions. The shape of the output is the same as in classical polarography when only the maximum current for each drop is used. The input pulse sequence is shown in Figure 16.13.

Kalousek polarography, also called the *Kalousek commutator method*, is an uncommon form of normal pulse polarography in which the interval of measurement of current and the interval of the pulse are coincident. The purpose of the Kalousek (S45, S46) switch is identical to that of the pulses in normal pulse

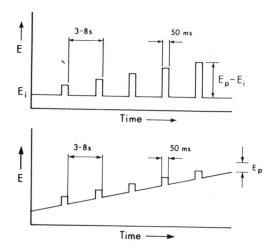

Figure 16.12   Normal pulse polarography: charging current due to pulses.

polarography. As analytical techniques the Kalousek methods offer no advantages over normal pulse polarography (S47), but for distinguishing between reversible and irreversible systems and for mechanistic studies, they have some value. The apparatus of Ishibashi and Fujinaga (S48–S50) operates in the same way as the Kalousek methods.

16.3.2   Differential Pulse Polarography

Differential pulse polarography (DPP) is the most widely used form of pulse polarography. In differential pulse polarography, a slow continuous DC scan is used as in classical polarography. On this slow voltage scan is superimposed a series of voltage pulses, themselves of constant magnitude rather than of increasing magnitude as in normal pulse polarography. The current is sampled twice—once just *prior to* the application of the pulse, and once at the *end of* the pulse—and the *difference* between these is recorded as the output. In the differential pulse mode of the P.A.R. Model 174A, the voltage pulse is 57 ms in length, and current sampling is averaged both over the 16.7 ms just before and the final 16.7 ms of the pulse; the difference between these two averaged samples serves as the recorded current. In other instruments, the instantaneous current before, and at the end of, the pulse is measured. The applied potential is shown in Figure 16.14.

The height of the pulses is constant and can usually be selected by the polarographer. Typical ranges of pulse heights are 5 to 100 mV. Within the limit that ΔE covers only the rising part of the ordinary polarographic wave, the magnitude of the signal output increases with ΔE. The resolution of closely spaced waves is improved, however, with smaller pulse heights and the polarographer must make an empirical tradeoff between sensitivity and resolution.

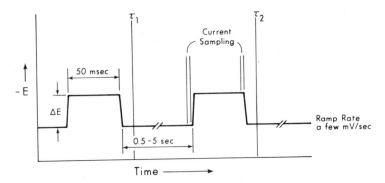

Figure 16.13 Differential pulse polarography: time sequence. Sampling intervals are shown in relation to drop and pulse times. Differential pulse mode of P.A.R. Model 174A. (Courtesy of Princeton Applied Research.)

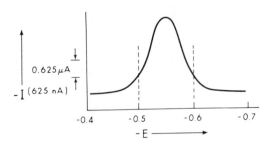

Figure 16.14 Differential pulse polarography: response curve. Typical differential pulse polarogram of a single reversible species.

The advantages of differential pulse polarography over classical polarography are similar to, but greater than, those of normal pulse polarography. The polarograms occur in differential form as peaks rather than waves as shown in Figure 16.15. For example, a solution of 1.3 × 10⁻⁵ mol/dm³ chloroamphenicol gives a peak current of about 800 nA, and the peak width is about 80 mV at half-maximum (50-mV pulse). The values of the peak currents, which are proportional to the concentration of electroactive species, are for a reversible system some 7 to 10 times greater than the diffusion currents measured in classical polarography. The charging current is drastically reduced because the charge of the electrode is changed only by the difference in potential that is the pulse height; thus the current flowing will be much smaller. More correctly, the *observed* charging is much smaller because only differential charging is observed. Hence the signal-to-noise ratio is increased, and determinations at lower concentrations are possible, down to about 10⁻⁸ mol/dm³. The major areas of application have been in trace analysis for heavy metal ions (lead, cadmium) and in analysis for biologically active materials such as antibiotics. The sensitivity of differential pulse polarography for lead and cadmium is comparable to that of atomic absorption methods, as is its precision; for speciation and molecular analysis, the electrochemical method is preferable.

Supporting electrolytes must be used in differential pulse polarography as in other electroanalytical methods, but the concentrations used can be smaller because the concentrations or electroactive ions are smaller and the currents passed are less. Thus determinations of 10^{-6} to 10^{-7} mol/dm^3 are possible in 10^{-4} mol/dm^3 supporting electrolyte (S47); this is an advantage in trace analysis since introduction of impurities from the supporting electrolyte is therefore less. Reagent-grade KCl has only 0.0001% heavy metals (as lead), but this is 1×10^{-6} mol/mole KCl or 1×10^{-7} mol/dm^3 in a 0.1-mol/dm^3 supporting electrolyte, enough to render trace analysis for lead in this medium questionable.

One problem in differential pulse polarography is that the height and shape of the peak is a sensitive function of the reversibility of the electrochemical system being studied. The presence of kinetic complications, whether charge-transfer or reaction kinetics, will reduce the height of the peak observed, and any factor that affects these kinetics will therefore affect the peak height also.

16.3.3 Square-Wave Polarography

Square-wave polarography originated with Barker in 1952 (S51, S52) and is comparatively uncommon because it is the most instrumentally complex of the variants of polarography. However, it can achieve the highest degree of sensitivity. It is useful at concentrations down to about 40 nmol/dm^3, and hence has some real analytical interest. Its only competition in sensitivity comes from differential pulse polarography, which in many ways it closely resembles, and which is instrumentally somewhat simpler. Its sensitivity for irreversible processes is less by at least an order of magnitude than its sensitivity for reversible processes. It is used primarily in direct trace analysis and is of little use in mechanistic studies.

In square-wave polarography, a square-wave voltage (typically 225 Hz, 30 mV peak-to-peak) is superimposed upon the slow DC scan voltage of classical polarography. The AC current arising from this signal is measured as in AC polarography. It is normal for the current to be measured late in drop life, in the manner of tast polarography, to suppress violent oscillations due to the dropping of the DME.

The output of the square-wave polarographic circuit is a plot of I_{AC}, or sometimes the maximum value of the square-wave current as measured on an oscilloscope, as a function of the slow linear DC sweep voltage E_{DC}. The resulting output closely resembles the output of AC polarography and differential pulse polarography for the same solution. Each peak corresponds to a redox process. The potential of the peak is identical to the $E_{1/2}$ of classical polarography for a

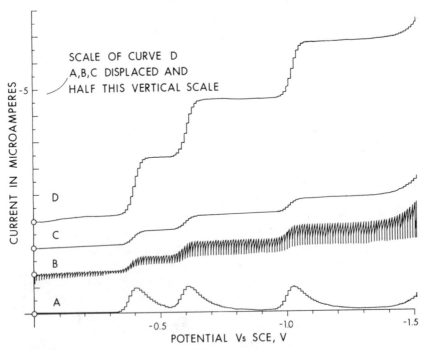

Figure 16.15 Comparison of modes of polarography. Aqueous solution 0.1 mmol/dm³ in Pb(II), Cd(II), Zn(II) and 10 mmol/dm³ in KCl. Curve A, differential pulse; B, classical polarography; C, tast polarography; D, normal pulse polarography. Done using P.A.R. Model 174A, drop time 2 s, scan rate 5 mV/s. Vertical scale of D is exaggerated by a factor of 2 to show the stepwise nature of the rise. Curve origins as indicated (author's laboratory).

reversible process, and will also be symmetrical. The greatest response is obtained with reversible processes, the detection limit being about 4×10^{-8} mol/dm³ for a two-electron reduction. For irreversible processes the response is less, the detection limit probably being about 1 μmol/dm³, and the peaks are no longer symmetric.

Square-wave, AC, and differential pulse polarography share the advantage of being able to detect small concentrations of a substance B that is reduced at a potential more negative than some substance A when A is present in much greater concentration than B. A difference of 10^5 in concentration ($c(A) = 100,000c(B)$) can be resolved given reasonable separation of half-wave potentials. Moreover, the separation of processes having a separation of only 50 mV in half-wave potentials is possible when their concentrations are comparable. Classical polarography cannot do nearly as well.

16.4 THEORY OF PULSE TECHNIQUES

The theory of pulse techniques derived by Barker (S50) is accessible in the papers of Osteryoung and co-workers (S51, S52). It suffices to deal with only the condition in which measurements are made at initial potentials where no reaction occurs and at potentials where the current is limited by diffusion alone.

16.4.1   Normal Pulse Polarography

For a cathodic process alone the current is given by the explicit solution for current of the Heyrovsky–Ilkovic equation (14.7):

$$I = I_{d,c}/(1 + P) \tag{16.14}$$

where the parameter P is given by

$$P = zF \exp (E - E_{1/2})/RT \tag{16.15}$$

Insertion of the Cottrell equation gives:

$$I = zFAD^{1/2}c\pi^{-1/2}t^{-1/2}/(1 + P) \tag{16.16}$$

At potentials negative enough to be on the diffusion-limited plateau the value of the parameter P approaches zero, so for normal pulse polarography:

$$I_{\text{lim}} = zFAD^{1/2}c\pi^{-1/2}t^{-1/2} \tag{16.17}$$

where t is the elapsed time between pulse application and current measurement. The Ilkovic equation, here used without explicit time dependence of area, gives for the classical polarographic diffusion-limited current:

$$I_{d} = (7/3)^{1/2}zFAD^{1/2}\pi^{-1/2}\tau^{-1/2} \tag{16.18}$$

For normal pulse polarography the increase in sensitivity is

$$I_{\text{lim}}/I_{d} = \tau^{1/2}/(7/3)^{1/2}t^{1/2} \tag{16.19}$$

This is a factor of 6 to 7 under the usual polarographic conditions, as is found experimentally (S53). Anodic processes give the same result. Normal pulse polarographic waves otherwise appear the same as those of classical polarography.

16.4.2   Differential Pulse Polarography

When the pulse amplitude ΔE is sufficiently small, less than RT/zF, the derivative of Eq. (16.16) is approximately equal to its differential:

$$\Delta I = z^2F^2AD^{1/2}c\,\Delta E\,P/RT\pi^{1/2}t^{1/2}(1 + P)^2 \tag{16.20}$$

The maximum value of ΔI is found at a potential E where $P = 1$, which for a reversible cathodic process is

$$E_p = E_{1/2} + \Delta E/2 \qquad\qquad (16.21)$$

This potential is always positive of $E_{1/2}$; the reverse relation is found for a reversible anodic process. When the process is not reversible, the relation is more complex (S54).

At this potential the value of ΔI is

$$\Delta I_p = z^2 F^2 A D^{1/2} c \Delta E / 4 R T \pi^{1/2} t^{1/2} \qquad\qquad (16.22)$$

Consideration of this relationship and those for the width of the peak shows (S55) that complete separation of two one-electron reversible processes whose half-wave potentials differ by 100 mV or less is not possible by any differential method since the minimum half-width of the peak for such a process exceeds 90 mV. With increasing ΔE the half-width of all differential pulse polarographic peaks increases in an approximately linear relationship. As the value of z increases the peaks become both higher and narrower. Peak height increases with pulse amplitude up to about 200 mV for a one-electron process and about 100 mV for a three-electron process, then becomes constant since the pulse now covers the entire rising portion of the polarographic wave.

16.5   AC TECHNIQUES

Alternating current (AC) techniques are those in which a periodic waveform, generally of low amplitude, is superimposed upon a DC potential applied to an electrochemical cell. The periodic waveform applied need not be a simple sine wave, although the sine wave is by far the most common form used for both theoretical and practical reasons. Since any periodic waveform can be expressed as a Fourier series of different frequencies and phase angles, AC techniques include all electroanalytical measurements in which periodic waveforms are used. The pulse techniques described in the preceding section are those in which the waveform (pulse) is aperiodic, since only one pulse is generally applied to each drop. AC techniques are used in both polarography and voltammetry as well as in modes where no sweep is applied. In polarographic work the AC signal can be applied, and the resultant signal measured over a short time interval late in drop life, in the manner of tast polarography. Care must then be taken that the interval of measurement follows the onset of AC application by enough time that any effects due to the onset have become negligible, so that the theory of periodic rather than aperiodic waveforms can be applied. The synchronization of signal and drop fall is necessary as in tast polarography.

In all AC methods it is convenient to deal with the observed phenomena in terms of electrical analogue models. Much of the theory of AC electroanalytical methods is based on AC circuit relations, an introduction to which was given in Chapter 2. Measurement methods include bridges and more elaborate sampling circuits. Since both amplitude and phase relationships, as well as frequency, must be recorded to give the complete information in a Fourier pattern, the complete analysis of complex waveforms requires computer assistance. In most applications, simpler equipment is used with concomitant loss of information.

The electrical analogue model of the cell described in Part 1 was a simple series circuit containing one resistive element, the cell resistance, and one capacitive component, the double-layer capacitance. The cell resistance is the resistance of the cell as seen by an external measuring circuit, and is the sum of the resistances of the electrodes and leads (usually negligible, except in the case of capillaries filled with mercury) and the resistance due to ionic transport, which is the reciprocal of the conductance. The double-layer capacitance is much larger than the capacitance due to the electrodes and leads, and the latter are usually neglected.

16.5.1   Faradaic Impedance

The method of Faradaic impedance in its modern form originated with D. C. Grahame and J. E. B. Randles (S53–S56) in the 1950s, although its theory dates from the work of Warburg at the beginning of the century (S57, S58). A cell containing no redox couple is placed in one arm of an impedance bridge, a small-amplitude AC signal of the desired frequency is applied, and the resistive and capacitive components of the impedance are obtained by null balance of the bridge. The resistive component is assigned as the cell resistance, and the capacitive component is assigned as the double-layer capacitance. The redox couple in question is then added to the cell, and the new null balances for the resistive and capacitive components of the impedance are obtained. The difference in the impedance caused by the presence of the redox couple constitutes the *Faradaic impedance* due to the presence of a Faradaic process. Both components of the redox couple must be present to obtain poising of the electrode (at the potential given by the Nernst equation for the appropriate activities of the species present) and to prevent rectification of the applied AC voltage by the absence of a Faradaic process capable of going in the appropriate direction. A solution containing only the oxidized form would block anodic Faradaic current and one containing only the reduced form would block cathodic Faradaic current.

Warburg visualized the addition of what is now known as the *Warburg impedance* Z_w as a path parallel to the double-layer capacitance to account for the charge passage due to the Faradaic reaction. An impedance can be taken as either a series or a parallel equivalent *RC* circuit; the Warburg impedance is usually taken as a series combination of a Warburg resistance R_w and a Warburg capacitance C_w.

The effect of the additional pathway of the Warburg impedance depends upon the nature of the Faradaic reaction. The simplest case is reversible charge transfer and diffusion-limited mass transfer of both reduced and oxidized species to the electrode, the same diffusion coefficient being used for both species. The resistive and capacitive components of the Warburg impedance are then identical:

$$R_w = 1/\omega C_w \tag{16.23}$$

and the phase angle θ of the output current from the Warburg impedance path will be 45°. The Warburg impedance Z_w is the vectorial sum of its two components, and its magnitude is thus equal to the square root of the sum of the squares of their magnitudes. The phase angle of the output current from the purely capacitive pathway via C_{dl} is 90° so that phase-sensitive detection can separate the capacitive double-layer impedance from the Warburg impedance of the Faradaic process.

16.5.2   Fournier Effect

The impedance presented by an electrochemical cell to an AC waveform is never completely adequately described by an analogue electrical circuit however complex. In an electrical sense the cell acts as a *nonlinear circuit element,* an element whose response is not always proportional to the input voltage signal. Fournier (S59) superimposed an AC potential of 50 to 100 mV, 10 to 1000 Hz on a DC potential and observed the direct current output, which appears as a normal classical polarogram. It appears with slightly shifted half-wave potential when the process is not reversible, and the potential of crossover with a normal polarogram depends upon the applied frequency. For a reversible process, the potential of crossover is the classical polarographic half-wave potential, and the potential of crossover is independent of frequency because the cell can then be approximated by a linear circuit element over a small voltage range (S60, S61). The Fournier effect has no modern interest.

16.5.3   Faradaic Rectification

In 1950 Doss and Agarwal (S62) applied a small AC voltage, through a transformer, to a cell containing a large auxiliary platinum electrode and a platinum microelectrode. A capacitor was placed in the microelectrode circuit to block the flow of direct current. The DC potential measured between the microelectrode and a reference electrode was found to depend upon the redox system in the cell and upon the frequency and amplitude of the applied AC voltage. This effect, initially called the *redoxokinetic potential,* is now known as *Faradaic rectification* (S63).

Faradaic rectification has been employed somewhat more widely than the Fournier effect; the review of Agarwal (S64) gives an effective summary. The redoxokinetic potential is, at low frequencies, proportional to the square root of the frequency. The theoretical equations permit calculation of the transfer coefficient α and the heterogeneous rate constant k^0_h from measurements of the redoxokinetic potential over a range of frequencies, and Faradaic rectification has sometimes been employed for this purpose. It has no significant advantages, or use, in analytical applications.

16.5.4 Classical AC Polarography

Most of the remaining AC methods, whether polarographic or voltammetric, are known under the name of AC polarography, and no clear terminology has evolved to distinguish them. The simplest such technique, *classical AC polarography*, is a variant of polarography (G3, G5, G7) in which a small AC potential of roughly 10 to 20 mV at 60 to 100 Hz is superimposed upon the usual slow ramp potential of classical polarography. No drop timer is needed for the DME. The usual direct current recorder is replaced by an AC recorder, and the DC component of the current is ignored. The AC polarogram thus consists of a plot of the alternating current against the DC slow ramp potential.

The theory of AC polarography has been worked out for many complex cases (G6, G7), but its primary use is in the determination of species undergoing reversible redox processes at a dropping-mercury electrode. Its output generally consists of one peak per electrochemical step or process, as shown in Figure 16.16. The height of the peak I_p is directly proportional to the bulk concentration of electroactive species, and hence quantitative as well as qualitative analysis is possible.

AC polarography gives output in the form of peaks for three different types of electrochemical phenomena. The first class is called *AC polarographic peaks* or *Faradaic peaks*. These peaks are due to the reduction or oxidation of an electrochemically active substance. The electrochemical process may be simple and reversible, or it may be complicated by slow reaction kinetics or by charge transfer kinetics of the electrode reaction. The second class is called *tensammetric peaks*. These peaks are due to reversible potential-dependent adsorption–desorption processes at the electrode surface. They involve movement of charge because the process changes the double-layer capacitance of the interface, and they also appear at a reproducible potential. No oxidation–reduction (Faradaic) process is involved in the formation of a tensammetric peak. Tensammetric peaks are of some analytical value in that they permit analysis for materials that adsorb or desorb on electrodes but that are neither oxidizable nor reducible. The third class is called *transition peaks*. These peaks are due to an electron transfer that produces a surface-active material whose subsequent adsorption on to the electrode surface gives rise to a tensammetric peak. An example would be the discharge of chloride ion on mercury in which a surface film of Hg_2Cl_2 is formed. Peaks of this third type are of little use in analysis.

Peaks that are due to Faradaic processes are of analytical importance because the observed alternating current is considerably greater than the direct current, and hence greater sensitivity can be obtained in AC polarography. Such peaks are also of interest in electroanalytical reaction studies of a fundamental nature.

The apparatus required for AC polarography can be quite simple. It consists of an AC generator, which can be either a transformer or a signal generator, in series

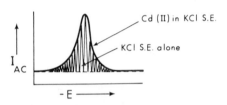

Figure 16.16 AC polarography: response curve. Typical polarograms of supporting electrolyte (KCl) and SE containing Cd(II), which is reversibly reduced. The envelope is drawn in a darker line, while the actual wave observed is the series of oscillations shown. The baseline, which is not at zero alternating current, is obtained for the KCl supporting electrolyte alone.

with the DC polarizing unit of a normal polarograph. The alternating current can be monitored using the AC ranges of a vacuum tube voltmeter or by a strip-chart recorder equipped to filter out the DC component. Sophisticated commercial units based on lock-in amplifiers are now on the market. These permit the measurement of the phase angle of the alternating current with respect to the AC voltage as well as the peak height and potential. The phase angle is a useful diagnostic in AC polarography.

16.5.5   Oscillopolarography

In oscillopolarography (G1, G2), a small alternating current (0.1–1 mA) is applied to the cell, and the resulting derivative of the potential dE/dt is observed as a function of the potential. This is a restricted meaning of oscillopolarography accepted by IUPAC (S65) which corresponds to the *alternating current oscillographic polarography* of Heyrovsky (G1); other areas included under oscillographic polarography by Heyrovsky are included under other titles in this chapter. The original meaning of oscillographic polarography—use of an oscilloscope as a measurement system—has been found to be too broad a classification to be useful.

The oscillographic responses for typical processes are given in Figure 16.17. The indentations in the circle or ellipse correspond to potentials at which reduction or oxidation occurs. For a reversible process both will occur at the same potential, but for an irreversible process there will be a voltage separation between them as shown on the bottom curve of Figure 16.17.

An alternative procedure for displaying oscillographic polarography curves is shown in Figure 16.18. Here the voltage output is applied to the vertical plates of an oscilloscope while the horizontal time sweep of the oscilloscope is triggered by the alternating current applied. The resultant peak form shows indents at the potentials for oxidation and reduction. For a reversible process the indents occur at the same potential, but for an irreversible process they show a separation in voltage, as in the lower curves of Figure 16.18. For convenience, the measurements are usually made late in drop life in the manner of tast polarography. Such techniques have special uses but are not widely employed outside Czechoslovakia (S66, S67).

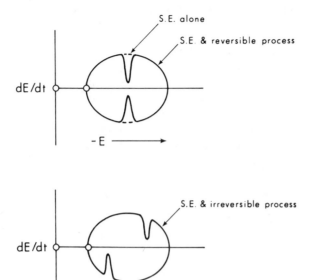

Figure 16.17 Oscillopolarography: response curves. The curve observed for a single reversible process is on the top, that observed for a single irreversible process is on the bottom. A curve for the supporting electrolyte alone shows no indentation (dashed line, top curve).

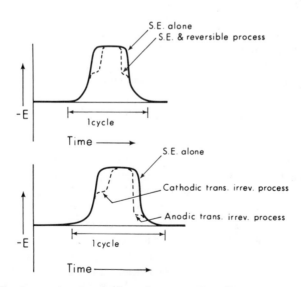

Figure 16.18 Oscillopolarography: alternate form of response curves. Upper curves are those observed for a reversible process, lower curves are those observed for an irreversible process. Curves for supporting electrolyte alone are solid, those for supporting electrolyte plus electroactive species are dashed.

16.6   THEORY OF AC TECHNIQUES

The parameters that are the independent variables in AC polarography are the AC frequency ω and the DC voltage E_{DC} along the slow voltage sweep. All the other parameters are dependent variables, including the peak current I_p, the peak potential E_p, the phase angle, and the peak width (normally taken as the full peak width at half the peak maximum). The peak width is usually sufficient to characterize the peak shape.

The solution of the differential equations of AC polarography requires the use of the Laplace transform. Smith (G7) has given a sound theoretical development.

16.6.1   Reversible Processes

The AC polarographic response for a reversible system is actually the derivative of the DC polarographic response. The fundamental harmonic alternating current for a reversible reduction process observed at either a stationary or expanding planar electrode is (G7)

$$I(\omega t) = z^2 F^2 Ac(\omega D)^{\frac{1}{2}} \,\Delta E \, \sin(\omega t + \pi/4)/4RT \, \cosh^2(j/2) \tag{16.24}$$

The parameter j is defined as

$$j = zF(E_{DC} - E_{\frac{1}{2}})/RT \tag{16.25}$$

The current $I(\omega t)$ is the alternating current observed, ΔE is the amplitude of the applied AC sine voltage, and ω is its frequency, usually in radians per second. The alternating current is a maximum when E_{DC} equals $E_{\frac{1}{2}}$, because $\cosh^2(j/2)$ is unity when j is zero. The peak or maximum current is then:

$$I_p = z^2 F^2 Ac(\omega D)^{\frac{1}{2}} \Delta E \, \sin(\omega t + \pi/4)/4RT \tag{16.26}$$

The potential at which the current has a maximum, E_p, for a simple reversible process is independent of the AC frequency and equal to $E_{\frac{1}{2}}$, the polarographic half-wave potential, as can be seen from the preceding equation. It can also be shown from this equation that the width of the peak will be independent of the AC frequency and equal to $90/z$ mV for a simple reversible process, that the phase angle is constant over the entire AC polarographic peak at $\pi/4$ radians or 45° and is independent of the AC frequency, and that the value of the maximum current I_p will be directly proportional to the square root of the AC frequency. Thus I_p will in principle, go to zero as the frequency decreases, and a plot of I_p against the square root of ω should be linear and have a zero intercept.

Rearrangement of the definition of j leads to

$$E_{DC} = E_{\frac{1}{2}} + RTj/zF \tag{16.27}$$

which upon appropriate substitution yields

$$E_{DC} = E_{\frac{1}{2}} + (2RT/zF) \ln [(I_p/I)^{\frac{1}{2}} - ((I_p - I)/I)^{\frac{1}{2}}] \tag{16.28}$$

This equation indicates that a linear plot of E_{DC} against the logarithm of the term in square brackets should be obtained whose intercept should be $E_{\frac{1}{2}}$ and whose slope should be $118/z$ mV at 25°C.

The above comparatively simple theory can be applied only if the AC voltage imposed is small, less than $16/z$ mV peak-to-peak. If it is larger than this there will be a significant contribution to the observed current from the higher harmonic components as well as the fundamental, while for low applied AC voltages only the fundamental need be considered.

16.6.2   Effect of Charge-Transfer Kinetics

If charge-transfer kinetics limit the rate of current flow, the observed AC peak current is generally reduced; in the case of a totally irreversible reaction, the AC response may be so small as to be invisible. Charge-transfer processes that are intermediate between reversible and totally irreversible are called *quasi-reversible*.

When slow charge-transfer kinetics are present, the peak potential E_p will generally vary with the AC frequency. For a reaction that is slow but not totally irreversible, the variation with AC frequency will be observed only when α is not equal to 0.5; for a totally irreversible reaction, the peak potential will shift cathodic by $30/\alpha z$ mV at 25°C for every 10-fold increase in ω. The value of the peak current I_p will depend upon drop time and potential; anodic of a "crossover" potential it increases, cathodic of this potential it decreases, with drop time. If the reaction is totally irreversible then I_p is inversely proportional to the square root of the drop time. In addition, I_p is not linear with the square root of ω. Depending upon the value of k^0_h, linearity with the square root of ω may be noted at low frequencies (the system approaches the reversible case). The peak current will approach a limit as ω increases, although this occurs at higher frequencies as k^0_h increases and for systems that are *not* totally irreversible may be beyond the experimentally accessible range. For reactions that *are* totally irreversible the limit is generally observable. Finally, the phase angle will vary with both ω and E_{DC}. The cotangent of the phase angle will increase linearly with the square root of ω, approaching unity as low frequency as shown in Figure 16.19. From the slope of a plot of the cotangent of the phase angle against the square root of the AC frequency the rate constant k^0_h can be extracted if α and D are obtained from these or other measurements. The value of the cotangent of the phase angle is a maximum at

$$E_{DC} = E_{\frac{1}{2}} + RT/zF \ln[\alpha/(1 - \alpha)] \tag{16.29}$$

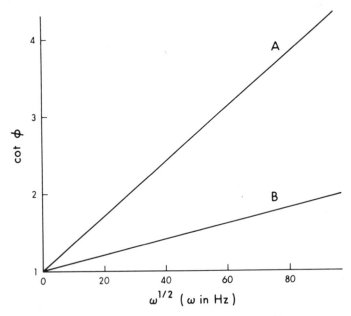

Figure 16.19 AC polarography: change transfer kinetics. Curves A and B are experimental plots observed for different reactions or for the same reaction under different conditions. Curve A, $k^0_h = 2 \times 10^{-4}$ m/s; B, $k^0_h = 1 \times 10^{-3}$ m/s; in both, the phase angle is measured at the half-wave potential.

which is at $E_{1/2}$ when α is 0.5. As the reaction becomes totally irreversible, the phase angle becomes very small; a phase angle much less than $45°$ is a useful criterion for totally irreversible reactions.

A major area of application of AC polarography has been the study of reactions in which the electrochemical charge transfer step is quasi-reversible.

16.6.3   Effect of Reaction Kinetics

Similar behavior of the dependent variables of AC polarography is observed when the kinetics of chemical reactions rather than those of charge transfer itself are involved (G6). In the case of adsorption, additional tensammetric peaks are observed.

16.7   STRIPPING TECHNIQUES

Stripping techniques are those techniques in which the potential of an electrode is initially held for a relatively long period of time (the *deposition step*) during which material moves from the solution to the electrode; no measurements are made during the deposition step. The material on the electrode may then be allowed to assume an equilibrium distribution or configuration at some other potential (the *equilibration step*). Again, no measurements are made. In a subsequent *measurement step*, the material on or in the electrode is determined. The significance of this procedure is that the time for the deposition step may be much greater than the time for the measurement step. If the coulombs used in both steps are the same, the current in the measurement step should then be much greater than the current in the deposition step. The effect of this is to significantly increase the sensitivity of the method since the current in the measurement step need no longer be limited by the rate of diffusion of material to the electrode surface.

Stripping techniques must involve three steps: a *deposition step* (preconcentration step), an *equilibration step,* and an *analysis step.* In the deposition step, the species of analytical interest is reduced or oxidized onto or into an electrode from the bulk of the solution to be analyzed. In the equilibration step, the deposition of new material is halted while the material in or on the electrode assumes a new and equilibrium distribution. In the analysis step, the deposited material is oxidized or reduced back into solution during some electroanalytical procedure, usually voltammetry, and is thereby determined.

Quantitative analysis using SV requires attention to each of the three steps above. The testing of stripping procedures by the method of standard addition (the usual method of testing them) is not sufficient to evaluate matrix effects, which may drastically and adversely affect the results obtained.

The only advantage of stripping methods such as anodic stripping voltammetry is the ability to determine much lower concentrations, 10^{-8} to 10^{-9} mol/dm^3, than can be determined by any other electrochemical method. Determination of two or more separate species is possible in principle and sometimes in practice. The major disadvantage of anodic stripping voltammetry is that it only works well for metals that form amalgams. Materials that are reduced to forms insoluble in mercury can be tackled with platinum electrodes but less satisfactorily. Metals that do form amalgams may form intermetallic compounds with each other or with mercury, which will inhibit their subsequent stripping. Stripping voltammetry (SV) is the common name for the entire class of modern electroanalytical stripping methods, all of which will be discussed under this heading. The one and only significant advantage of stripping techniques over other electroanalytical methods is their sensitivity, which is greater than that of any other electroanalytical method. For this reason it is used for trace or ultratrace analysis. Detection limits are comparable with nonflame atomic absorption methods for those elements detectable by both methods, but the number of elements determined by SV methods is less than for atomic absorption. Anodic stripping (ASV) methods have been used for Ag, As, Au, Ba, Bi, Cd, Cu, Ga, Ge, Hg, In, K, Mn, Ni, Pb, Pt, Rh, Sb, Sn, Tl, and Zn, but practical analyses are limited to Ag, Au, Bi, Cd, Cu, Hg, In, Pb, Rh, Sn, Tl, and Zn; most of the emphasis has been on Cd, Cu, Pb, Sn, Tl, and Zn. Stripping methods (G8, G9) are generally used for elemental analysis; speciation requires particular conditions or applications.

16.7.1   Electrode Materials

Electrodes used in SV work may be inert solid materials such as gold, platinum, or carbon. The latter may be either a graphite rod or, more commonly, either glassy carbon or a wax-impregnated graphite electrode (WIGE) since the latter give a much smoother surface. Solid electrodes may be of different geometry. The rotating disk electrode and rotating ring-disk electrode are desirable in fundamental studies since their hydrodynamic equations have been quantitatively derived, but simpler and varied geometries are more commonly used, especially stationary disks and cylinders. Noble metals, especially platinum, have been used but carbon electrodes are becoming steadily more common because their potential range is superior and they do not interact with deposited material. Glassy carbon and pyrolytic graphite are superior, owing to their less porous surface, but are more difficult to shape into electrode form and are usually used as disk surfaces. Noble metals can be used in any form; tubular electrodes are convenient at the outputs of liquid chromatographic columns.

Alternatively, the electrode may be a liquid, virtually invariably mercury, of which two major types are used in SV. The most popular type is the hanging-mercury-drop electrode (HMDE), which consists of a drop of mercury hanging in the solution to be analyzed. The HMDE can be prepared in two ways. One is by extrusion of a mercury drop from a capillary fed from a reservoir by a micrometer-driven syringe. Alternatively, mercury drops obtained by extrusion or from a conventional DME can be attached to an inert metal support—generally amalgamated platinum, although gold or silver have also been used. This latter method is considerably less convenient, and has the disadvantage that the contact or support metal dissolves in mercury with deleterious effects on the analytical results obtained. The capillary extrusion type uses the mercury thread as its contact, so is not subject to this contamination, but does suffer from the reverse problem since material deposited in the HMDE may diffuse into the capillary thread, leading to contamination of subsequent extruded drops. This difficulty can be easily overcome by extruding two or three additional drops of mercury between the drops used for analysis. A variation of the HMDE, the sitting-mercury-drop electrode (SMDE), also uses a mercury drop and a vertical support, but the drop now sits on the support rather than hangs from it. There appears to be little advantage to this variation.

The second major type of electrode based upon mercury is the mercury thin-film electrode (MTFE), which consists of a thin (1–100-μm) film of mercury on a solid support. The film may be cathodically deposited either separately before deposition or, if the deposition is not adversely affected by the presence of Hg(II), as part of the deposition step itself by adding sufficient Hg(II) to the solution being analyzed. If the latter procedure is employed the mercury deposition can be at the same potential as deposition of the material to be analyzed for, or at a potential somewhat less cathodic followed by the usual deposition procedure.

16.7.2   Other Materials Considerations

Since the only advantage of SV lies in its ability to measure very small quantities of material and the method is otherwise inferior to more direct electroanalytical determination procedures, SV is applied to the analysis of very dilute solutions. Although in some applications such as those involving natural waters, sample volume is effectively unrestricted, samples of a few cubic centimeters suffice and are universally used. Many of the problems found in the applications of SV pertain to the solution and container rather than to SV itself; handling of solutions of concentration 10^{-6} mol/dm^3 or below presents some unusual problems not encountered in millimolar solutions. These problems arise in three areas container materials, the base electrolyte, and the cell atmosphere.

Glass, the usual laboratory container material, has two major disadvantages in work with dilute solutions: glass surfaces adsorb many substances, including metal ions, and alkali metal ions and other ions can be leached from glass, especially by basic solutions. Vessels of quartz, or of Teflon or other plastic, if well leached, are better than glass, although some adsorption will still occur. Glass vessels can be improved by siliconizing the walls. The problem of contamination from the container materials becomes increasingly serious as the solution concentration decreases below 1 mmol/m^3 (μmol/dm^3).

The base electrolyte in SV is normally aqueous. The water should be purified by distillation, preferably double distillation with the addition of alkaline permanganate to decompose organic impurities. The base electrolyte also normally contains an inorganic acid, base, or salt. The purity of even analytical reagent-grade materials is often not high enough to prevent serious interference with SV determinations from traces of heavy metals or transition metals in the inorganic material used. The most efficient way of dealing with this problem is exhaustive electrolysis of concentrated solutions of the inorganic material using a mercury cathode of large area. The potential of the mercury electrode should be controlled at a potential more negative than the potential that will be used in the deposition step of the particular SV determination to ensure that all of the impurities that might be deposited are removed.

The choice of base electrolyte depends upon the specific analysis. A sufficient concentration of base electrolyte is one 2 orders of magnitude greater than that of the material of interest, and thus a lower concentration of base electrolyte is normal in SV than in classical polarography. If the sample is taken into solution by mineral acid or fusion, or if the sample already contains sufficient electrolyte (as seawater, for example, does), addition of more inert electrolyte may not only be unnecessary but deleterious, since additional trace impurities may be introduced with it.

Treatment of the solutions to be analyzed with strong mineral acids or with ozone is useful in removing complexing agents whose stable complexes with metal ions would prevent their cathodic deposition. Ozone has the advantage that its use does not add electrolyte to the solution.

The cell atmosphere is almost invariably purified nitrogen, and the same gas is used to remove dissolved oxygen from the base electrolyte. The interference of oxygen in stripping voltammetry is more severe than in classical polarography, not only because the concentrations are generally lower in SV but also because dissolved oxygen can reoxidize the deposited material, decreasing the efficiency of the deposition step in a manner that is not reproducible. One possible problem in the nitrogen system is flexible plastic tubing, which is *not* impermeable to oxygen—glass or copper is preferable.

16.7.3   Deposition Step

The deposition step in SV is the step in which the material being analyzed for is deposited onto and into the electrode from the bulk of the solution to be analyzed. When mercury electrodes of either type are used, the deposition step may include deposition into, as well as deposition onto, the surface of the electrode. When solid inert electrodes are used, deposition is only on the surface of the electrode. The solution from which the material is deposited need not be the same solution in which the analysis step is carried out, but it is almost invariably so since physical transfer of either solution or electrode adversely affects the analytical results. Deposition may be quantitative, in that essentially all of the desired material may be concentrated into the electrode from the bulk solution, but this is time-consuming, and most applications of SV do not use quantitative deposition. It is only necessary that the quantity deposited be directly proportional to the concentration in the bulk solution, which can be achieved without quantitative deposition.

Electrochemical reactions may be governed by quite complex and frequently unknown reaction kinetics, but for SV purposes most such reactions can be dealt with quite simply as three successive processes: mass transport of the electroactive species to the electrode, charge transfer between the electroactive species and the electrode, and steps subsequent to charge transfer. This latter category of processes may include mass transport of products from the electrode surface either in the bulk of the electrode or in the bulk of the electrolyte solution.

In stripping voltammetry, one of the most significant parameters under the control of the analyst is the potential at which the electroactive species is deposited onto or into the electrode. This potential is normally set at a value on a polarographic limiting-current plateau, which is a potential sufficiently cathodic or anodic that the rate of charge transfer is essentially infinitely fast and the current is limited by mass transport, diffusive or convective. Under these conditions the deposition step may be considered as being controlled by mass transport, charge transfer may be neglected, and subsequent steps may be regarded as generally undesirable complications to be minimized. It is possible to control the current rather than the potential of electrodeposition, but this alternative has no significant practical use.

The deposition step is by far the longest in the determination of substances by SV. The sensitivity of the method is directly proportional to the ratio of the deposition time to the analysis time because, ideally, the coulombs used in deposition are identical in number to those used in stripping. The time is reduced by increasing the rate of mass transport by stirring. Stirring is achieved by an external stirring bar or by rotating the electrode. The *collection efficiency* increases with increasing deposition time, but the increase is an inverse exponential one, and increasing the deposition time can therefore achieve only a limited increase in collection efficiency. Collection efficiency depends upon the deposition potential, the degree of stirring, and the type of electrode employed.

16.7.4 Equilibration Step

An equilibration step is often omitted, and in a practical analysis is not strictly necessary, but quantitative or theoretical studies on mercury do require it since a uniform concentration in the mercury or in the solution is necessary to permit solution of the equations. When the deposition is only on the surface of the electrode, equilibration is less necessary.

16.7.5 Analysis Step

Analysis can be carried out by a variety of electroanalytical techniques, but the two that are by far the most common are fast linear-sweep voltammetry and differential pulse voltammetry. The sensitivity of DPASV is greater than that of linear-sweep ASV because the analysis method is more sensitive.

The voltammetric technique used for the stripping procedure need not be normal fast-sweep voltammetry. Other methods, such as differential pulse voltammetry (S68, S69), phase-sensitive AC techniques (S70, S71), and square-wave techniques have been used. Applications of the technique have been extensively reviewed (G8, S72–S74).

One of the problems in ASV on mercury, and to a much lesser extent on solid electrodes, is the formation of intermetallic compounds between two of the components being analyzed for or between a component being analyzed for and another material present in the matrix deposited with it, regardless of whether or not the second material is later stripped. The effect of this appears when the matrix changes significantly. The deposition step is not affected, but the analysis step may not detect all of the deposited material if some of it is in the form of an intermetallic compound.

16.7.6 Cathodic Stripping Voltammetry

Cathodic stripping voltammetry is possible when an *anodic* reaction can form an insoluble compound with the electrode material. This usually requires a mercury electrode, although silver electrodes can be used for some species such as halides. The reference electrodes cannot introduce halide ions, and thus either a Hg_2SO_4

/Hg in saturated K_2SO_4 reference electrode or KNO_3 isolation salt bridges must be used. For halides, the addition of methanol or acetonitrile may be necessary to reduce the solubility of the halide compounds. Cathodic stripping is from the surface rather than the bulk of the mercury electrode.

Cathodic stripping voltammetry has been used to analyze for S, Se, Cl⁻, Br⁻, I⁻ and SCN⁻, as well as certain thioureas.

16.7.7   Stripping Voltammetry with Collection

Blaedel (S75) has suggested use of stripping voltammetry with collection. When two successive tubular platinum electrodes are used at the output of a liquid chromatographic column, the sought-for constituents can be reduced (plated) on to the first electrode and collected there, then oxidized off in a brief pulse and analyzed by the second (downstream) electrode. The method has been applied to analysis of copper, lead, and cadmium in tap water.

16.8   REFERENCES

General

G1   J. Heyrovsky and J. Kuta, *Principles of Polarography*, Academic Press, New York, 1966. Good general text with emphasis on theory. Does not cover all advanced techniques.

G2   R. Kalvoda, *Techniques of Oscillographic Polarography*, Elsevier, Amsterdam, 1965.

G3   B. Breyer and H. H. Bauer, *Alternating Current Polarography*, Wiley-Inerscience, New York, 1963.

G4   B. B. Damaskin, *The Principles of Current Methods for the Study of Electrochemical Reactions*, G. Mamantov, Trans., McGraw-Hill, New York, 1967.

G5   P. Delahay, *New Instrumental Methods in Electrochemistry*, Wiley-Interscience, New York, 1954, 437 pp.

G6   E. R. Brown and R. F. Large, Chapter VI in *Techniques of Chemistry, Vol. I, Part IIA*: Electrochemical Methods, A. Weissberger and B. W. Rossiter, Eds., Wiley-Interscience, New York, 1971.

G7   D. E. Smith, "AC Polarography and Related Techniques: Theory and Practice," *in Electroanalytical Chemistry, Vol. 1*, A. J. Bard, Ed., Marcel Dekker, New York, 1966.

G8   F. Vydra, K. Stulik, and E. Julakova, *Electrochemical Stripping Analysis*, Ellis Horwood/John Wiley, New York, 1976, 283 pp.

G9   K. Z. Brainina, *Stripping Voltammetry in Analysis*, Halsted Press/John Wiley, New York, 1974, 222 pp. Translated from the Russian edition of 1972. Most of the book consists of detailed methods for the analysis of specific ions by stripping voltammetry.

G10   D. D. Macdonald, *Transient Techniques in Electrochemistry*, Plenum, New York, 1977, 329 pp. Unified theoretical presentation at a highly mathematical level.

Specific

S1   D. N. Hume, *Anal. Chem.* **36**, 200R (1964).

S2   D. N. Hume, *Anal. Chem.* **38**, 261R (1966).

S3   D. N. Hume, *Anal. Chem.* **40**, 174R (1968).

S4   R. S. Nicholson, *Anal. Chem.* **42**, 130R (1970).

S5   R. S. Nicholson, *Anal. Chem.* **44**, 478R (1972).

S6   P. T. Kissinger, *Anal. Chem.* **46**, 21R (1974).

S7   P. T. Kissinger, *Anal. Chem.* **48**, 17R (1976).

S8   W. R. Heineman and P. T. Kissinger, *Anal. Chem.* **50**, 166R (1978).

S9   W. H. Reinmuth, *Anal. Chem.* **36**, 211R (1964).

S10 W. H. Reinmuth, *Anal. Chem.* **38**, 270R (1966).

S11 W. H. Reinmuth, *Anal. Chem.* **40**, 185R (1968).

S12 D. K. Roe, *Anal. Chem.* **44**, 85R (1972).

S13 D. K. Roe, *Anal. Chem.* **46**, 8R (1974).

S14 D. K. Roe and P. Eggiman, *Anal. Chem.* **48**, 9R (1976).

S15 D. K. Roe, *Anal. Chem.* **50**, 9R (1978).

S16 E. Wahlin and A. Bresle, *Acta Chem. Scand.* **10**, 935 (1956).

S17 A. Bresle, *Acta Chem. Scand.* **10**, 943, 947, 951 (1956).

S18 A. W. Elbel, *Z. Anal. Chem.* **173**, 70 (1960).

S19 P. O. Kane, *J. Polarogr. Soc.* **8**, 10 (1962).

S20 J. E. B. Randles, *Trans. Faraday Soc.* **44**, 3241 (1950).

S21 A. Sevcik, *Coll. Czech. Chem. Commun.* **13**, 349 (1948).

S22 M. M. Nicholson, *J. Amer. Chem. Soc.* **76**, 2359 (1954).

S23 W. T. de Vries and E. Van Dalen, *J. Electroanal. Chem.* **6**, 490 (1963).

S24 P. Delahay, *J. Amer. Chem. Soc.* **75**, 1190 (1953).

S25 H. Matsuda and Y. Ayabe, *Z. Elektrochem.* **59**, 494 (1955).

S26 R. S. Nicholson and I. Shain, *Anal. Chem.* **36**, 706 (1964).

S27 R. S. Nicholson and I. Shain, *Anal. Chem.* **37**, 178 (1965).

S28 R. S. Nicholson and I. Shain, *Anal. Chem.* **37**, 190 (1965).

S29 R. P. Frankenthal and I. Shain, *J. Amer. Chem. Soc.* **78**, 2969 (1956).

S30 W. H. Reinmuth, *J. Amer. Chem. Soc.* **79**, 6358 (1957).

S31 W. H. Reinmuth, *Anal. Chem.* **32**, 1891 (1960).

S32 W. H. Reinmuth, *Anal. Chem.* **33**, 185 (1961).

S33 W. H. Reinmuth, *Anal. Chem.* **33**, 1793 (1961).

S34 W. H. Reinmuth, *Anal. Chem.* **34**, 1446 (1962).

S35 R. D. DeMars and I. Shain, *J. Amer. Chem. Soc.* **81**, 2654 (1959).

S36 S. Piekarski and R. N. Adams, "Voltammetry with Stationary and Rotated Electrodes," in *Techniques of Chemistry, Vol. I. Part IIA*: Electrochemical Methods, A. Weissberger and B. W. Rossiter, Eds. , Wiley-Interscience, New York, 1971, pp. 531ff.

S37 R. H. Wopschall and I. Shain, *Anal. Chem.* **39**, 1514 (1967).

S38 R. H. Wopschall and I. Shain, *Anal. Chem.* **39**, 1535 (1967).

S39 R. H. Wopschall and I. Shain, *Anal. Chem.* **39**, 1535 (1967).

S40 J. E. B. Randles, *Trans. Faraday Soc.* **44**, 322 (1947).

S41 J. E. B. Randles, *Trans. Faraday Soc.* **44**, 327 (1947).

S42 G. C. Barker and A. W. Gardner, *Z. Anal. Chem.* **173**, 79 (1960).

S43 E. P. Parry and R. A. Osteryoung, *Anal. Chem.* **36**, 1366 (1964).

S44 H. E. Keller and R. A. Osteryoung, *Anal. Chem.* **43**, 342 (1971).

S45 M. Kalousek, *Coll. Czech. Chem. Commun.* **13**, 105 (1948).

S46 M. Kalousek and M. Ralek, *Coll. Czech. Chem. Commun.* **19**, 1099 (1954).

S47 W. F. Kinard, R. H. Philp, and R. C. Propst, *Anal. Chem.* **39**, 1556 (1967).

S48 M. Ishibashi and F. Fujinaga, *Bull. Chem. Soc. Japan* **23**, 261 (1950).

S49 M. Ishibashi and F. Fujinaga, *Bull. Chem. Soc. Japan* **25**, 68, 238 (1952).

S50 F. Fujinaga, *J. Electrochem. Soc. Japan* **23**, 588 (1955).

S51 G. C. Barker, in **Progress in Polarography, Vol. 2**, P. Zuman, Ed. , Wiley-Interscience, New York, 1962, p. 411.

S52 G. C. Barker, *Anal. Chim. Acta* **18**, 118 (1958).

S53 J. E. B. Randles, *Disc. Faraday Soc.* **1**, 11 (1947).

S54 J. E. B. Randles, *Trans. Faraday Soc.* **48**, 828 (1952).

S55 J. E. B. Randles and K. W. Somerton, *Trans. Faraday Soc.* **48**, 937 (1952).

S56 J. E. B. Randles and K. W. Somerton, *Trans. Faraday Soc.* **48**, 951 (1952).

S57 E. Warburg, *Ann. Phys.* **67**, 493 (1899).

S58 E. Warburg, *Ann. Phys.* **69**, 125 (1901).

S59 M. Fournier, *J. Chim. Phys.* **49**, C183 (1952).

S60 G. S. Buchanan and R. L. Werner, *Aust. J. Chem.* **7**, 239 (1954).

S61 G. S. Buchanan and R. L. Werner, *Aust. J. Chem.* **7**, 312 (1954).

S62 K. S. G. Doss and H. P. Agarwal, *J. Sci. Ind. Res. (India)* **9B**, 280 (1950).

S63 K. B. Oldham, *Trans. Faraday Soc.* **53**, 80 (1957).

S64 H. P. Agarwal, "Faradaic Rectification Method and its Applications to the Study of Electrode Processes," *Electroanalytical Chemistry,* Vol. 7, 1974, pp. 161–271.

S65 *Pure Appl. Chem.* **45**, 81 (1976).

S66 J. Heyrovsky, *Coll. Czech. Chem. Commun.* **18**, 749 (1953).

S67 J. Heyrovsky, *Anal. Chim. Acta* **12**, 600 (1955).

S68 M. S. Krause and L. Ramaley, *Anal. Chem.* **41**, 1365 (1969).

S69 G. D. Christian, *J. Electroanal. Chem.* **23**, 1 (1969).

S70 H. Siegerman and G. O'Dom, *Amer. Lab.* **4**, 59 (1972).

S71 J. B. Flato, *Anal. Chem.* **44**, 75A (1972).

S72 E. Barendrecht, *in Electroanalytical Chemistry, Vol, 2,* A. J. Bard, Ed. , Marcel Dekker, New York, 1967.

S73 I. Shain, "Stripping Analysis," in *Treatise on Analytical Chemistry,* Part I, Vol. 4, Chapter 50, I. M. Kolthoff and P. J. Elving, Eds. , Wiley-Interscience, New York, 1965.

S74 T. P. Copeland and R. K. Skogerboe, *Anal. Chem.* **46**, 1257A (1974).

S75 G. W. Schieffer and W. J. Blaedel, *Anal. Chem.* **50**, 99 (1978).

16.9   STUDY PROBLEMS

16.1 Analysis for lead in blood by DPASV gave a 100-nA peak current for a 700-ppb concentration of lead (7.0×10^{-8} mol/dm^3). The working curve is linear. The deposition time was 3 min at –0.7 V, and the differential scan (at 4 mV/s and modulation amplitude 50 mV) ran from there through the peak (potential at –0.38 V, half-maximum width 100 mV). Normal levels of lead in blood are 150 to 400 ppb; a peak of 30 nA can be distinguished from baseline. Can this method reach the bottom of the normal range? If not, how should it be modified to do so?

16.2 A solution containing 1.0 mmol/dm^3 of Pb(II), Cd(II), and Zn(II) in aqueous 0.1-mol/dm^3 KNO$_3$ buffered with acetate gave three polarographic waves, each of height 4.7 μA. Differential pulse polarography using 50-mV pulses gave three symmetric peaks of height 80.0 μA; their widths at half-maximum were 75 mV. Peak potentials were −0.467, −0.634, and −1.047 V against SCE. Sketch both of these curves on the same graph, indicating the half-wave potentials. Are they identical to the peak potentials? Why or why not?

CHAPTER
17
CHRONOMETHODS

Those electroanalytical methods in which the dependence of an electrical signal on time is measured are collectively known as the *chronomethods*. The electrical parameter measured may be current (chronoamperometry), in which the process is initiated by a potential step, or it may be potential (chronopotentiometry), in which the process is initiated by a current step. Alternatively, either of these time-dependent signals may be initiated by injection of a charge pulse rather than a step function. The methods are then referred to as *coulostatic* or *charge-step* methods, and the time-dependent signal follows the course of a relaxation toward an equilibrium or pseudoequilibrium state. Chronocoulometry, in which charge is followed as a function of time, is derivable from either chronoamperometric or, less frequently, chronopotentiometric studies; it often presents the data in more easily interpretable form. The biennial reviews in *Analytical Chemistry* cited at the beginning of the preceding chapter also cover these techniques.

17.1   CHRONOAMPEROMETRY

Chronoamperometry is that class of electroanalytical measurements in which the current that flows through the indicating electrode is measured as a function of time, while the excitation signal, usually a voltage, is either constant or some simple function of time. Only a few of the possible variants of chronoamperometry have analytical significance. When used without further qualification, *chronoamperometry* denotes the current–time curve for an indicating electrode that is stationary in unstirred solution. The mass transfer to such an electrode will occur by diffusion, and the resulting current–time curve will be an exponential decrease of current with time. Chronoamperometry in this general sense is of no significance, but the useful techniques of *potential-step chronoamperometry* are often called by the simpler general name.

17.1.1   Potential-Step Chronoamperometry

In potential-step chronoamperometry, the potential of the electrode is discontinuously stepped *from the open-circuit potential* to some other final potential. This technique has no advantages over double potential-step chronoamperometry and some significant disadvantages in that the open-circuit potential of the cell may be neither stable nor suitable for the study of the desired process. Chronoamperometry used in this single-step mode in which the potential is stepped from the equilibrium or open-circuit potential to a potential at which a reaction takes place is also known as the potentiostatic method for the measurement of the kinetics of charge-transfer reactions (S1–S4). Details of these and other methods for study of reaction rates have been reviewed by Delahay (S5).

17.1.2   Double Potential-Step Chronoamperometry

Double potential-step chronoamperometry is that technique in which a potential step is applied to the indicating electrode, and the observed current is measured as a function of time. The excitation signal and form of the observed output are as shown for a single simple reversible process in Figure 17.1. The technique is normally applied to a stationary electrode in unstirred solution. It has been applied to a DME, in which case it is called *polarographic chronoamperometry*. Polarographic chronoamperometry is uncommon and has no advantage over double potential-step chronoamperometry at an HMDE. The potential step is applied to the drop in the last stages of drop life, when the rate of change of drop area with time is small. Use of polarographic chronoamperometry requires synchronization of the application of the potential step and the recording of the current–time curve with the drop time.

In double potential-step chronoamperometry, which in deference to current practice will henceforth be called simply "chronoamperometry," the independent variables or controlled parameters are the two potentials, the initial potential E_I and the final potential E_F. The other independent variables are those of the solution, such as concentration and pH. The dependent variables, or measured parameters, are the shape of the curve, the amplitude of the curve, and the dependence of the curve on the value of E_F. This last dependence is a most useful diagnostic and should never be overlooked.

In chronoamperometric measurements, the initial potential E_I is normally set sufficiently anodic (or cathodic) that the reduction (or oxidation) process being observed does not occur to any measurable extent. The effect of the application of the chronoamperometric double-step signal will then be, in the absence of any other electroactive species, simply to charge the double-layer capacitance to the final potential E_F from the potential E_I. The charging current that flows will then decay from infinity, exponentially as the double-layer capacitance is charged. The observed finite limit is because of the speed of response of the recording device used, as is the subsequent very rapid initial exponential decay. The useful part of the chronoamperogram is the section *subsequent* to this rapid excursion.

Chronoamperometry has little to no direct analytical significance, but it is useful for measurements of diffusion coefficients, rates of electrode processes, adsorption parameters, and rates of coupled chemical reactions. Other methods

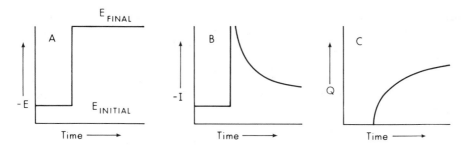

Figure 17.1   Chronoamperometry and chronocoulometry. (*A*) Excitation potential step; (*B*) chronoamperometric response; (*C*) chronocoulometric response, which is the integral of the chronoamperometric response. The curves in (*B*) and (*C*) would be obtained with a single reversible process.

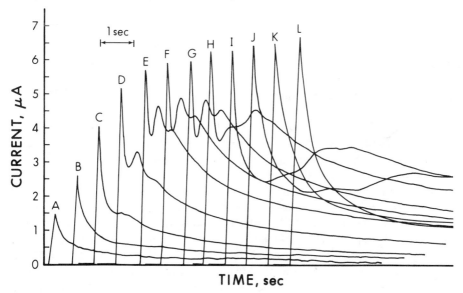

Figure 17.2   Chronoamperometry. Curves for the Tl(I)/Tl(Hg) reduction in $Ca(NO_3)_2 \cdot 4H_2O$. Time scale arbitrarily displaced for each curve shown. Potential step (against Ag(I)/Ag reference) applied: A, –0.90 V; B, –0.92 V; C, –0.94 V; D, –0.96 V; E, –0.98 V; F, –0.99 V; G, –1.00 V; H, –1.02 V; I, –1.04 V; J, –1.06 V; K, –1.08 V; L, –1.10 V. Concentration of Tl(I) is $4.36 \times 10^{-3} M$. Area of HMDE, 0.0176 cm²; temperature, 50°C. The initial potential was set at –0.2 V before each run [from Bansal and Plambeck (S6)].

also exist for the study of all of these, of which cyclic voltammetry is probably the most powerful. Since most of the plots used in analysis of chronoamperometric curves involve some function of current against the square root of time, and since measurements for more than a few seconds are not possible due to the interference of convection, an oscilloscope is normally used to permit measurements of response at very short times. A rapid chart recorder can also be used.

Chronoamperometric measurements performed on an HMDE, which is the most popular electrode used for them, should be carried out on a new drop of mercury each time. When this is done, a sequence of chronoamperometric curves taken at different potentials is comparable to a polarogram obtained manually at those potentials, and the current at any desired time on the chronoamperometric curve corresponds to the polarographic maximum current that would be observed for a DME of that drop time. The correspondence is not exact, and is more comparable to that of normal pulse polarography, since the reducible or oxidizable substance in the immediate vicinity of the electrode surface has not been depleted by electrolysis prior to the application of the chronoamperomeric potential step. A polarographic plot of this type is often useful in qualitative interpretation of chronoamperometric data.

A series of chronoamperometric curves obtained at different values of E_F in the author's laboratory (S6) is shown in Figure 17.2. The peaks appearing in these curves are due to nucleation in the deposition of Tl(I) on the HMDE employed. The polarograms constructed from these chronoamperograms and other similar sets taken in solutions of different Tl(I) concentration are shown in Figure 17.3. The peaks on these polarograms are due to the nucleation process rather than to convection, the usual cause of polarographic maxima.

17.1.3   Cyclic Chronoamperometry

This method is called *double-potential-step chronoamperometry* by some authors (G3). In cyclic chronoamperometry, the potential step is first made in such a direction that a reduction (say) can occur, and then at some time t the potential is stepped again, this time to a value at which the reverse reaction (say, oxidation) can occur. The potential of the second step usually is, but need not be, the same as the initial potential. The excitation signal and chronoamperometric response are shown in Figure 17.5. Measurements are made from the start of the first potential step to well beyond t the time at which the potential is again stepped, so that both the forward and the reverse directions of the electrochemical processes can be studied. The major advantages of this method are that it is quite sensitive to adsorption of the species formed in the initial step and to chemical reactions involving either the oxidized or the reduced form. The theory has been developed by Schwarz and Shain (S7).

17.1.4   Chronoamperometric Theory

The fundamental equation of chronoamperometry, which was derived in Chapter 13, is the *Cottrell equation*:

$$I_d = zFAcD^{1/2}\pi^{-1/2}t^{-1/2} \tag{17.1}$$

for linear diffusion control of current to a stationary planar electrode. Cylindrical or spherical diffusion will give rise to similar but more complex expressions (S8). The Cottrell equation applies immediately in chronoamperometry if the final potential applied is sufficient to immediately force the concentration of electroactive species at the electrode surface to zero. In most experiments this is

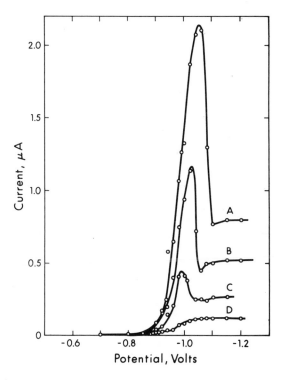

Figure 17.3 Chronoamperometry: constructed polarograms. HMDE, area 0.0176 cm². Concentration of Tl(I) in Ca(NO₃)₂•4H₂O: A, 4.36 × 10⁻³M; B. 2.56 × 10⁻³M; C, 1.3 × 10⁻³M; D, 4.59 × 10⁻⁴M; temperature, 50°C. Current was measured 10 s after applying the potential step [from Bansal and Plambeck (S6)].

not the case, because a lesser applied potential is used. The resulting time-dependent surface concentrations for a single reversible process are (G1)

$$c^0{}_{Ox} = c_{Ox}[D_{Ox}/D_{Red}]^{1/2} \exp[zF(E - E^0/RT]/(1 + [D_{Ox}/D_{Red}]^{1/2} \exp[zF(E - E^0/RT]) \qquad (17.2)$$

$$c^0{}_{Red} = c_{Red}(D_{Ox}/D_{Red})^{1/2}/(1 + [D_{Ox}/D_{Red}]^{1/2} \exp[zF(E - E^0)/RT]) \qquad (17.3)$$

rather than zero for a reduction process, if the activity coefficients of the species involved are taken as unity. The current–time dependence curve is then (G1)

$$I = zFAc_{Ox}D_{Ox}{}^{1/2}\pi^{-1/2} \, t^{-1/2}/(1+(D_{Ox}/D_{Red})^{1/2} \, \exp[zF(E-E^0)/RT]) \qquad (17.4)$$

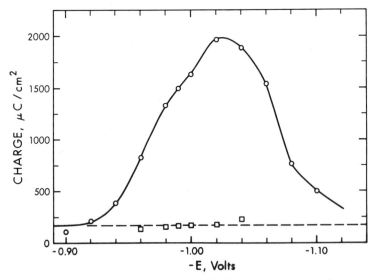

Figure 17.4   Integration of chronoamperometric data. ○, Q(total) in microcoulombs per square centimeter for charge of all process to 14 s from step application; □, Q(nucl) in microcoulombs per square centimeter for charge of nucleation–nuclei growth step alone. Dashed horizontal line at 171.2 $\mu C/cm^2$ represents theoretical charge required to form monolayer of thallium metal. Area of electrode 0.0176 cm^2, HMDE [from Bansal and Plambeck (S6)].

This equation reduces to the Cottrell equation in the limit of sufficiently negative potential, as expected, since at sufficiently negative potential the concentration of electroactive species at the electrode surface is forced to zero at once. Note that as the potential E, which for a chronoamperometric experiment is the final potential E_F, increases, the current increases also. For a reversible process, this is the only effect of stepping the potential to, in the case of a reduction, successively more negative values.

17.1.5   Effect of Reaction Kinetics

The effect of a kinetic process preceding the electrochemical step is to reduce the observed current to below the value predicted by the Cottrell equation or its modifications. The rate constant for a preceding reaction, or of charge transfer itself, can be obtained from suitable plots of such data.

17.1.6   Effect of Adsorption

The effect of adsorption, if strong, or of other processes such as nucleation is the introduction of peaks on the current–time curve as shown in Figure 17.2. Integration of the peak between minima can yield the charge associated with the process responsible for the peak. In the case of thallium reduction, integration of the curves of Figure 17.2 leads to the values plotted in Figure 17.4.

17.2   CHRONOCOULOMETRY

Chronocoulometry is very similar to chronoamperometry both in the method used and in the information that can be obtained from it. The excitation signal, or potential step, is the same as in chronoamperometry, and the measured signal is the charge flow as a function of time. This is identical to the integral of the chronoamperometric signal as shown in Figure 17.1. The chronocoulometric curve contains no more information than does the chronoamperometric one, although its interpretation is sometimes simpler. The theoeretical treatment of chronoamperometry applies also to chronocoulometry.

17.2.1   Potential-Step Chronocoulometry

Potential-step chronocoulometry is identical to potential-step chronoamperometry except for the integration effect of reading charge rather than current. Like potential-step chronoamperometry, it is rarely used since there is generally no advantage in stepping from the open-circuit potential.

17.2.2   Double Potential-Step Chronocoulometry

The use of double potential-step chronocoulometry, like the use of the analogous double potential-step chronoamperometry, is of considerable value in studies of adsorption or of kinetics or of influencing electrode reactions. The theory has been developed by Christie (S9). It is a method of choice for adsorption measurements (G3). The response is shown in curve C of Figure 17.5.

17.2.3   Cyclic Chronocoulometry

This technique, like its counterpart cyclic chronoamperometry, is useful in studies of mechanism and rates of reactions involved with the charge-transfer process. The technique was introduced by Anson (S10) and its theory has been developed by Christie, Osteryoung, and Anson (S11). Like double potential-step chronocoulometry, it finds its primary use in the study of processes in which adsorption is involved (G3).

17.3   CHRONOPOTENTIOMETRY

Chronopotentiometry is that class of electroanalytical methods in which the potential of the indicating electrode is measured as a function of time while the current that passes through it is controlled at a value either constant or known as a function of time. The analytically useful variants of chronopotentiometry are those in which the potential is monitored after either a current-step initiation signal or a charge-step initiation signal. Current-step chronopotentiometry is often denoted by the simpler general name.

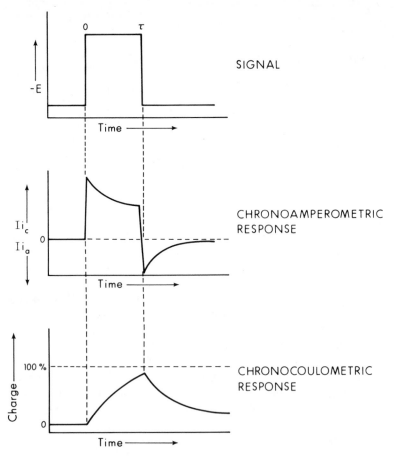

Figure 17.5 Chronoamperometry and chronocoulometry. Cathodic reaction-reversal responses. Curve A (upper), excitation signal for both techniques; curve B (middle), chronoamperometric response; curve C (lower), chronocoulometric response.

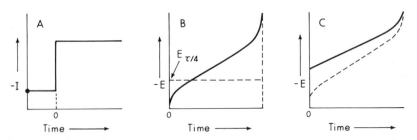

Figure 17.6 Chronopotentiometry: excitation signal and responses. Responses shown are for a single reversible (center) and a single irreversible (right) process.

Current-step chronopotentiometry is that form of chronopotentiometry in which the current is stepped from zero to an arbitrary constant value and the potential of the electrode is then measured as a function of time as shown in Figure 17.6. The theory of chronopotentiometric measurements was worked out by Sand (S12) at the turn of the century, but received little attention for some 50 years until Gierst and Juliard (S13) reintroduced it to modern chemists. The technique received close attention from 1953 through about 1970, and, while it is still significant, recent interest has declined somewhat.

17.3.1   Chronopotentiometric Theory

The conditions of chronopotentiometry are usually derived theoretically under the conditions that the current is constant and that mass transport is by semi-infinite linear diffusion to a planar electrode of fixed area. Under these conditions a series of diffusion gradients is established with time as shown in Figure 17.7. When nonplanar electrodes such as a cylindrical wire or a spherical HMDE are used for chronoamperometric measurements, cylindrical or spherical diffusion theory should be used to take into account the curvature of the electrode (S14, S15). The usual initial time and infinite time boundary conditions for reduction are

$$c_{Red,x,0} = 0, c_{Ox,x,0} = c_{Ox}$$

$$c_{Red,x,t} \rightarrow 0 \text{ when } x \rightarrow \text{infinity}$$

$$c_{Ox,x,t} \rightarrow c_{Ox} \text{ when } x \rightarrow \text{infinity}$$

These relations lead to differential equation solved by Sand (S12) and others, and in complete form by Karaoglanoff (S16), for values of $c_{Red,x,t}$ and $c_{Ox,x,t}$. These relationships are general. From them, values of $c^0_{Red,t}$ and $c^0_{Ox,t}$ can be derived. Insertion of these values into the Nernst equation for both oxidized and reduced forms at the electrode surface yields (G1)

$$E = E_{1/2} + RT \ln[(c_{Ox} - Pt^{1/2})/Pt^{1/2}]/zF \tag{17.5}$$

where the parameter P is given by

$$P = 2I/\pi^{1/2}zFAD_{Ox}^{1/2} \tag{17.6}$$

And $E_{1/2}$ is given by the usual definition of the polarographic half-wave potential $E_{1/2}$ (whether or not the activity coefficients of the ions are taken as unity, as has been done in this derivation).

Theoretically, the observed potential would go to infinity whenever the numerator of the logarithmic term went to zero. This will happen when the time t has the specific value τ given by

$$\tau = c_{Ox}^2/P^2 \tag{17.7}$$

Solution of the above equation for c_{Ox} and insertion of that value into Eq. (17.5) yields:

$$E = E_{1/2} + RT/zF \ln[(\tau^{1/2} - t^{1/2})/t^{1/2}] \tag{17.8}$$

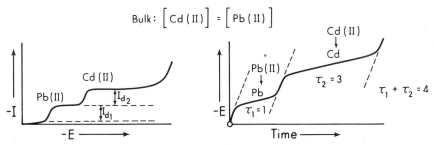

Figure 17.7   Chronopotentiometry: concentration gradients.

This equation is exactly analogous to the Heyrovsky–Ilkovic equation, which gives the electrode potential as a function of the cathodic limiting current and the actual current for the same reduction case. In polarography, plots of $\log[(I_{d,c} - I)/I]$ against potential give straight lines of slope $2.303\ RT/zF$; chronopotentiometric plots of $\log[(\tau^{1/2} - t^{1/2})/t^{1/2}]$ also yield the same information. The relationship between a polarogram and a chronopotentiogram is shown graphically in Figure 17.8.

The potential will not actually increase to infinity at times greater than the transition time τ, although the theory for a single process so predicts. The potential actually increases only until some process other than the reduction of Ox occurs at the electrode, such as reduction of the solvent or supporting electrolyte. In anodic chronopotentiometry the potential increases, at times greater than τ, to a limit set by other oxidizable species in solution or by the oxidation of the solvent or supporting electrolyte.

In the slightly simpler case when only the concentration of the oxidized form is considered, as by Sand (S12), the physical picture can be explained more clearly. Assuming Nernstian response of the electrode, the electrode potential E will be dependent only upon the instantaneous surface concentration of the oxidized species c^0_{Ox}. This in turn is affected only by the rate of diffusion of the species and the applied current, thus:

$$c^0_{Ox} = c - 2It^{1/2}/zFA(\pi D)^{1/2} \tag{17.9}$$

where t is the time, I is the applied constant current, and the remaining symbols have their usual significance. If and only if $c^0 = 0$, the remaining terms are equal. The value of the time at which c^0 becomes zero is called the *transition time* τ, after Butler and Armstrong (S18, S19). At this time, from the above equation, one obtains the *Sand equation*:

$$I\tau^{1/2} = zFAc(\pi D)^{1/2}/2 \tag{17.10}$$

The Sand equation is the chronopotentiometric counterpart of the Ilkovic equation. The Sand equation can be rearranged to a form that has on the left the term $\tau^{1/2}/c$, which is known as the *transition time constant*. At the time $t = \tau/4$, analysis of the time-dependence equation (17.9) shows that the surface

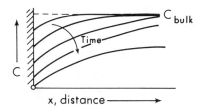

Figure 17.8   Comparison of polarography and chronopotentiometry. In both cases, the curves are shown for equimolar solutions of Cd(II) and Pb(II).

concentration is exactly one-half the bulk concentration. Hence $E(\tau/4)$, the potential of this point, should correspond to the polarographic half-wave potential $E_{1/2}$ for reductions at mercury electrodes because the logarithmic term is then zero.

The equation of a chronopotentiometric wave can be obtained by insertion of concentration values into the Nernst equation:

$$E = E(\tau/4) + RT/zF \ln[(\tau^{1/2} - t^{1/2})/t^{1/2}] \tag{17.11}$$

which will hold for the shape of a reversible reduction wave such as that for reduction of Cd(II). The reversible equation is rarely used in practical chronopotentiometric studies, unlike the plot of electrode potential against $\log[(I_d - I)/I]$, which is of considerable practical use in polarography.

For a totally irreversible reduction the equation obtained (S17) is similar:

$$E = E^{0\prime} + RT/\alpha zF \ln[(\tau^{1/2} - t^{1/2})/t^{1/2}] \tag{17.12}$$

where

$$E^{0\prime} = E^0 + (RT/\alpha zF) \log (2k^0{}_h/\pi D_{Ox}{}^{1/2}), \tag{17.13}$$

so that $E^{0\prime}$ includes the rate constant of the electrochemical reaction.

If two electrochemical processes can occur at significantly different potentials, as in a solution containing both Cd(II) and Pb(II), then two chronopotentiometric waves or plateaus will be obtained, just as two separate waves will be observed polarographically. However, polarography is generally better at resolving such waves because of its slower rate of change of potential with time and the more precise readout available on a strip-chart recorder as opposed to an oscilloscope.

In polarography, two electrochemical processes such as those described above will give two waves, the first of which will continue while the second begins; that is, the second wave will be superimposed upon the first as shown in Figure 17.8. The diffusion-limited current of the second wave is obtained by subtraction of the diffusion-limited current of the first wave from the observed total diffusion current in polarography. The situation in chronopotentiometry is exactly analogous, because it is $\tau^{1/2}$ that is analogous to I_d in being directly proportional to the bulk concentration of the diffusing species. The chronopotentiogram contains two waves as shown in Figure 17.8.

Multiple waves will also arise when a single species can be reduced in two successive steps at two different potentials, such as $Cu(II) \rightarrow Cu(I) \rightarrow Cu$. When the number of electrons in each of the two steps is the same, $\tau_2 = 3\tau_1$. This relation is independent of the concentration of the species since the same material is reduced in both steps. If the number of electrons is not the same in both steps, the general relation is (S19):

$$\tau_1[((z_1+z_2)/z_2)^2] - \tau_1 = \tau_2 \qquad (17.14)$$

The result of reversal of current flow after transition is a unique and highly useful diagnostic tool. Suppose that a chronopotentiometric reduction is carried out and the current is reversed at the end of the transition time. If the reduced product were insoluble, as in the reduction lead(II) to lead on platinum, then all of it that was reduced would still be on the electrode to be reoxidized, and (if the forward and reverse currents are equal) then the forward and reverse transition times would be equal. If, however, the reduced product were soluble *in either the electrode or the electrolyte*, some of it would diffuse away from the electrode and not be available for reoxidation, so the forward transition time must be greater than the reverse transition time. Berzins and Delahay (S20) have shown that τ_{fwd} = $3\tau_{rev}$ if the diffusion coefficients of the oxidized and reduced forms are equal.

Complications can occur in chronopotentiometric studies as in other methods; these would be slow charge-transfer kinetics of the electrode reaction, slow reaction kinetics of an associated chemical reaction, or adsorption of reactants and/or products.

The effect of adsorption on chronopotentiometry is to increase the value of $I\tau^{1/2}$ over that expected (S21, S22). A quantitative analysis of the data depends on the choice of model—that is, whether the reaction of the adsorbed material is before, during, or after the reaction of the diffusing material. The symptom of such behavior is that the value of the diffusion coefficient D calculated from the Sand equation increases with increasing current.

The value of the chronopotentiometric constant should be independent of the current if the reaction is diffusion-controlled. If, however, there is a slow chemical kinetic step in the reaction, it will limit the rate. Thus for a slow kinetic step or equilibrium preceding the electrochemical step it will be found that the chronopotentiometric constant decreases with increasing current as shown in Figure 17.9. From the slope of plots such as Figure 17.9 a value of the reaction

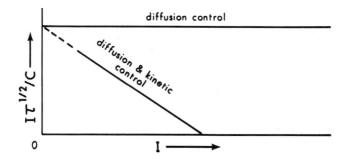

Figure 17.9   Chronopotentiometry: kinetic and diffusion control.

rate constant can be determined. Chronopotentiometric studies are normally carried out in milliseconds, while polarography is carried out in multiple seconds (the drop time), so that if a process is shown to be diffusion-controlled by chronopotentiometry it is so by polarography as well, but the converse is not true.

For example, the totally irreversible reduction of aqueous Ni(II) to Ni(Hg) was studied polarographically by Orlemann and Sanborn (S23), who obtained the polarographic curves shown in Figure 17.10. These curves were explained on the basis that at high concentrations of $NaClO_4$ the reduction is followed by a disproportionation step:

$$Ni(II) + e^- \rightarrow Ni(I)$$

$$2Ni(I) \rightarrow Ni(II) + Ni(Hg)$$

A later chronopotentiometric study of Gierst (S24) obtained the curves shown in Figure 17.10 in the same sodium perchlorate medium. From these plots Gierst deduced that the actual reaction mechanism is probably

$$Ni^{2+}(aq) \rightarrow Ni^{2+}(aq\text{-}x),$$

$$Ni^{2+}(aq\text{-}x) + 2e^- \rightarrow Ni(Hg)$$

The polarographic difficulty is due to a kinetic complication involving desolvation of the Ni(II) ion rather than Ni(I) disproportionation.

17.3.2   Chronopotentiometric Practice

The analytical uses of chronopotentiometry are similar to, but considerably more restricted than, those of polarography. Chronopotentiometry is more often used to determine diffusion coefficients, or for systems where a DME does not work, owing to the potential required or, at high temperature, mercury volatilization. It can be used for measurements of adsorption, but is inferior to chronocoulometry for this purpose (S29). The potential range accessible is limited only by the solvent

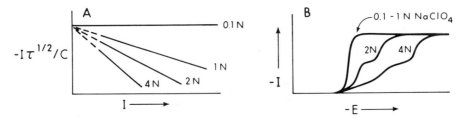

Figure 17.10   Reduction of aqueous Ni(II). (*A*) Polarography; (*B*) chronopotentiometry.

or by the material used to fabricate the chronopotentiometric indicator electrode. There are, however, limits on the usable times. The short time limit is the time required to charge the double-layer capacitance, about 1 ms. The long time limit is the time when convective mixing begins to become effective. In aqueous solutions, this may be up to 100 s, although 10 s is more usual, while in fused salts only about 5 s can be obtained. As to the current, no limits are imposed by the nature of the technique, but the currents employed must be chosen to give times within the above limits. The usual current range is about 0.1 to 1000 A/m² (cgs: 0.01 – 100 mA/cm²).

The chronopotentiometric cell must consist of at least two compartments, one for the auxiliary electrode and one for the indicating electrode; more often there are three compartments, with the reference electrode separated also. Electrodes used as reference electrodes are the SCE (aqueous) or Ag⁺/Ag type (nonaqueous). The auxiliary electrode is generally a large platinum flag or a large mercury pool, while the indicating electrode is a small platinum flag or wire, a small mercury pool, or an HMDE. For short transition times, the electrode shape does not matter greatly. For long transition times, it is best to have a vertical diffusion column to inhibit convection. Only for transition times greater than 10 s is this recommended. A simple vertical diffusion column is easily made by placing mercury in the bottom of an open-top glass tube in the solution. Electrical contact is made by sealing a small piece of platinum wire through the glass.

Chronopotentiometric instrumentation consists of a source of constant current, a potential-recording device, and a switching circuit. The potential measuring device is sometimes a strip-chart recorder. The alternative, and most common, potential-measuring device is the oscilloscope, which must have an external trigger circuit. The switching circuitry is necessary to start the current and the recording device simultaneously. It is preferable to start the recording device just before the current step in order to record the open-circuit potential. This can be done with a pushbutton hand switch, a mercury-wetted relay (better) or more complex electronic circuitry.

The measurement of transition times in chronopotentiometry has been a matter of considerable controversy and several graphical methods have been advocated (S25–S29). None has achieved universal acceptance. The writer advocates the method shown in Figure 17.11. A line is constructed through the

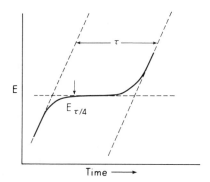

Figure 17.11   Measurement of chronopotentiometric transition times. Sloping lines should be parallel.

rising portion of the wave *following* the horizontal or slightly rising voltage plateau. A second line *parallel to the first* is then constructed from the initial rise of the chronopotentiogram so as to include the entire voltage plateau. In general, both of these lines will exhibit a considerable slope. A line *parallel to the horizontal axis*, which in general will not be parallel also to the voltage plateau, is constructed through these lines such that the distance from the leading line to the experimental plateau is one fourth the distance between the leading and the following lines. This point of intersection with the experimental curve, extrapolated to the vertical axis, gives $E(\tau/4)$; the distance between the two parallel lines, measured along the horizontal time axis, gives the transition time τ.

None of these methods reliably removes the contribution to the chronopotentiometric τ value arising from the charging of the double layer. Computational methods for doing so, together with some theoretical justification for their use, can be applied to data obtained by these graphical procedures. Such methods do yield significantly improved data (S30–S34), but, in general, the precision of chronopotentiometric data are sufficiently inferior to that obtainable by other chronomethods or by voltammetric techniques, making its use in precise quantitative analysis unwise.

17.3.3   Charge-Step Chronopotentiometry

Charge-step chronopotentiometry is a systematic name for what are commonly known as *coulostatic methods.* These were developed in 1962 by Delahay (S35–S37) and Reinmuth (S38–S40) after a suggestion by Barker. Their analytical utility (S41) is less than their utility in the study of rapid electrode processes (S42–S45) and adsorption kinetics (S46, S47) for which they were originally developed.

In a coulostatic method, an electrode is perturbed from a condition in which no Faradaic reaction is taking place to a condition in which a Faradaic reaction does take place by injection of a known amount of charge into the capacitance of the double layer. The injection is carried out as rapidly as possible by discharge of a capacitor charged to a known voltage or by application of a short-duration current pulse from an appropriate pulse generator. The charge source is then decoupled from the cell, either physically or by diodes, so that the injected charge can only be removed from the double layer by the Faradaic reaction being studied. The time dependence of the potential then tracks the relaxation of the process back toward the initial state. For analytical purposes, the charge step must be large enough that the Faradaic process is limited only by diffusion, and the bulk concentration of a species involved in the Faradaic reaction is then determinable; for kinetic studies this is not required. The coulostatic methods differ from current-step chronopotentiometry in that current is not imposed by an external circuit while potential measurements are being made.

Interpretation of coulostatic results are easier if the number of coulombs passed is followed as a function of time, but this is not possible by direct measurement. Since the current flow does not involve an external source, it is also not directly accessible by an external circuit, and the coulombs cannot be obtained by a current integration method. If the double layer capacitance were independent of the potential of the electrode then the potential change would be simply related to the current flowing by the equations for discharge of a parallel-plate capacitor (Chapter 2); but it is not, and calibration of the cell for the dependence of C_{dl} on E over the potential range used is necessary. This can be done in the absence of the Faradaic reaction by application of pulses of different size, since in the presence of the usual supporting electrolyte the structure of the double layer will be similar in the presence and in the absence of the Faradaic reaction.

17.3.4   Double Charge-Step Chronopotentiometry

This technique has been suggested by Daum (S48, S49) although under different names, including *double charge-step chronocoulometry*. It differs from charge-step chronopotentiometry in that successive charge pulses of opposite sign are applied to the electrode, and the relaxation in both directions is recorded. Appropriate calibration of the potential dependence of the double-layer capacitance is used to obtain data in chronocoulometric form, although the actual experimental data are chronopotentiometric.

17.3.5   AC Chronopotentiometry

Alternating-current chronopotentiometry, or AC chronopotentiometry, is performed in the same manner as DC chronopotentiometry except that an alternating *current* of small and constant amplitude is superimposed upon the direct electrolysis current. The resulting potential of the indicating electrode is the algebraic sum of two components: a direct component, which is identical to the DC voltage observed in ordinary chronopotentiometry, and an alternating component. The changes in the *alternating* component of the potential of the indicating electrode are recorded as a function of time. This technique was suggested and some of the theory developed by Takemori *et al.* (S50); the first actual application was by Bansal and Plambeck (S51) who employed a lock-in amplifier to separate the AC and DC components. The technique has received very little use because of the high level of instrumental complexity required. Its only advantage over DC chronopotentiometry is slightly better definition of some transition times.

17.4 REFERENCES

General

G1 P. Delahay, *New Instrumental Methods in Electrochemistry*, Wiley-Interscience, New York, 1954, 437 pp.

G2 D.G. Davis, *in Electroanalytical Chemistry: A Series of Advances*, A. J. Bard, Ed., Marcel Dekker, New York, 1966, p. 157. Best introduction to chronopotentiometry.

G3 R. W. Murray, "Chronoamperometry, Chronocoulometry, and Chronopotentiometry," *in Physical Methods of Chemistry,* Part IIA, A. Weissberger and B. W. Rossiter, Eds., Wiley-Interscience, New York, 1971, pp. 591ff.

G4 Z. Galus, *Fundamentals of Electrochemical Analysis*, Ellis Horwood/John Wiley, New York, 1976, 520 pp.

G5 D. D. Macdonald, *Transient Techniques in Electrochemistry*, Plenum, New York, 1977, 329 pp. Unified theoretical presentation at a highly mathematical level.

Specific

S1 H. Gerischer and W. Vielstich, *Z. Phys. Chem. (Frankfurt)* **3**, 17 (1955).

S2 W. Vielstich and H. Gerischer, *Z. Phys. Chem. (Frankfurt)* **4**, 10 (1955).

S3 C. A. Johnson and S. Barnartt, *J. Electrochem. Soc.* **114**, 1256 (1967).

S4 C. A. Johnson and S. Barnartt, *J. Phys. Chem.* **71**, 1637 (1967).

S5 P. Delahay, "Study of Fast Electrode Processes," **in Advances in Electrochemistry and Electrochemical Engineering,** Vol. 1, P. Delahay, Ed., Wiley-Interscience, New York, 1961.

S6 N. P. Bansal and J. A. Plambeck, *J. Electrochem. Soc.* **124**, 1036 (1977).

S7 W. M. Schwarz and I. Shain, *J. Phys. Chem.* **69**, 30 (1965).

S8 N. P. Bansal and J. A. Plambeck, *Can. J. Chem.* **56**, 155 (1978).

S9 J. H. Christie, *J. Electroanal. Chem.* **13**, 79 (1967).

S10 F. C. Anson, *Anal. Chem.* **38**, 55 (1966).

S11 J. H. Christie, R. A. Osteryoung, and F. C. Anson, *J. Electroanal. Chem.* **13**, 236 (1967).

S12 H. J. S. Sand, *Phil. Mag.* **1**, 45 (1901).

S13 L. Gierst and A. Juliard, *J. Phys. Chem.* **57**, 701 (1953).

S14 D. H. Evans and J. E. Price, *J. Electroanal. Chem.* **5**, 77 (1963).

S15 B. J. Welch, H. C. Gaur, A. K. Adya, and R. K. Jain, *Indian J. Chem.* **14A**, 150 (1976).

S16 Z. Karaoglanoff, *Z. Elektrochem.* **12**, 5 (1906).

S17 J. A. V. Butler and G. A. Armstrong, Proc. Roy. Soc. (London) **A139**, 406 (1933).

S19 J. A. V. Butler and G. Armstrong, *Trans. Faraday Soc.* **30**, 1173 (1934).

S19 P. Delahay and T. Berzins, *J. Amer. Chem. Soc.* **75**, 2486 (1953).

S20 T. Berzins and P. Delahay, *J. Amer. Chem. Soc.* **75**, 4205 (1953).

S21 S. V. Tatwawadi and A. J. Bard, *Anal. Chem.* **36**, 2 (1964).

S22 H. A. Laitinen and L. M. Chambers, *Anal. Chem.* **36**, 5 (1964).

S23 R. H. Sanborn and E. F. Orleman, *J. Amer. Chem. Soc.* **78**, 4852 (1956).

S24 L. Gierst and H. Hurwitz, *Z. Elektrochem.* **64**, 37 (1960).

S25 P. Delahay and G. Mamantov, *Anal. Chem.* **27**, 478 (1955).

S26 R. P. Buck, *Anal. Chem.* **35**, 1853 (1963).

S27 P. Delahay and C. C. Mattax, *J. Amer. Chem. Soc.* **76**, 874 (1954).

S28 R. W. Laity and J. D. E. McIntyre, *J. Amer. Chem. Soc.* **87**, 3306 (1955).

S29 J. J. Lingane, *Anal. Chem.* **39**, 485 (1967).

S30 W. H. Reinmuth, *Anal. Chem.* **33**, 485 (1961).

S31 A. J. Bard, *Anal. Chem.* **35**, 340 (1963).

S32 J. J. Lingane, *J. Electroanal. Chem.* **1**, 379 (1960).

S33 D. H. Evans, *Anal. Chem.* **36**, 2027 (1964).

S34 W. T. DeVries, *J. Electroanal. Chem.* **17**, 31 (1968).

S35 R. S. Rodgers and L. Meites, *J. Electroanal. Chem.* **16**, 1 (1968).

S36 P. Delahay, *Anal. Chim. Acta* **27**,90 (1962).

S37 P. Delahay, *Anal. Chem.* **34**, 1267 (1962).

S38 P. Delahay, *Anal. Chem.* **34**, 1662 (1962).

S39 W. H. Reinmuth, *Anal. Chem.* **34**, 1272 (1962).

S40 W. H. Reinmuth and C. E. Wilson, *Anal. Chem.* **32**, 1509 (1960).

S41 W. H. Reinmuth and C. E. Wilson, *Anal. Chem.* **34**, 1159 (1962).

S42 P. Delahay and Y. Ide, *Anal. Chem.* **34**, 1580 (1962).

S43 P. Delahay, *J. Phys. Chem.* **66**, 2204 (1962).

S44 P. Delahay and A. Aramata, *J. Phys. Chem.* **66**, 2208 (1962).

S45 P. Delahay and D. M. Mohilner, *J. Amer. Chem. Soc.* **84**, 4247 (1962).

S46 P. Delahay and D. M. Mohilner, *J. Phys. Chem.* **66**, 959 (1963).

S47 P. Delahay, *J. Phys. Chem.* **67**, 135 (1963).

S48 P. H. Daum, *Anal. Chem.* **45**, 2276 (1973).

S49 P. H. Daum and M. L. McHalsky, *Anal. Chem.* **52**, 340 (1980).

S50 Y. Takemori, T. Kambara, M. Senda, and I. Tachi, *J. Phys. Chem.* **61**, 968 (1961).

S51 N. P. Bansal and J. A. Plambeck, *J. Electroanal. Chem.* **78**, 205 (1977).

17.5 STUDY PROBLEMS

17.1 Explain why a peak appears in a chronoamperometric plot when strong specific adsorption is present. Do you expect it when the reactant is strongly adsorbed, when the product is strongly adsorbed, or in both of these cases?

CHAPTER 18

AMPEROMETRY AND AMPEROMETRIC TITRATIONS

Amperometry is that class of electroanalytical measurements in which the applied potential is held constant and the current that flows is a function of concentration, time, and other variables. The time-dependent methods of chronoamperometry have already been discussed. Time-independent amperometric measurements are of significance in two areas, *amperometric sensors* and *amperometric titrations*. The amperometric response is to concentration rather than activity in both areas.

18.1 AMPEROMETRIC SENSORS

An *amperometric sensor* consists of at least two electrodes placed in a medium whose contents are to be investigated; a constant potential is applied between them, and the resulting current is measured. The most common configuration is that of one indicator electrode and one reference electrode. Amperometry with two indicator electrodes is also possible when two indicator electrodes, usually identical, are placed in the same solution, and amperometric sensors of this type are also possible.

Amperometric sensors of themselves have only limited selectivity. That selectivity is provided by the adjustment of the potential at which the indicator electrode is held and provides good discrimination against species reduced (oxidized) at potentials significantly negative (positive) of the potential at which the desired species is reduced or oxidized. Amperometric sensors cannot of themselves discriminate against species that are more easily reduced (oxidized) than the desired species.

Amperometric sensors are commonly employed for the measurement of dissolved oxygen (G3), and commercial apparatus is available for this purpose (Beckman). These sensors consist of a small volume of electrolyte containing a noble metal or other microelectrode and a reference electrode. A potential that maintains the microelectrode on the diffusion-limited plateau of oxygen reduction is provided by an external circuit that applies a potential of about 0.8 V between the microelectrode and the reference electrode (usually SCE). A diaphragm permeable to oxygen separates the sensor from the external electrolyte whose oxygen concentration is being monitored. The oxygen content of the sensor electrolyte equilibrates with the oxygen content of the external electrolyte, and the microelectrode is the amperometric sensor for oxygen content of the sensor electrolyte.

A useful amperometric sensor is the tubular platinum electrode developed by Blaedel, Olson, and Sharma (S1) for use in flowing streams of electrolyte. Such an electrode can be coated with mercury (S2) and is then suitable for work at more negative potentials because of the higher overvoltage of hydrogen. Considerable development of flow-through electrodes and electrodes covered with only a thin film of electrolyte, as sensors at the output of liquid chromatography columns, has taken place.

18.2 AMPEROMETRIC TITRATION

Although in principle any nonequilibrium electroanalytical analytical method could be used to follow the course of a titration, in practice the only method so used is voltammetry (and its subclass polarography), in which the potential is held at some fixed value on a diffusion-limited plateau, and the current is observed during the course of a titration. Such an *amperometric titration* can be based on any stoichiometric chemical reaction, be it acid–base, precipitation, redox, or complex formation, in which the concentration of any soluble species can be followed voltammetrically. That is, the electrode potential is set on a diffusion plateau for some species that is either generated or removed during the titration, and the electrode current is used to follow the course of the titration.

The essential reference work on amperometric titrations is the book by Stock (G1). More recent references are available in the biennial reviews of the same author (S3–S11), which also include other types of titrations.

Straight-line plots are the most desirable type of titration curve; they can be obtained when (a) the potential is that of a diffusion plateau; (b) there is no migration current, that is, a supporting electrolyte is used; (c) the dilution is negligible or corrected for; and finally, (d) there is no incompleteness of reaction.

Since the concentration of a reducible (or, rarely, oxidizable) ion is being followed over the course of a titration, various forms of titration curves can be obtained depending upon the potential at which the electrode is used and the nature of the titration reaction. If the potential of an amperometric titration were set on the diffusion plateau for the reduction of Pb(II), the curves would be as shown in Figure 18.1. The curve shown in Figure 18.1C, which is obtained if one reducible species is titrated with another, can give slightly more accurate endpoints.

Thus far amperometry has been dealt with as a derivative of voltammetry, in the sense of carrying out an amperometric tiration using a single polarized

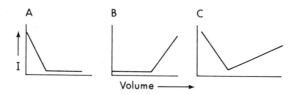

Figure 18.1 Amperometric titration curves (one polarized electrode). Curve A, titration of reducible species with nonreducible species; B, titration of nonreducible species with reducible species; C, titration of reducible species with reducible species. Titration of nonreducible species with nonreducible species cannot be followed amperometrically.

indicator electrode with the current flowing through either an unpolarizable reference electrode (two-electrode cell configuration) or an electrode of large area whose potential is not measured (three-electrode cell configuration). It is possible to use a two-electrode system with two identical (usually platinum) indicator electrodes in an amperometric titration. A small constant potential difference is applied between them. In this case the two electrodes switch function as electrodes at which the current is limited in the course of the titration. The amperometric curves obtained are somewhat more complex, as shown in Figure 18.2. This is obtained for the simplest possible case, in which one reversible electroactive couple is titrated with another.

This technique, known as *amperometric titration with two polarized electrodes*, has the advantage of slightly simpler cell configuration and simpler instrumentation. Its precision is not significantly less than that of amperometric titration with a single polarized electrode when one or both of the redox couples are reversible, the concentrations are relatively high, perhaps about 0.1 mol/dm³, and the sample is well-defined. If these conditions are not met the methods that use two polarized electrodes will have significantly worse precision and may have serious error in end-point determination; the use of the single-polarized-electrode method is fundamentally more sound and is to be preferred.

Automated titration units or analysis systems often employ amperometric methods, either to follow the course of a titration or to follow the rate of a chemical reaction when the analytical method is that of a rate determination. The electrodes used in these methods are usually platinum, although gold is also used, and may vary in form between wires and disks (rotated or stationary) or in some cases rings around tubes through which the liquid to be analyzed flows.

18.2.1 Amperometric End-point Indicators

Amperometric end-point indicators must be used if neither the titrant nor the sought-for constituent that reacts with it is reducible under amperometric conditions. An amperometric indicator is a reducible or oxidizable species that reacts with the titrant sufficiently less strongly than the sought-for constituent that its concentration changes only after all the sought-for constituent has been titrated. An example is the complexometric titration of Ca^{2+} with

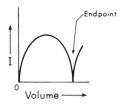

Figure 18.2   Amperometric titration curve (two polarized electrodes). Curve for titration of one reversibly reducible species with another reversibly reducible species. The small potential difference between the two electrodes is constant during the titration.

ethylenediaminetetraacetic acid, symbolized Y^{4-}, using Zn^{2+} as the indicator ion. The potential is set at a value at which Zn^{2+} is reduced, and Ca^{2+} is not. The resulting curve is as shown in Figure 18.3.

The amperometric indicator ion must be both more easily reduced and less strongly complexed than the ion whose end point it will be used to indicate. Addition of iodate and iodide to titration of base with strong acid, or addition of nitrite to titration of halides with silver ion, are also examples of amperometric end-point indicators, although the operation of these is somewhat different.

18.2.2   Rotating Platinum Electrodes

A rotating platinum-wire electrode is made by sealing a platinum wire through the side of a small, closed-end, Pyrex tube. The tube is held vertically in a chuck and rotated at about 600 revolutions per minute so that the wire sweeps out a small volume of solution. Contact to the platinum wire is made by filling the tube with mercury into which the external connecting wire is inserted. Although the *rotating platinum-wire electrode* (RPE) tends to be avoided in direct voltammetry since mass transport to its surface is ill-defined and the surface is not renewed, it is often used in amperometric titrations. It has the advantage over the DME that it is 50 to 100 times more sensitive because the stirring of the rotating wire reduces the thickness of the diffusion layer. Another advantage is that no charging current is present, and, there is somewhat less error. Again as compared with the DME, the disadvantages of the RPE are, first, that there is a low hydrogen overpotential on platinum and hence less cathodic potential range available, and second, that the surface is not renewed.

18.2.3   Comparison with Other Methods

Amperometric titrations have one advantage over the otherwise comparable potentiometric titrations—the electrochemical couple(s) involved in an amperometric titration need not be reversible, but in a potentiometric titration at least one of them must be. The comparison with conductometric titration is considerably more favorable. As a neutral point, a supporting electrolyte must be present in amperometric titrations and must not be in conductometric titrations.

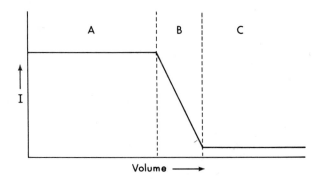

Figure 18.3 Titration curve using amperometric indicator. In region A, Ca^{2+} is being titrated and the concentration of Zn^{2+} remains unchanged. In region B, Zn^{2+} is being titrated, and in region C both have been titrated so the current remains at the low residual level. The observed current in all regions is residual or for reduction of Zn^{2+} alone, since Ca^{2+} is not reducible at this potential.

The advantage of amperometric titrations arises in selectivity. Amperometry is more selective in that at the potential chosen some ions will not be reducible and thus will not contribute to the current, while all ions will contribute to the conductance.

As compared with classical polarography, the amperometric titration is more precise, since its accuracy depends only on the stoichiometry, and 0.01% precision is possible; polarography is limited to about 1% to 3% precision. The amperometric titration curve is independent of the capillary characteristics (except at the end-point itself), whereas the polarographic wave is not. A minor advantage is that the apparatus in amperometric titrations can be simpler since no recording device is necessary.

18.3 ROTATING DISK ELECTRODES

Rotating disk electrodes can be classified as amperometric devices because they are normally operated at constant potential and used as current sensors. Unlike most other amperometric devices, current is limited by convection rather than diffusion. If a horizontal disk electrode in a homogeneous solution is rotated, mass transfer to such an electrode can occur by means of three driving forces: diffusion, convection, and ionic migration. The last of these can be effectively removed by a suitable concentration of supporting electrolyte. Mass transfer to such a disk electrode is therefore due to both convection and diffusion. The equation for the reduction current limited by mass transfer to a rotating disk electrode is (S12–S14) the *Levich equation*:

$$I_{lim} = -zFAcD^{2/3}\omega^{1/2}/1.61\nu^{1/6} \qquad (18.1)$$

where ω is the angular velocity of rotation of the disk and ν is the kinematic viscosity of the solution. This is identical to the Nernst diffusion-layer equation for the limiting current when

$$\delta = 1.61 D^{1/3} \nu^{1/6} / \omega^{1/2} \tag{18.2}$$

It is then seen that the thickness of the diffusion layer increases with the viscosity of the solution and decreases with increasing speed of rotation of the electrode.

Rotating disk electrodes, or RDEs, possess the great advantage over virtually all other indicating electrodes (except the DME) that a stable *steady state* is achieved, dependent upon the bulk concentration of the electroactive species and the rate of rotation of the disk. Such electrodes can be used for quantitative analysis, but the complexity of fabrication of the RDE is sufficiently great that it is not often employed in this way, although quantitative verification of the Levich equation has been accomplished. The shape of the waves obtained is, for a reversible charge transfer and for that only, identical to the shape of a classical polarographic wave.

18.3.1   Effect of Charge-Transfer Kinetics

When the rate of charge transfer need be considered, a more complex form of the Levich equation must be used that includes this rate:

$$I_{\text{lim}} = -zFADc / [(1.61 D^{1/3} \nu^{1/6} / \omega^{1/2}) + (D/k_h)] \tag{18.3}$$

In the limiting case of rapid charge-transfer kinetics this equation reduces to the simpler form of the Levich equation already given. This will occur at low rotation speeds and high values of k_h; a value of k_h greater than 10^{-3} m/s is sufficient at any practical rotation rate (S15, S16).

As the rotation rate increases or the value of k_h decreases, the effect of the charge-transfer term in the denominator of the Levich equation becomes dominant, and in the limit of slow charge transfer the Levich equation simplifies to

$$I_{\text{lim}} = -zFAk_h c \tag{18.4}$$

A value of k_h less than 10^{-6} m/s is low enough for this limit to be reached at all practical rotation speeds (S17).

18.3.2   Effect of Reaction Kinetics

The mechanisms in which the kinetics of chemical reactions can affect the currents observed at a RDE are those in which a chemical step can precede the electron transfer—the CE and ECE mechanisms. In 1958 Koutecky and Levich (S17) quantitatively treated the effect of reaction kinetics in a CE mechanism upon the current observed at an RDE. The preceding reaction was considered to be an equilibrium between an electrochemically inactive substance and a species reducible at the RDE, with their diffusion coeofficients assumed to be equal. The equilibrium constant K is the ratio between the forward rate constant k_f and the reverse or back rate constant k_b. The result of this treatment, in the form of an explicit solution for the limiting current observed, is a form of the *Koutecky-Levich equation*:

$$I_{lim} = -zFAD^{2/3}\omega^{1/2}c/1.61\nu^{1/6}(1 + 0.621\omega^{1/2}D^{1/6}\nu^{-1/6}K^{-1}(k_f + k_b)^{-1/2}) \tag{18.5}$$

which includes as c the sum of the bulk concentrations of the active and inactive species. This equation reduces to the Levich equation as the effect of the reaction kinetics becomes negligible. This will occur when the equilibrium constant is large because the solution species are then all in the form of the reducible species c.

The most useful form of the Koutecky–Levich equation is

$$I_{lim}/\omega^{1/2} = [I_d/\omega^{1/2}] - [D^{1/6}I_{lim}/1.61\nu^{1/6}K(k_f + k_b)^{1/2}] \tag{18.6}$$

In this equation I_{lim} is the actual current limited by the reaction kinetics, and I_d is the current that would be measured in the absence of any kinetic limitation and that is therefore limited only by diffusion. A linear plot should be obtained for $I_{lim}/\omega^{1/2}$ against I_{lim}. The slope of the line gives the kinetic parameter $K(k_f + k_b)^{1/2}$, and the intercept gives the diffusion-limited current. This form of the Koutecky–Levich equation was used to interpret the data of Vielstich and Jahn (S18). Additional studies have been reviewed by Riddiford (S19); more recent work, classified by the type of mechanism involved, is discussed by Galus (G7).

18.3.3   Rotating Ring-Disk Electrodes

A considerable advance upon the rotating disk electrode was made in 1959 when Frumkin and Nekrasov (S20, S21) developed a rotating ring–disk electrode (RRDE) in which the disk is surrounded by a ring electrode. This electrode, set flush with the disk, is separated from it by a small gap filled with insulating material so that the ring electrode and the disk electrode can be held at different potentials. Since the direction of convectively driven flow is outward uniformly from the center of the disk and since the ring–disk spacing is uniform, the time for an intermediate generated electrochemically at the disk to travel across the thin gap to the ring electrode whose potential is set to detect it is controllable by adjusting the speed of rotation of the RRDE; increasing the speed of rotation will reduce the time required to cross the gap, and thus adjustment of the three independent parameters of disk potential, ring potential, and rotation speed, with observation of the currents at generating disk and detecting ring, gives considerable information about the existence and duration of lifetime of electrochemically generated intermediates. A useful introduction can be found in the book of Albery and Hitchman (G4); additional material can be found in the briefer review of Piekarski and Adams (S22).

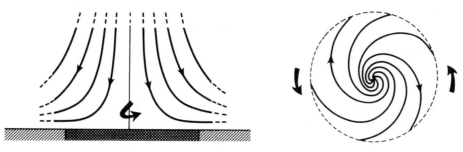

Figure 18.4   Rotating disk electrode: convective flow (after Levich).

The rotating disk electrode and its ring–disk derivative have been employed comparatively infrequently despite their significant advantages in the detection of short-lived intermediate species, because of the experimental difficulties associated with the construction and use of such an assembly. The electrode must be mounted vertically and rotated by a synchronous motor. Careful fabrication and alignment of the vertical bearing is essential to prevent vibration of the electrode in either the horizontal or vertical direction. Such vibration would, by increasing the convection in a manner unpredictable by theoretical equations, make quantitative use of the electrode impossible. The electrode must be of reasonable size, about 1 cm^2 for a disk (and considerably less for a ring, which must be comparatively thin) to avoid significant anomalous effects at the edge of the disk, and must possess a smooth surface to avoid disturbing the laminar flow, which is as shown in Figure 18.4.

The study of metal anodic dissolution, in particular that of indium, has been the subject of several rotating disk electrode studies. Losev (S23) employed a rotating indium disk and a stationary solid indium electrode in his studies, while Kiss (S24) used an RRDE with an indium disk and a noble metal ring in a solution containing In(III); in the latter configuration the ring was used as a potentiometric sensor for lower-valent states of indium. Miller and Visco (S25) used solid indium or indium amalgam disk electrodes with gold or graphite rings, including a split-ring configuration in which the potentials of each half of the ring could be separately adjusted, thus permitting simultaneous monitoring of anodic and cathodic currents.

18.4   REFERENCES

General

G1   J. T. Stock, *Amperometric Titrations*, Wiley-Interscience, New York, 1965.

G2   L. Meites, *Polarographic Techniques,* 2nd ed., Wiley-Interscience, New York, 1964, 752 pp. Excellent discussion of polarography; also covers amperometric titration and, very briefly, other related techniques.

G3   I. Fatt, *Polarographic Oxygen Sensors*, CRC Press, West Palm Beach, Florida,, 1977, 292 pp. Discusses theory and practice of amperometric sensors for oxygen, with emphasis on physiological applications.

G4   W. J. Albery and M. L. Hitchman, *Ring–disc Electrodes*, Clarendon Press, Oxford, 1971, 175 pp.

G5   Yu. V. Pleskov and V. Yu. Filinovskii, Eds., *The Rotating Disc Electrode*, H. Wroblowa, Trans., Plenum Press, New York, 1976.

Specific

S1   W. J. Blaedel, C. L. Olson, and L. R. Sharma, *Anal. Chem.* **35**, 2100 (1963).

S2   T. O. Oesterling and C. L. Olson, *Anal. Chem.* **39**, 1543 (1967).

S3   J. T. Stock, *Anal. Chem.* **36**, 355R (1964).

S4   J. T. Stock, *Anal. Chem.* **38**, 452R (1966).

S5   J. T. Stock, *Anal. Chem.* **40**, 392R (1968).

S6   J. T. Stock, *Anal. Chem.* **42**, 276R (1970).

S7   J. T. Stock, *Anal. Chem.* **44**, 1R (1972).

S8   J. T. Stock, *Anal. Chem.* **46**, 1R (1974).

S9   J. T. Stock, *Anal. Chem.* **48**, 1R (1976).

S10   J. T. Stock, *Anal. Chem.* **50**, 1R (1978).

S11   J. T. Stock, *Anal. Chem.* **52**, 1R (1980).

S12   V. G. Levich, *Acta Physicochim. U.R.S.S.* **17**, 257 (1942).

S13   V. G. Levich, *Acta Physicochim. U.R.S.S.* **19**, 117, 133 (1944).

S14   V. G. Levich, *Physicochemical Hydrodynamics*, Prentice-Hall, Englewood Cliffs, N. J., 1962.

S15   D. Jahn and V. Vielstich, *J. Electrochem. Soc.* **109**, 849 (1962).

S16   Z. Galus and R. N. Adams, *J. Phys. Chem.* **67**, 866 (1963).

S17   J. Koutecky and V. G. Levich, *Zh. Fiz. Khim.* **32**, 1565 (1958).

S18   W. Vielstich and D. Jahn, *Z. Elektrochem.* **64**, 43 (1960).

S19   A. C. Riddiford, *in Advances in Electrochemistry and Electrochemical Engineering*, Volume 4, P. Delahay, Ed., Interscience, New York, 1966.

S20   A. N. Frumkin and L. N. Nekrasov, *Dokl. Akad. Nauk SSSR* **126**, 115 (1959).

S21   A. N. Frumkin, L. N. Nekrasov, V. G. Levich, and J. Ivanov, *J. Electroanal. Chem.* **1**, 84 (1959/60).

S22 S. Piekarski and R. N. Adams, "Voltammetry with Stationary and Rotating Electrodes," *in Physical Methods of Chemistry,* Part IIA, A. Weissberger and B. W. Rossiter, Eds., Wiley-Interscience, New York, 1971, pp. 531ff.

S23 A. P. Pchelnikov and V. V. Losev, *Zashch. Metall.* **1**, 482 (1965).

S24 L. Kiss, *Magy. Kem. Folyoirat* **72**, 191 (1966).

S25 B. Miller and R. E. Visco, *J. Electrochem. Soc.* **115**, 251 (1968).

18.5 STUDY PROBLEMS

18.1 In acid sulfate media, the reaction of V(IV) with V(II) to give V(III) is quantitative and can be used for amperometric titration of V(IV). Against SCE, V(II) gives an anodic wave of half-wave potential –0.51 V, and V(IV) gives a cathodic wave of half-wave potential –0.85 V. Select a suitable potential for the amperometric titration of 10 cm³ of 1.0 mmol/dm³ V(IV) with 2.0 mmol/dm³ V(II), and draw the amperometric titration curve obtained at that potential.

18.2 Compute the effect on the thickness of a Nernst diffusion layer at a disk electrode as the speed of the disk goes from 0 (stationary) to 100 to 1000 to 10,000 rpm (revolutions per minute).

18.3 Reversible polarographic reduction of Fe(III) occurs at a half-wave potential of –0.74 V against SCE in 1.0 mol/dm³ aqueous KF solution at pH 7. The resulting Fe(II) may react in solution with a certain species X, not reducible at this potential, to regenerate Fe(III) and give the electrochemically inactive product Y. Discuss the use of an RRDE to obtain the homogeneous rate constant of the reaction Fe(II) + X → Fe(III) + Y, including in your answer the potentials at which the two electrodes should be operated and the reasons for selecting these values.

APPENDIX
1
ABBREVIATIONS AND SYMBOLS

ABBREVIATIONS

ASV	anodic stripping voltammetry
CSV	cathodic stripping voltammetry
DME	dropping-mercury electrode
DPASV	differential pulse anodic stripping voltammetry
EDTA	ethylenediaminetetraacetic acid
HMDE	hanging-mercury-drop electrode
ISE	ion-selective electrode
LSV	linear-sweep voltammetry
MFE	mercury-film electrode
MTFE	mercury thin-film electrode
NCE	normal calomel electrode
NHE	normal hydrogen electrode
OTE	optically transparent electrode
RDE	rotating disk electrode
RPE	rotating platinum electrode
RRDE	rotating ring–disk electrode
SCE	saturated calomel electrode
SCV	staircase voltammetry
SHE	standard hydrogen electrode
SMDE	sitting-mercury-drop electrode
SV	stripping voltammetry
TFE	thin-film electrode
WIGE	wax-impregnated graphite electrode

TABLE A1.1   Symbols for Physical Quantities (Italic)

Letter	Quantities	Letter	Quantities
A	area	a	chemical activity
B		b	
C	capacitance	c	concentration; speed of light
D	diffusion coefficient	d	diameter
E	potential difference	e	electron charge
F	Faraday constant	f	frequency
G	free energy; conductance	g	
H	enthalpy	h	height; Planck constant
I	current; ionic strength	i	
J	flux	j	current density
K	constant (equilibrium)	k	constant (rate); Boltzmann constant
L	inductance	l	length
M	molarity	m	molality
N	Avogadro number	n	amount of substance
O		o	
P	power	p	pressure
Q	electrical charge	q	heat
R	resistance; molar gas constant	r	radius
S	entropy	s	
T	temperature (K)	t	time; temperature(°C); transference number
U	internal energy	u	mobility
V	volume	v	
W		w	work
X	mole fraction; reactance	x	distance
Y	admittance $(1/Z)$	y	activity coefficient
Z	impedance	z	charge number

TABLE A1.2 Symbols for Units and Unit Prefixes (Roman)

Letter	Unit (Prefix)	Letter	Unit (Prefix)
A	ampere	a	(atto-)
B		b	
C	coulomb	c	(centi-)
D		d	(deci-)
E	(exa-)	e	2.71828...
F	farad	f	(femto-)
G	(giga-)	g	gram
H	henry	h	hour (hecto-)
I		i	
J	joule	j	
K	kelvin	k	(kilo-)
L		l	
M	(mega-)	m	meter (milli-)
N	newton	n	(nano-)
O		o	
P	(peta-)	p	(pico-)
Q		q	
R		r	
S	siemens	s	second
T	(tera-)	t	
U		u	
V	volt	v	
W	watt	w	
X		x	
Y		y	
Z		z	

TABLE A1.3 Physical Quantities (Greek)

Letter	Quantity	Letter	Quantity
A		α	degree of dissociation; transfer coefficient
B		β	
Γ	surface excess	γ	activity coefficient
Δ	difference	δ	thickness of (Nernst) layer
E		ϵ	permittivity
Z		ζ	
H		η	overpotential
Θ		θ	phase angle
I		ι	
K		κ	conductivity
Λ	molar conductivity	λ	molar ionic conductivity
M		μ	chemical potential; (micro-)
N		ν	kinematic viscosity
Ξ		ξ	
O		o	
Π		π	3.14159...
P		ρ	density; resistivity
Σ	summation	σ	surface tension
T		τ	time (drop, transition, etc.)
Υ		υ	stoichiometric number
Φ	inner electric potential	ϕ	a parameter
X		χ	surface electric potential
Ψ	outer electric potential	ψ	a parameter
Ω	ohm	ω	angular velocity
		∂	partial symbol

TABLE A1.4 Potential Symbols

E	general symbol for potential, also used for the real or observed potential
E'	a potential defined specifically in the text; used as a notation to simplify equations
$E(0)$	zero-current potential observed when no current is drawn from the electrode. Throughout Part II, all usage of E is usage of $E(0)$.
E^0	standard potential; the activities of the species involved in the potential-determining couple are unity
$E^{0'}$	formal potential; the activities of the species involved in the potential-determining couple are fixed but not unity
$E_{1/2}$	polarographic half-wave potential; differs from a formal potential for the couple only in the ratio of the square roots of the diffusion coefficients of the oxidized and reduced species, if the polarographic reduction is reversible.
$E_{\tau/4}$	chronopotentiometric quarter-time potential. For a reversible process taking place on a mercury electrode, this is equal to the polarographic half-wave potential.
E_{pzc}	potential of an electrocapillary maximum of the electrode in a solution; taken as the potential of zero net charge on the electrode
E_p	potential of a current peak observed in some nonequilibrium electroanalytical technique
$E_{p/2}$	half-peak potential; potential at which the current observed in some nonequilibrium technique is halfway between the baseline or residual current and the peak current. If the peak width is measured it should be measured at this point.
E_{DC}	total DC component of a potential having both DC and AC components
E_{AC}	total AC component of a potential having both AC and DC components

TABLE A1.5   Current Symbols

I — general symbol for current, also used for the net or observed current

I^0 — standard current, more often called the *exchange current*. This is the current flowing when no net current is drawn from the electrode, when $(E = E(0))$

I_c — cathodic (negative) component of the current

I_a — anodic (positive) component of the current. Net current is the algebraic sum of the cathodic and anodic components, and exchange current is the absolute magnitude of either of them when they are equal.

I_{lim} — current limited by some process or processes

I_d — current limited by diffusion of species alone; a subclass of I_{lim}

I_{surf} — current limited by available electrode surface; a subclass of I_{lim}

I_{DC} — total DC component of a current having both AC and DC components

I_{AC} — total AC component of a current having both AC and DC components

$I(\psi t)$ — an AC component current whose frequency in hertz is $\psi/2\pi$; may not be the only or total AC component of the current

I_p — current at maximum (peak) value obtained by some nonequilibrium electroanalytical procedure

TABLE A1.6   Equilibrium Constant Symbols

K	general symbol for equilibrium constants
K_a	general symbol for acid ionization constants; a subclass of K_d
K_d	general symbol for dissociation constants
K_f	general symbol for formation constants. For the same equilibrium the formation and dissociation constants are reciprocals of each other
K_n	stepwise equilibrium constant for the nth step. The type of equilibrium constant must be made clear by additional letter subscripts if it is not clear from the context.
K_s	solubility product constant; a subclass of K_d
$K_{i,j}$	potentiometric selectivity coefficient for ion j over ion i, used with ion-selective electrodes
β	cumulative formation constant for metal ion complex

TABLE A1.7   Other Constant Symbols

k	general symbol for a constant, often a rate constant; in homogeneous media unless otherwise indicated. This symbol is also used for the Boltzmann constant.
k_h	a constant that refers to a heterogeneous reaction, most often one taking place on the surface of an electrode
$k_{a,h}$	heterogeneous rate constant for an anodic (oxidation) electrode reaction
$k_{c,h}$	heterogeneous rate constant for a cathodic (reduction) electrode reaction
k^0_h	standard heterogeneous rate constant observed at the zero-current potential of the couple to which it refers
k_f	rate constant for forward reaction
k_b	rate constant for backward or reverse reaction
$k_{X,i}$	constant of Henry's law, mole fraction scale
$k'_{x,i}$	empirical constant used with Henry's law, mole fraction scale; $k'_{m,i}$ and $k'_{c,i}$ are used for the molality and general concentration scale constants.

APPENDIX
2
PHYSICAL DATA

TABLE A2.1 Fundamental Physical Constants[a]

Constant[b]	Symbol	Value ± Uncertainty	SI Unit
Avogadro	N	$6.02252 \pm 0.00028 \times 10^{23}$	mol^{-1}
Boltzmann	k	$1.38054 \pm 0.00018 \times 10^{-23}$	J/K
Faraday	F	$9.64870 \pm 0.00016 \times 10^4$	C/mol
Planck	h	$6.6256 \pm 0.0005 \times 10^{34}$	J s
gas	R	8.3143 ± 0.0012	J/K mol
electron charge	e	$1.60210 \pm 0.00007 \times 10^{-19}$	C
"ice-point"	T(ice)	273.1500 ± 0.0001	K
speed of light	c	$2.997925 \pm 0.000003 \times 10^8$	m/s

[a]These data are a partial set of the values of the fundamental constants recommended by IUPAC [*Pure and Applied Chemistry* **9**, 453 (1964)] and are used throughout this text. The estimated uncertainty given here is three times the standard deviation.

[b]The speed of light given is that in vacuum. The "ice-point" temperature is the temperature of equilibrium of solid and liquid water saturated with air at a pressure of 1 atm; subtraction of this temperature from an absolute temperature gives the Celsius temperature. The Faraday constant is the product of N and e, the gas constant is the product of N and k.

TABLE A2.3 Limiting Molar Conductivities of Ions[a]

Cation	λ^0_i (mS m²/mol)	Anion	λ^0_i (mS m²/mol)
H^+	34.99(8)	OH^-	19.84
Li^+	3.87(0)	F^-	5.54
Na^+	5.012	Cl^-	7.639
K^+	7.354	Br^-	7.818
Rb^+	7.785	I^-	7.68(8)
Cs^+	7.73(0)	N_3^-	6.9
Ag^+	6.19(3)	NO_3^-	7.150
Tl^+	7.47	ClO_3^-	6.46
NH_4^+	7.35(9)	BrO_3^-	5.57(7)
$CH_3NH_3^+$	5.87(5)	IO_3^-	4.05(6)
$(CH_3)_2NH_2^+$	5.190	ClO_4^-	6.73(9)
$(CH_3)_3NH^+$	4.72(7)	IO_4^-	5.45(8)
$(CH_3)_4N^+$	4.49(4)	ReO_4^-	5.50(0)
$(C_2H_5)_4N^+$	3.26(8)	HCO_3^-	4.45(2)
$(C_3H_7)_4N^+$	2.34(3)	$HCOO^-$	5.46(2)
$(C_4H_9)_4N^+$	1.94(8)	CH_3COO^-	4.09(2)
$(C_5H_{11})_4N^+$	1.74(9)	CH_2FCOO^-	4.44(1)
$(CH_3)_3C_6H_5N^+$	3.46(7)	CH_2ClCOO^-	4.22(2)
$HOCH_2CH_2NH_3^+$	4.22(5)	CH_2BrCOO^-	3.92(4)
Be^{2+}	9.0	CH_2ICOO^-	4.06(2)
Mg^{2+}	10.62	CH_2CNCOO^-	4.34(4)
Ca^{2+}	11.91	$CH_3CH_2COO^-$	3.58

Cation	λ^0_i (mS m^2/mol)	Anion	λ^0_i (mS m^2/mol)
Sr^{2+}	11.90	$CH_3(CH_2)_2COO^-$	3.26
Ba^{2+}	12.73	$C_6H_5COO^-$	3.24(0)
Cu^{2+}	10.73	SO_4^{2-}	16.01
Zn^{2+}	10.57	$C_2O_4^{2-}$	14.84
Co^{2+}	11.0	CO_3^{2-}	13.87
Pb^{2+}	13.91	$Fe(CN)_6^{3-}$	30.28
La^{3+}	20.92	$P_3O_9^{3-}$	25.09
Ce^{3+}	20.95	$Fe(CN)_6^{4-}$	44.22
Pr^{3+}	20.89	$P_4O_{12}^{4-}$	37.4(6)
Nd^{3+}	20.83	$P_2O_7^{4-}$	38.3(8)
Sm^{3+}	20.56	$P_3O_{10}^{5-}$	54.5
Eu^{3+}	20.35	$[Co(NH_3)_6]^{3+}$	30.58
Gd^{3+}	20.20	$[Co_2(trien)_3]^{6+}$	41.2(4)
Dy^{3+}	19.69	$[Ni_2(trien)_3]^{4+}$	21.0(1)
Ho^{3+}	19.90		
Er^{3+}	19.78		
Tm^{3+}	19.63		
Yb^{3+}	19.69		

[a]Values given are those selected by R. A. Robinson and R. H. Stokes (*Electrolyte Solutions*, 2nd ed., Butterworths, London, 1959), recalculated to SI units by the author. Figures in parentheses are less accurate and should be taken as ±5. Values for dilute aqueous solutions at 25°C.

TABLE A2.4 Standard Reduction Potentials in Aqueous Acid[a]

Electrode Couple	E^0 (V)	dE/dT (mV/K)
$Na^+ + e^- \rightarrow Na$	-2.714	-0.772
$Mg^{2+} + 2e^- \rightarrow Mg$	-2.363	+0.103
$Al^{3+} + 3e^- \rightarrow Al$	-1.662	+0.504
$Zn^{2+} + 2e^- \rightarrow Zn$	-0.7628	+0.091
$Fe^{2+} + 2e^- \rightarrow Fe$	-0.4402	+0.052
$Cd^{2+} + 2e^- \rightarrow Cd$	-0.4029	-0.093
$Sn^{2+} + 2e^- \rightarrow Sn$	-0.136	-0.282
$Pb^{2+} + 2e^- \rightarrow Pb$	-0.126	-0.451
$2H^+ + 2e^- \rightarrow H_2(SHE)$	±0.0000	±0.000
$Cu^{2+} + e^- \rightarrow Cu^+$	+0.153	+0.073
$AgCl + e^- \rightarrow Ag + Cl^-$	+0.2223	-0.653
$Hg_2Cl_2 + 2e^- \rightarrow 2Cl^- + 2Hg$	+0.2676	-0.317
$Cu^{2+} + 2e^- \rightarrow Cu$	+0.337	+0.008
$Cu^+ + e^- \rightarrow Cu$	+0.521	-0.058
$I_2 + 2e^- \rightarrow 2I^-$	+0.5355	-0.148
$Hg_2SO_4 + 2e^- \rightarrow 2Hg + SO_4^{2-}$	+0.6151	-0.826
$Fe^{3+} + e^- \rightarrow Fe^{2+}$	+0.771	+1.188
$Ag^+ + e^- \rightarrow Ag$	+0.7991	-1.000
$Br_2(aq) + 2e^- \rightarrow 2Br^-$	+1.087	-0.478
$O_2 + 4H^+ + 4e^- \rightarrow 2H_2O$	+1.229	-0.846
$MnO_2 + 4H^+ + 2e^- \rightarrow Mn^{2+} + 2H_2O$	+1.23	-0.661
$Cl_2 + 2e^- \rightarrow 2Cl^-$	+1.3595	-1.260

[a]Data at 25°C taken from A. J. DeBethune, T. S. Licht, and N. Swendeman, *J. Electrochem. Soc.* **106**, 616 (1959). The isothermal temperature coefficient given in this table is the derivative dE/dt of the isothermal cell in which one of the electrodes is the SHE. For a discussion of the thermal temperature coefficient see the original reference.

TABLE A2.5 Standard Reduction Potentials

in Aqueous Bas[a] Electrode Couple	E^0 (V)	dE/dT (mV/K)
$Mg(OH)_2 + 2e^- \rightarrow Mg + 2OH^-$	−2.69	−0.074
$Al(OH)_3 + 3e^- \rightarrow Al + 3OH^-$	−2.30	−0.06
$Cr(OH)_3 + 3e^- \rightarrow Cr + 3OH^-$	−1.48	−0.11
$Zn(OH)_2 + 2e^- \rightarrow Zn + 2OH^-$	−1.245	−0.131
$ZnCO_3 + 2e^- \rightarrow Zn + CO_3^{2-}$	−1.06	−0.293
$Fe(OH)_2 + 2e^- \rightarrow Fe + 2OH^-$	−0.877	−0.19
$2H_2O + 2e^- \rightarrow H_2 + 2OH^-$	−0.828	+0.037
$Cd(OH)_2 + 2e^- \rightarrow Cd + 2OH^-$	−0.809	−0.143
$FeCO_3 + 2e^- \rightarrow Fe + CO_3^{2-}$	−0.756	−0.422
$Co(OH)_2 + 2e^- \rightarrow Co + 2OH^-$	−0.73	−0.193
$2SO_3 + 3H_2O + 4e^- \rightarrow S_2O_3^{2-} + 6OH^-$	−0.58	−0.27
$Fe(OH)_3 + e^- \rightarrow Fe(OH)_2 + OH^-$	−0.56	−0.27
$PbCO_3 + 2e^- \rightarrow Pb + CO_3^{2-}$	−0.506	−0.423
$S + 2e^- \rightarrow S^{2-}$	−0.447	−0.06
$S_4O_6 + 2e^- \rightarrow 2S_2O_3$	+0.08	−0.24
$Pt(OH)_2 + 2e^- \rightarrow Pt + 2OH^-$	+0.15	−0.273
$O_2 + 2H_2O + 4e^- \rightarrow 4OH^-$	+0.401	−0.809
$MnO_4^- + 2H_2O + 3e^- \rightarrow MnO_2 + 4OH^-$	+0.588	−0.907

[a]Data at 25°C taken from A. J. DeBethune, T. S. Licht, and N. Swendeman, *J. Electrochem. Soc.* **106**, 616 (1959). The isothermal temperature coefficient given in this table is the derivative *dE/dt* of the isothermal cell in which one of the electrodes is the SHE. For a discussion of the thermal temperature coefficient see the original reference.

TABLE A2.6   Table of Atomic Weights, 1975
(Scaled to the Relative Atomic Mass $^{12}C = 12$)

The atomic weights of many elements are not invariant, but depend on the origin and treatment of the material. The footnotes elaborate the types of variation to be expected for individual elements. The values given here apply to elements as they exist naturally on earth and to certain artificial elements. When used with due regard to the footnotes they are considered reliable to ± 1 in the last digit or ± 3 when followed by an asterisk. Values in parentheses are used for certain radioactive elements whose atomic weights cannot be quoted precisely without knowledge of origin; the value given is the atomic mass number of the isotope of that element of longest known half-life.

Name	Symbol	Atomic Number	Atomic Weight
Actinium[a]	Ac	89	227.0278
Aluminum	Al	13	26.98154
Americium	Am	95	(243)
Antimony	Sb	51	121.75*
Argon[b,c]	Ar	18	39.948*
Arsenic	As	33	74.9216
Astatine	At	85	(210)
Barium[c]	Ba	56	137.33
Berkelium	Bk	97	(247)
Beryllium	Be	4	9.01218
Bismuth	Bi	83	208.9804
Boron[b,d]	B	5	10.81
Bromine	Br	35	79.904
Cadmium[c]	Cd	48	112.41
Cesium	Cs	55	132.9054
Calcium[c]	Ca	20	40.08
Californium	Cf	98	(251)
Carbon[b]	C	6	12.011
Cerium[c]	Ce	58	140.12
Chlorine	Cl	17	35.453
Chromium	Cr	24	51.996
Cobalt	Co	27	58.9332
Copper[b]	Cu	29	63.546*
Curium	Cm	96	(247)
Dysprosium	Dy	66	162.50*
Einsteinium	Es	99	(254)
Erbium	Er	68	167.26*
Europium[c]	Eu	63	151.96

Fermium	Fm	100	(257)
Fluorine	F	9	18.998403
Francium	Fr	87	(223)
Gadolinium[c]	Gd	64	157.25*
Gallium	Ga	31	69.72
Germanium	Ge	32	72.59*
Gold	Au	79	196.9665
Hafnium	Hf	72	178.49*
Helium[c]	He	2	4.00260
Holmium	Ho	67	164.9304
Hydrogen[b]	H	1	1.0079
Indium[c]	In	49	114.82
Iodine	I	53	126.9045
Iridium	Ir	77	192.22*
Iron	Fe	26	55.847*
Krypton[c,d]	Kr	36	83.80
Lanthanum[c]	La	57	138.9055*
Lawrencium	Lr	103	(260)
Lead[b,c]	Pb	82	207.2
Lithium[b,c,d]	Li	3	6.941*
Lutetium	Lu	71	174.97
Magnesium[c]	Mg	12	24.305
Manganese	Mn	25	54.9380
Mendelevium	Md	101	(258)
Mercury	Hg	80	200.59*
Molybdenum	Mo	42	95.94
Neodymium[c]	Nd	60	144.24*
Neon[d]	Ne	10	20.179*
Neptunium[a]	Np	93	237.0482
Nickel	Ni	28	58.70
Niobium	Nb	41	92.9064
Nitrogen	N	7	14.0067
Nobelium	No	102	(259)
Osmium[c]	Os	76	190.2
Oxygen[b]	O	8	15.9994*
Palladium[c]	Pd	46	106.4
Phosphorus	P	15	30.97376
Platinum	Pt	78	195.09*
Plutonium	Pu	94	(244)

Polonium	Po	84	(209)
Potassium	K	19	39.0983*
Praseodymium	Pr	59	140.9077
Promethium	Pm	61	(145)
Protactinium[a]	Pa	91	231.0359
Radium[a,c]	Ra	88	226.0254
Radon	Rn	86	(222)
Rhenium	Re	75	186.207
Rhodium	Rh	45	102.9055
Rubidium[c]	Rb	37	85.4678*
Ruthenium[c]	Ru	44	101.07*
Samarium[c]	Sm	62	150.4
Scandium	Sc	21	44.9559
Selenium	Se	34	78.96*
Silicon	Si	14	28.0855*
Silver[c]	Ag	47	107.868
Sodium	Na	11	22.98977
Strontium[c]	Sr	38	87.62
Sulfur[b]	S	16	32.06
Tantalum	Ta	73	180.9479*
Technetium	Tc	43	(97)
Tellurium[c]	Te	52	127.60*
Terbium	Tb	65	158.9254
Thallium	Tl	81	204.37*
Thorium[a,c]	Th	90	232.0381
Thulium	Tm	69	168.9342
Tin	Sn	50	118.69*
Titanium	Ti	22	47.90*
Tungsten	W	74	183.85*
Uranium[c,d]	U	92	238.029
Vanadium	V	23	50.9414*
Xenon[c,d]	Xe	54	131.30
Ytterbium	Yb	70	173.04*
Yttrium	Y	39	88.9059
Zinc	Zn	30	65.38
Zirconium[c]	Zr	40	91.22

[a] Element for which the value given is that of the radioisotope of longest half-life.

[b] Elements for which known variations in isotopic composition in normal terrestrial material prevent a more precise atomic weight being given; values should be applicable to any "normal" material.

[c] Element for which geological specimens are known in which the element has an anomalous isotopic composition, such that the difference in atomic weight of the element in such specimens from that given in the table may exceed considerably the implied uncertainty.

[d] Element for which substantial variations from the value given can occur in commercially available material because of inadvertent or undisclosed change of isotopic composition.

APPENDIX
3
OBSOLETE ELECTRICAL UNITS

Electrical measurements sufficiently precise to be of current use were not made prior to the twentieth century, by which time the metric system was well established. In 1910, the International Committee on Weights and Measures established the so-called international units, adopted by the major European and North American countries. The volt was considered the derived unit and was left to be derived from Ohm's law. Operational definitions were established for the ampere and the ohm, as follows:

> The international ampere is the unvarying direct current that, when passed through an aqueous solution of silver nitrate under certain specified conditions, will deposit silver at the rate of 0.00111800 g/s.
>
> The international ohm is the resistance offered to an unvarying direct current at 0°C by a uniform column of mercury 106.300 cm long whose mass is 14.4521 g. (The cross-sectional area of such a column is very close to 1 mm².)

In 1946, agreement was achieved on the relation of the international ampere and the international ohm to the absolute ampere (operationally defined in terms of the force produced by the current in a current balance) and the absolute ohm (operationally defined in terms of inductance and frequency), which are related to the more fundamental units of mass, length, and force. As a consequence, the previously agreed-upon "international" units were abandoned in 1948 in favor of the "absolute" or SI units. The relationships between the units are:

> one international ohm = 1.00049 absolute ohms
> one international ampere = 0.99985 absolute ampere, and consequently
> one international volt = 1.00034 absolute volts

These factors should be used in measurements employing instruments made prior to 1948. For instruments whose calibration is traceable to the U.S. National Bureau of Standards, as most precise instruments manufactured in North America were, use the factors that 1 international ohm = 1.000495 Ω and 1 international volt = 1.000330 V. Instruments manufactured since 1948 are normally calibrated in absolute units, which are identical in previous metric systems and SI.

INDEX